SMART DATA

SMART DATA
Enterprise Performance Optimization Strategy

JAMES A. GEORGE
JAMES A. RODGER

With a Foreword by David E. Chesebrough,
President, Association for Enterprise Information (AFEI)

A JOHN WILEY & SONS, INC., PUBLICATION

Library of Congress Cataloging-in-Publication Data

George, James A., 1948-
  Smart data : enterprise performance optimization strategy / James A. George, James A. Rodger; with a foreword by David E. Chesebrough.
      p. cm. - - (Wiley series in systems engineering and management)
  Includes bibliographical references and index.
  ISBN 978-0-470-47325-2
  1. Information resources management – Quality control. 2. Database management – Quality control.
  I. Rodger, James A., 1952- II. Title.

  T58.64.G43 2010
  658.4'013 - - dc22

                                                                2009031431

Printed in the United States of America

10 9 8 7 6 5 4 3 2 1

To my wife, Maureen, for her lifelong inspiration and support for all my endeavors, as well as for preparing the graphics for this book

James A. George

To Mom, Dad and Keiko-san: Domo Arigato Gozaimasta

James A. Rodger

Contents

Foreword

Information is the lifeblood of civilization. It has always been so. Many thousands of years ago information was transmitted primarily as the spoken word. Inevitably, ways of recording information were created so that it could be retained and passed on. Thus humans evolved their ability to convey not just data but their thoughts and meanings, their knowledge, through recorded information. The desire to unlock knowledge and put it to use is a compelling impetus for human advancement, and while development of alphabets and writing forms moved on, the literacy of the common person surely lagged behind. Ancient recorded information was likely barely accessible only to a very few in the upper classes. Enter the first wave of mass production of information. The advent of printing made information widely available and, by the end of the 15th century, an estimated 15 million books had been printed on presses. But information still spread slowly down through the strata of society. Even in America during the mid-19th century, books were difficult to access in the rural parts of our young and expanding country. During his youth, Abraham Lincoln sometimes walked miles just to find a book. Fast forward just one century and libraries dot the landscape from shore to shore. In the mid-20th century books were widely available, and most people could read them. Information was available to everyone. But the country was poised to enter an era that would produce a literal explosion of information, unlike anything ever witnessed by humankind. Advancing technology spawned radio, movies, television, computers, the Internet, and the World Wide Web. With technology, everyone could not only get information but could also publish.

Information is ubiquitous. Its ever-increasing volume overwhelms our enterprises and our lives. The question today is not *"Can I get the data?"* The real question is *"What is the correct data?"* In this book its authors, James A. George and James A. Rodger, have taken on a topic imperative for our time. We are awash in data. It comes at use from every conceivable direction. We cannot absorb it or even grasp the extent of it. Government and business struggle to deal with burgeoning volumes of data.

George and Rodger, in creating this book, have placed themselves in the company of Frederic Taylor, W. Edwards Deming, and Peter Drucker. Making business organizations more effective and efficient has always been a fundamental goal of management science. The authors, in defining smart data concepts and strategy, have put us on the road to making information technology matter again. They persuasively address the problem of "data whose life cycle often outlasts the information technology (IT) paradigm under which it was created." Their work sets a pattern that helps us see how to survive the immutable effects of Moore's Law and paves the way for optimizing enterprise performance in the digital age. Its effect will be to usher in a new age of agility and resilience. Instead of choking the arties of our enterprises, information will once again propel our advancement.

DAVID E. CHESEBROUGH
President, Association for Enterprise Information

Preface

This book is written for enterprise executives and addresses the single most important activity for which they are ultimately responsible—optimization of performance. Whether you are an executive of government or commercial enterprise, or any other organization, your primary aim is to maximize return from limited resources to accomplish the unique mission and goals of your enterprise. Optimizing performance means applying scarce resources to business processes under constraint and transforming them into highest yield and best use outcomes by managing people and enabling technology.

Our aim in writing *Smart Data* is to contribute to the optimization of enterprise performance with a strategy that will vastly improve enterprise data resource management reflected in more efficient and cost effective information technology (IT) support that produces high impact results for executive and operations management users. "Smart data" describes data that have been engineered to have certain superior characteristics and that is a product of state-of-the-art data engineering discipline such that it is interoperable and readily exchangeable among qualified members of an enterprise user community. Smart data is an executive strategy and tool for exacting higher performance across the enterprise. It operates by leveraging state-of-the-art infrastructure and enterprise services.

Smart Data covers the following points:

1. The context is optimizing enterprise performance, whereby enterprise data resource management is owned by executives and operations management users and not relegated to enabling technologists and information systems specialists.

2. More efficient and cost effective IT support performance is managed and evaluated by executives and users and directly correlated with results.

3. High impact results are engineered from the beginning through collaboration between executives, operations management, and enabling technologists.

4. Measurements are meaningful and expressed in ways that tie directly to optimizing enterprise performance.

5. A strategy and detailed plan is needed that includes roles, responsibilities, and accountability—in the case of government that is understandable by ultimate citizen consumers; in the case of commerce that is understandable by employees, stakeholders, partners, investors, and, ultimately customers.

Included in the target audience are other executive specialists upon whom the CEO depends, such as Chief Information Officer, Chief Technology Officer, and Chief Performance Officer, among all other executives who are unique to your organization type.

There are six elements to this book: Introduction and Chapters 1–4 are the foundation for content presented to the CEO or Department Secretary as the CEO's Smart Data Handbook (Chapter 5). Each chapter element is approximately equal in length. An executive may read and reference Chapter 5 without reading the balance of the book as it is a standalone product. However, a performance advantage will be developed through the reader's complete grasp of topics presented in the entire text.

Introduction presents a *Comprehensive Overview* of smart data and smart data strategy.

- Chapter 1—Context: The case and place for smart data strategy—presents the big picture including critical terms.
- Chapter 2—Elements—presents the essentials for optimizing performance in the enterprise.
- Chapter 3—Barriers—discusses how to optimize performance and how to overcome barriers to optimum performance.
- Chapter 4—Visionary Ideas: Technical enablement—presents enabling technologies to help implement recommendations.
- Chapter 5—CEO's Smart Data Handbook—sums up the book into actionable form for the executive audience.

All enterprises exist for a purpose that is expressed in mission and value statements, goals and objectives, otherwise summarized into business plans. Once desired outcomes are identified, leaders can organize resources into functions. Functions identify the work that needs to be done to produce outcomes. How the work is accomplished is defined as processes where process activities constitute proprietary differentiation.

Proprietary differentiation or unique ways of accomplishing things is achieved through a variety of means that begins with creative leadership:

- Selecting the right customers to service
- Attending to all constituents and specific needs (government)
- Selecting the right things to do
- Organizing activities and designing work
- Attributing activities with application of a superior mix of people and technology
- Applying scarce resources in an optimal manner
- Structuring the balance of consequences such that doing the right things the right way is rewarded and deviations are dissuaded
- Ensuring that customers receive valuable results
- Assuring stakeholders that the enterprise is performing optimally

Data is at the heart of each of these management activities and that is why we focus on data as a principal contributor to optimizing performance.

In today's global performance environment, we are cognizant of economic and political factors that affect the creation and flow of capital and materials moving commercial and government enterprise. We address these issues to the extent that our focus and subject relates, as they do indeed, though we resist the temptation to deviate from our immediate focus that is bounded by the terms enterprise performance optimization and smart data.

We employ two types of case studies to support and illustrate our ideas: (1) scientific/academic and (2) anecdotal/programs. The scientific/academic examples tend to support the technical aspects of smart data whereas the anecdotal/programs are more philosophical and qualitative. All examples are intended to give real world life to our ideas.

We ask our readers to work with us to discover the transferability of our examples horizontally, across different types of enterprises, functions, and structures as they truly apply broadly. We ask you to understand that our efforts to develop, innovate, and apply our ideas are constrained by the opportunities that we have to work with client organizations. Therefore they may not be perfect examples, though they are surely honest ones.

DIFFERENT VIEWPOINTS: EXECUTIVE AND INFORMATION TECHNOLOGIST

This text is written from different viewpoints:

1. Government: James George is a government consultant who also has extensive commercial enterprise management experience and is also a business process improvement consultant.
2. Academia: Dr. James Rodger is a college professor and government and commercial consultant who worked with George on the *Battlefield Backorder Initiative (B3I)* and *Supplier Response Improvement Program (SRI)*

for the U.S. Department of Defense. Dr. Rodger is an expert in the subject of decision support and management information systems.

The B3I/SRI project involved analyzing millions of backorder military parts records to determine attributes that could be used to predict, prevent, and mitigate perpetual shortages plaguing actual war combatants. The effort produced predictive methodology and algorithms that illustrate the applications of a part of our vision for smart data and smart data strategy. Former DOD client (retired), Donald Hall, collaborated with George in creating the *Service-Oriented Enterprise (SOE)* paradigm and notion of "smart data."

RECENT EXPERIENCES AND USING THIS BOOK

Some contributions to the effort are from recent experiences that have produced pioneering achievements in enterprise integration included:

- Creating the SOE paradigm for the U.S. Department of Defense
- Developing data exchange capability leveraging international standards for neutral translation
- Developing the Integrated Data Strategy (IDS) for DOD with roots in ideas developed by aerospace contractors
- Harmonizing electronic business standards and practices in the aerospace and defense industry
- Pioneering uses of metadata
- Applying artificial intelligence methods to decision making
- Publishing case studies on business process redesign
- Creating algorithms to solve complex business problems
- Generating program coding to enable problem solving

This text employs comparisons and contrasts to differentiate the unique characteristics and value from the smart data paradigm and is written with the following ideas in mind:

- Directly addresses the responsibilities of Mr./Ms. Secretary and CEO
- Compares/contrasts *enterprise performance optimization* versus *enterprise integration*
- Makes data strategy an explicit element of enterprise performance optimization versus subordinating to infrastructure elements
- Compares/contrasts *government enterprise* versus *private enterprise, and collaboration*
- Compares *smart data exchange* versus *standardization*

This text makes smart data strategy explicit, understandable, and actionable to those entities capable of leading their unique contribution to enterprise performance. Proven examples are provided of how government and business enterprises can achieve their missions and goals more effectively so:

- Leaders discover unique performance improvement advantages from adopting smart data strategy
- Government enterprise executives improve ability to achieve results more quickly
- Commercial enterprise participants discover how to maximize demonstrated value to government enterprise participants and discover how to maximize demonstrated value to government enterprise as members of the supply chain as prime contractors, subcontractors, and suppliers
- All benefit from more cost-effective strategy that changes information management economics.

By surveying government and commercial enterprise customers for a number of years, and from having been directly engaged in related topics for more than 15 years, we observed the following:

- Deficiencies and omissions from government and commercial data strategy
- Advances in commercial enabling technology with gaps in application and implementation
- Continuous investment in information technology on a large scale
- Increasing pressure on government and private enterprise for better use of scarce resources as invested in information technology
- Improved commercial-off-the-shelf technologies that can accelerate adoption of smart data strategy
- Requirements for management and technical training

A survey of Amazon.com and other publication sources show the same publication titles appear when searching "data strategy" and "enterprise integration" with some notable attention to the subject as envisioned in *Adaptive Information* by Jeffrey T. Pollock and Ralph Hodgson. Pollock also worked on a government program where his concepts were applied to enterprise services and security architecture development for a major weapon systems program. This landmark publication is a technical foundation on which the complementary smart data strategy presented here is advanced.

Current publications are leveraged here by positioning and differentiating beyond the current literature so

- Commercial vendors champion our message.
- Enterprise leadership adopts our ideas as they become collaborative partners in future development.
- The book becomes a catalyst for change and improvement by providing *uniquely actionable* methods.

What makes the book "*uniquely actionable?*" Messages are organized by processes and targeted to specific audience segments in a collaborative context. Practical examples are designed to appeal strongly to executive interests and priorities in a strained and recovering economic environment.

WHO NEEDS THIS BOOK?

The audiences for this book are:

- Government management and technical enterprise executives
- Commercial enterprise participants in government enterprise: systems integrators and prime and subcontractors
- Technology providers
- Academics: graduate students in private institutions as well as the Defense Management University and others
- Members of professional associations and international standards organizations
- Congressional staff and governance organization, such as Office of Management & Budget and General Accounting Office
- Governance boards

At the core is the desire to make a difference in government and commercial enterprise performance. By advancing this smart data strategy that was developed as a product of discussions during development, we show it has worked effectively in the past. One of our first products on this journey is to produce a powerful strategy for power-filled enterprise leadership.

As a consultant in the early 1990s, George used the phrase "catalysts for change and improvement" in an article for *Business Engineering Newsletter* to characterize various initiatives embraced by management for a period of time, however fleeting, for the purpose of motivating organizations to change their behavior. Such catalysts have life cycles. Recollection about the Western Electric Hawthorne studies prompted this observation. George thought it humorous that the act of studying workers' environment, changing lighting conditions, adding and removing light bulbs and such, could positively affect human performance. It was the act of attention to workers that mattered most as concluded by landmark research.

Deliberate though unscientific observations reveals that management often grows tired of catalysts that manifest in programs, initiatives, projects, and slogans before they actually catch on. It takes time for people and organizations to grasp their intent, and by the time they do, management has moved on. They either move on physically to another station, or they simply change their minds.

The hazard from changing in midstream was revealed in studies performed by a former colleague, Michael Reiman Ph.D., who determined that attempts at implementing concurrent engineering strategy without following through to completion produced worse results than if the organization had not embarked upon change at all.

Often, to realize the benefit, organizations must confirm that they have the capacity for change and improvement that is a combination of capital, material, and intellectual resources, including management and technical. What capacity will it take to change from their present intent to actually adopting a smart data strategy? This book provides a construct for answering this question.

Of course, there are numerous examples in management science history about CEOs who lead with initiatives aimed at improving quality and service, and reducing costs. Often, CEOs are engineers who introduce a degree of science and math to their initiatives, principally aimed at desired metrics and outcomes. Today, the more popular catalysts do this such as "Six Sigma" and "Lean." Efforts such as "CMMI" aim at improvement from processes and continuous process improvement. "Balanced Score Card" addresses a mix of dynamics with corresponding elements in organization strategy.

The more complicated catalysts are, the least likely they are to become sustainable. They burn out too soon. One reason for this is organizations have limited capacity for change and improvement. Leaders must select strategies that can be accomplished by organizations within the bounds of certain resource constraints.

Our natural inclination is to change the subject every so often because we get bored. We sometimes confuse overcoming boredom with a quest for continuous improvement. A better way to regulate our inclination to move forward is to insist on accountability for milestone achievements as a prerequisite.

Commercial enterprise must pursue becoming best at what's new. Government enterprise must optimize enterprise performance and leverage the most from commercial trading partners. That is why we devote considerable attention to the sources of best technologies to determine what they offer to implementing smart data.

In government, it is interesting because leaders are encouraged to embrace best commercial practices and commercial-off-the-shelf technologies for which the metrics aim at maximizing profits for shareholders, which is quite different from government's desire to maximize service utility for citizens. Government leaders have predictably short tenures that accompany planning and budget cycles that exacerbate the short time in which they have to apply their talent.

The magnitude of problems and needs, and their associated life cycles are large and long respectively. Do the catalysts used by government executives possess characteristics best suited for their tentative tenure? We don't have all the answers to this question, although by being aware, we can factor this into our recommendations about strategy to improve performance.

JAMES A. GEORGE
JAMES A. RODGER

Arlington, Virgina
Hooversville, Pennsylvania
January 2010

Acknowledgments

A premiere data guru in our lifetime is Daniel S. Appleton, and his personal training and coaching provide a foundation from which to advance the focus on data and making it smart. The idea to write this book came from a government client, Donald Hall, former Chief at the Defense Logistics Agency, who consistently pushed for ways to leap beyond legacy systems and culture that prevent organizations from performing optimally. Wilbert Bailey, now at the Department of Treasury, supervised some important work. Broad support for improving performance from application of information technology is the Association for Enterprise Information and its president, David Chesebrough.

We would also like to acknowledge Parag C. Pendharkar Ph.D. for contributing quantitative insights into the cases and David J. Paper, Ph.D. for his qualitative inputs. These two mentors have provided valuable guidance and a deep impact upon our understanding of basic and applied research, in both decision sciences and information systems theory.

J. A. G.
J. A. R.

Introduction: A Comprehensive Overview

We shall not cease from exploration
And the end of all our exploring
Will be to arrive where we started
And know the place for the first time.
　　　　　　　　　—T. S. Eliot

The foundation for this books postulates that *smart data* and *smart data strategy* are essential elements for optimizing enterprise performance—the *most basic* executive responsibility. Smart data is a product of data engineering discipline and advanced technology. Making data smart enables a smart data strategy. Today, many organizations do not have a data strategy, much less a smart one, and the subject of data is not a part of executive's lexicon. Data should be at the top of executives' priorities when executing their responsibilities.

Joshua Cooper Ramo, Managing Director of Kissinger & Associates, advocates "resilience" [1] as the new strategic watchword. In computer technology terms, this means "the ability to provide and maintain an acceptable level of service in the face of faults and challenges to normal operation." [2]

U.S. President Barack Obama describes this as "fits and starts" for which adjustments are needed as better ways to improve things are discovered:

> We dug a very deep hole for ourselves. There were a lot of bad decisions that were made. We are cleaning up that mess. It's going to be sort of full of fits and starts, in terms of getting the mess cleaned up, but it's going to get cleaned up. And we are going to recover, and we are going to emerge more prosperous, more unified, and I think more protected from systemic risk.

Becoming more resilient and working through problems iteratively characterize the work ahead. Smart data and associated strategy are strategic to supporting these pursuits because nothing gets done without actionable data. Getting things done without the proper data slows down progress and makes management much less precise than it needs to be when resources are so constrained.

Government and commercial enterprise executives face a more complex world and a global economy; therefore, they require better strategies and the ability to plan, decide, solve problems, make sense, and predict with greater speed and confidence. Accessing and leveraging data is central to improving this ability.

While much attention is given to improving infrastructure for processing data and communicating globally, attention is deficient in improving the data itself: data quality, data characteristics, data exchange, and data management and security. Strategic focus be given to data because it will lead executives and their enterprises to more precise solutions and optimized performance.

Smart data is a consumer-driven, constituent-driven, investor-driven demand for higher enterprise performance from executives to use data smartly to manage more effectively; to make data smart through data engineering; to make enterprises smarter by adopting smart data strategy; to make infrastructure more responsive by employing smart data technology in an enterprise context.

On a recent visit to the doctor for a routine checkup, the doctor spoke about the effort to digitize patient records to meet a new national standard. He lamented that was difficult for the hospital to get disparate systems and sources of data to interface. He said the hospital pays hundreds of thousands of dollars for interfaces intended to achieve interoperability, although they are so brittle they rarely work for very long.

The doctor so clearly articulates the symptom of the problem, it surely must be pervasive among this and other professions, as well. The problem is a flawed data strategy that results in brittle interfaces being developed and maintained instead of a open and interoperable data exchange independent from applications.

The money spent on brittle interfaces would be better spent on direct patient care and lowering healthcare costs, for instance—one small practical example of where and how smart data can make a difference. Here, the doctor can't make a difference, but the head of the hospital can. The problem must be scoped and scaled, and understood by the enterprise level for action.

Smart data is our invention. The conclusion of this book presents our best description and notions but executives and information technologists must advance the concept, as we have done with specific examples and demonstrations. At a high level, smart data is applied as part of the systems engineering discipline. Yet, this subject is very much integral to enterprise management and therefore management and information science. Modern executives must have a hybrid command of these various disciplines.

There are six elements to this book:

The Introduction presents a *Comprehensive Overview* of smart data and smart data strategy.

Chapter 1, *Context*, presents the big picture including critical terms.

Chapter 2 presents the *Elements* essential to optimizing performance in the enterprise.

Chapter 3 discusses *Barriers* to optimizing performance and how to overcome them.

Chapter 4 presents *Visionary Ideas* including enabling technologies to help implement recommendations.

Chapter 5, called a *CEO's Smart Data Handbook*, presents an *actionable* summary format for the executive audience. An executive may read and reference this chapter without reading the balance of the book; however, a performance advantage is developed through the reader's complete grasp of topics presented here.

Whether you lead a commercial enterprise or a government enterprise, the one most important activity for which all CEOs and senior executives are responsible and accountable is *optimizing performance*. Optimizing performance means applying scarce resources to business processes under constraint, and transforming them into highest yield and best use outcomes by managing people and enabling technical mechanisms (technology).

All enterprises exist for a purpose, which is expressed as mission and value statements, goals and objectives, otherwise summarized into business plans. Once desired outcomes are identified, then leaders organize resources into functions. Functions identify the work that needs to be done to produce outcomes. How the work is accomplished is defined as processes, where process activities constitute proprietary differentiation.

Proprietary differentiation, or unique ways of accomplishing things, is achieved through a variety of means that begins with creative leadership:

- Selecting the right customers to service
- Attending all constituents and specific needs (government)
- Selecting the right things to do
- Organizing activities and designing work
- Attributing activities with application of a superior mix of people and technology
- Applying scarce resources in an optimal manner
- Structuring the balance of consequences such that doing the right things the right way is rewarded, and deviations are dissuaded
- Ensuring that customers receive valuable results
- Assuring stakeholders that the enterprise is performing optimally

Data is at the heart of each of these management activities which is why data as a principal contributor to optimizing performance is the focus of this book.

In today's global performance environment, economic and political factors affect the creation and flow of capital and materials moving commercial and government enterprise. These issues are addressed here to the extent that our focus and subject relates, as they do indeed; our immediate focus is bounded by the terms "enterprise performance optimization" and "smart data," as much as deviation from these terms is a temptation. Observing a new U.S. administration, President Obama advances a shift in government spending priorities that should result in a new definition of outcomes. President Obama's stated outcomes are [3]:

- Alternative energy production doubled in three years
- 75% of federal buildings modernized and improved
- 2,000,000 American homes made energy efficient
- 100% of all medical records computerized
- Schools equipped as twenty-first century classrooms
- Broadband access expanded
- Science, research, and technology invested made to advance medical break-throughs, new discoveries, new industries

This list can be vastly improved, and with more crispness added, by applying smart data thinking.

Smarter outcomes will be achieved from smarter data in the United States by achieving:

- Energy independence by a guaranteed certain date by the following means in specified percentage allocation.
- Full employment at specified % of unemployment by a guaranteed certain date.
- Increased manufacturing capability and production of more goods made in America by specified amount, by industry type with specified consumption and export targets.
- Increased border security at a specified percentage with a specified percent reduction in illegal immigration and specified deportation of illegal aliens.
- Military capacity to fight and win specified threats at specified levels with superiority.
- Access to healthcare of equal to or greater quality than currently available to all Americans at specified expense and in specified mix of public and private healthcare capacity.
- Specified amounts of investment in specified fields of research and development for the purpose of achieving specified goals and objectives.
- Reduced greenhouse gases by specified amount in specified time frame.

Now, you may say these lists require a lot more data to be specific, which is exactly correct. Let's take just one subject from Obama's list and compare it to the same subject with smart data thinking applied.

The Obama plan calls for "Alternative energy production doubled in three years." If alternative energy production is minuscule today compared with the demand, doubling it in three years may be insignificant or may be much less than what is needed in rate and volume of production to make a difference that realizes benefit for all.

Is not the desired outcome for the United States to achieve energy independence by a date certain by the various alternatives and allocated in specified percentages? Should not the quantities be based on known demand and facts about alternative development and availability, as well as competition?

Pressing for data facts and meaning is a part of the smart data strategy that begins with how the executive defines the outcomes. The Obama administration inherited obligations and has limits on discretion and department and agency processes and a government framework that is fixed by the U.S. Constitution. Knowing the precise starting position is essential.

What isn't fixed, and in fact is broken, is the financial system that provides capital for industry and government. Fixing this requires government and private sector collaboration among government and commercial enterprises in a global environment.

After taking office, President Obama immediately expressed dismay at the White House technology that is at least 10 years behind his usual functional level. His management toolkit is limited in desired functionality. While Obama projects a new executive skill set and charisma, he has said little about government performance data, and even the initial stimulus package intended to create 3 to 4 million jobs was described more qualitatively than quantitatively. This is symptomatic of a government operating without a strategic data focus. According to Lori Montgomery of the *Washington Post* [4]:

> In a new report that provides the first independent analysis of President Obama's budget request, the nonpartisan Congressional Budget Office predicted that the administration's agenda would generate deficits averaging nearly $1 trillion a year over the next decade— $2.3 trillion more than the president predicted when he unveiled his spending plan just one month ago.

Government is terrible at predicting revenues, expenses, deficits. . .everything!

Obama inherited this situation; but it is the administration's chance to do something about it by adopting a fundamental change in how it treats data as a strategic element in managing to optimize performance.

There are so many moving parts in an enterprise that it is unfathomable that modern executives would attempt to manage without techniques to keep track of them; but many do—a condition that is unacceptable in the highly automated world of the twenty-first century because resources are exceedingly scarce and risks are too high to operate intuitively.

For example, many top executives do not have a computer on their desk. They do not read or write their own e-mail. Instead, they ask a secretary to open their e-mail account, individually print a paper copy, and wait for a designated time until the top executive dictates each e-mail response. The secretary then sends out the dictated e-mail. This is not what Bill Gates meant whenever he wrote his book, *Business at the Speed of Thought*. This antiquated approach is counterintuitive to real-time decision making. Any top executive who answers e-mail only once or twice a day is by definition out of touch with reality. As an aside, this approach requires the secretary to work at the office on weekends, in an attempt to divert the deluge of e-mail responses that would ordinarily build up over this two-day hiatus. Heaven help whoever has to clean up the mess, whenever the top executive travels. Under these circumstances, perhaps she dictates the responses over her cell phone? This is not a smart data strategy worth emulating. Rather it is analogous to one aspect of the problems being addressed by this text, albeit the simplest.

One of the greatest barriers to realizing the full benefit of automation occurs when people insert themselves in unnecessary manual intervention, usually unintentionally. They need and want data to perform better, though it can't because they did not plan and prepare sufficiently for the moment of need. Incoming executives often inherit a data planning and preparation deficit situation.

For enterprises to perform optimally, executives must insist on better data planning, preparation, and engineering. Recent fiascos in commercial and government enterprise are largely a result of not paying attention to data that is essential to accurate anticipation and optimal management.

- Not paying attention to the quantitative impact of rules and regulations
- Not paying attention to the end result outcomes

The trial-by-error approach employed by the U.S. government to right the economy is a symptom of the absence of smart data strategy. The imprecision by which various stimulus packages were presented and pursued is symptomatic of flawed enterprise management for which outcomes are imprecisely defined and the relationship between stimulus and results is too ambiguous.

As Joshua Cooper Ramo said on MSNBC's *Morning Joe*, "There is no American strategy." This is not only true for the national government but for many U.S. companies that are so embroiled in the moment, trying to stay alive, causing investments in research and discovery to shrink therefore starving the future of opportunity and resilience. They are future data-starved.

A hierarchy of reasons exists why executives don't pay attention to essential data including:

- Governments and private sector financial institutions turning off regulations that would make performance data transparent
- Corporations and financial institutions electing market manipulation as a strategy to make money at the expense of the real economy and satisfying real consumer needs
- Corporations and governments failing to address data facts about resources and the environment that have direct bearing on goods and services production, and product and services characteristics being optimally suited to consumer needs
- Corporations, governments, and individuals ignoring data facts describing associated risks about their mutual transactions and responsibilities

- Corporations and governments not assessing their capacity for change and improvement that is surely limited
- U.S. government not accurately evaluating the competitive playing field in a global market environment with respect for other government competitive factors and relationship to commerce

A simple way for keeping track of essential things needed to optimize performance can be employed. One is a technique employed by the U.S. Department of Defense (DOD), some other government agencies, and commercial enterprises called IDEF.[1]

IDEF is a mature modeling technique that provides a standard for defining how work gets done so that processes can be compared as apples-to-apples. It is easy for executives to learn and employ this technique used to understand enterprise performance and to express expectations for improvement with sufficient precision. Having the ability to compare processes is essential to:

1. Changing from current performance to expected improvement,
2. Integrating processes either internally or externally, and
3. Enabling continuous improvement.

The IDEF process modeling technique is preferred because it allows executives to track the relationship of critical elements to activity performed by the enterprise to produce or accomplish its outcomes.

A number of techniques that are a part of a Business Process Modeling (BPM) and Six Sigma, for instance, are useful for answering certain questions about performance. That is the point: different modeling techniques address different needs. Our focus is the executive, although we must consider what enabling technologists need to support them.

Significant effort has been made by government and commercial enterprise to improve process designs, although too often, work is applied to modeling processes and changing processes with insufficient attention to measuring the results. In government, for instance, by the time a process is modeled, the executives have moved on never to see changes implemented, much less the outcome.

The Government Accountability Office (GAO) continuously reports problems and suggest ways to improve them, even though there is no mechanism to ensure that GAO recommendations are followed. Often, the horse is out of the barn by the time the GAO report is made. Lost investment of time and resources cannot be recovered. Smart data strategy can help the GAO leap beyond the current trap; that is, if the U.S. Congress also gains a grasp of smart data and the associated strategy.

---

[1] Wikipedia states IDEF (*Integration DEFinition*) is a family of modeling languages in the field of software engineering. They cover a range of uses from function modeling to information, simulation, object-oriented analysis and design and knowledge acquisition. These "definition languages" have become standard modeling techniques.

PREDICTIVE MANAGEMENT

The smart data discussion and strategy place greater emphasis on having the ability to predict, prevent, and mitigate problems before they happen; this is called *predictive management*. Foreseeing the future with a higher degree of certainty requires:

- Knowledge about history
- Knowledge about the present situation
- Knowledge about developing new technologies
- Knowledge about evolving future needs and trends
- Knowledge about competitive threats
- Knowledge about capacity for change and improvement

All of these things are addressed by data that describe how things get done and their associated metrics. In anticipating the future, data is needed that addresses possible scenarios for which probabilities are determined about their possible occurrence.

Data is made useful through applying various analytical techniques, methods, and algorithms. The smart data paradigm encourages executives to press for better data and best methods and algorithms to support their requirements.

Information technologists employ a host of techniques to design and develop complex systems and software. UML is a popular family. For executives, we return to IDEF because it is the simplest way to address critical elements that includes accounting for enterprise data in context of process performance. By applying IDEF process modeling and other data modeling techniques, executives can visualize an operating enterprise at various degrees of aggregation or decomposition (detail).

Other modeling techniques are employed as well and our approach is not to advocate any particular set, but we choose a certain set as it suites our aim for clarity. Some of the techniques used in this book may inspire you to investigate the tools that are or should be a part of your *enterprise performance optimization portfolio*, which is discussed in a later chapter.

At the start of his administration, George W. Bush issued *The President's Management Agenda* [5] in which he said: "Government likes to begin things—to declare grand new programs and causes. But good beginnings are not the measure of success. What matters in the end is completion: performance and results. Not just making promises, but making good on promises." There is merit to this, though the impatient executive may have missed an important aspect of management.

President Bush focused on outcomes where his predecessor President Clinton emphasized the importance of how things get done, or process. Daniel S. Appleton, a leading IT guru, often said that "the outcome has more to do with the process than the objective." That is, one can identify what needs to be done, but until one knows how, the desired outcome remains elusive. This view is collaborated by systems theory: Inputs are processed into outputs. Inputs are then changed by feedback, from the outputs, thus changing the process.

$$(\text{Inputs} \rightarrow \text{Process} \rightarrow \text{Outputs} \rightarrow \text{Feedback})$$

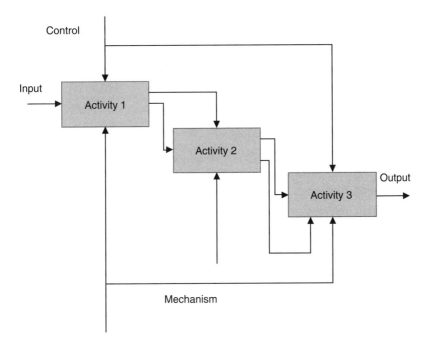

Figure 1 IDEF Inputs, Controls, Outputs, Mechanisms (ICOM).

President Obama stated that to achieve his vision, "I will encourage the deployment of the most modern communications infrastructure. In turn, that infrastructure can be used by government and business to reduce the costs of health care, help solve our energy crisis, create new jobs, and fuel our economic growth. And an Obama administration will ensure America remains competitive in the global economy." [6]

Smart data and smart data strategy are strategic contributions toward this aim. Emphasis here is on enabling mechanism and technology investments by government. Although still unexplained is how performance will be improved and certainly how explicit measurable outcomes will be defined. Enterprise board members, shareholders, and government constituents (citizens) must demand more from executives. Accountability and credibility should be measured by saying not only what will be accomplished, but how it will be accomplished. Answering "how-to" question is the truth detector for senior management in commercial and government enterprise.

Figure 1 considers the power and use of these simple terms: Inputs, Controls, Outputs Mechanisms. This technique can be employed to define all processes when processes are three to six activities. Each activity can be broken down into sub-activities if needed in defining very complex processes. Each of these terms will be defined here, although suffice it to say that a simple technique is needed to define and explain complex enterprise performance.

The DOD employed this technique to model the defense enterprise, which serves as a source of best practices in this regard. Application of such an approach can be employed to model parts of an enterprise or all of the enterprise whereby having a standard for modeling enables integration of individual pieces. Applying the technique to all or any part of the enterprise, executives can answer such questions as:

- What is the intended output or outcome?
- How will the outcome be produced?
- What is the process or activities?
- What resources are required as inputs?
- What people are required to perform the work?
- What technologies are required to perform the work?
- What are the controls and constraints on performance?
- At what rate, frequency, quality, cost, and schedule are work to be performed to deliver required outputs?

By attributing activities with cost and time metrics, executives can see how much cost and time is consumed by activities to produce intended results. In addition, quality attributes can be associated with activity such as volume, rate, frequency, and other specific measurements. Managing with smart data and smart data strategy enables transparent accountability.

Throughout this book we discuss government and commercial enterprise because they are inextricably linked. Some may wonder why we do not include consideration of various other global financial community players who greatly influence the availability of capital and other resources, and who act without transparency to manipulate markets; our answer is that this topic is outside the scope of this text.

However, in the instance of the Bernie Madoff case, (a $50 billion Ponzi scheme) whistleblower Harry Markopolos' testimony revealed that a CPA reading SEC filings would have noted:

1. Only one CPA was auditing an $80 billion company;
2. The CPA firm was too small and had no other clients; and
3. The straight-line ROI was implausible.

Markopolos said that the SEC regulators were too inexperienced to catch the problems and lacked incentive to do so. In this case, if the SEC had adopted a smart data strategy and equipped regulators with intelligent tools that red flagged this type of problem (and those much more complex), this scheme could not have been accomplished.

IDEF LEXICON FOR EXECUTIVES

The following definitions are consistent with other authoritative references:

Activities. Actions performed by people and machines, usually in combination, to produce certain outcomes, products, or accomplishments. Activities are used to define and describe processes.

Inputs. Resources such as capital and materials that are consumed or transformed by the activities that manifest in value-added outcomes with associated accounting for time, cost, and quality.

Controls. Constraints on activities imposed by sources of regulation. Sources of regulation may be identified as organizations that issue controlling laws, regulations, policies, guidance, plans, budgets, and schedules that define expectations for how work is done, and their accomplishments and attributes. Controls may be self-imposed as in the case of business rules, for instance or may be externally imposed as in the case of laws and regulations.

Outputs. All results that may be identified as accomplishments, products, assets, cost, and time consumed. Outputs can be expressed as positive values or negative values. In the production of goods and services, outputs will be a combination of positives and negatives with the difference being gain, profit, or value.

Mechanisms. Methods or procedures that enable work to be done as defined by processes that include people and machines, i.e., associated jobs and organizations as well as specific equipment, hardware, and software (infrastructure).

The precision with which executives and managers lead organizations to produce outcomes represents the enterprise brand and quality of management. Management is responsible for designing the work, needed by the enterprise, to achieve prescribed outcomes.

Mature organizations comprise a collection of processes, people (organizations or bureaucracy), and technologies, otherwise known as systems, that evolved to their present state. The collection of historical elements is referred to as legacy. Legacy assets have varying degree of usefulness. Some legacy assets have long life cycles; while others have short ones. Life cycles may be extended through upgrade and change. Legacy consumes resources to sustain and maintain it. Legacy elements with depleted usefulness may represent targets for retirement or for footprint reduction.

This book was envisioned a year before the Obama inauguration in order to address the following main idea and supporting points:

Main Idea

Smart Data describes data that have been engineered to have certain superior characteristics which are a product of state-of-the-art data engineering discipline so that the data are interoperable and readily exchangeable among qualified members of an enterprise user community. Smart data is an executive strategy and tool for exacting higher performance enterprise-wide. It operates leveraging state-of-the art infrastructure and enterprise services. Smart data contributes to optimizing enterprise performance by providing a strategy that will vastly improve enterprise data resource management reflected in more efficient and cost effective IT support, producing high impact results for executive and operations management users.

Supporting Points

1. Optimization of enterprise performance whereby enterprise data resource management is owned by executives and operations management users and not relegated to enabling technologists and information systems specialists.
2. More efficient and cost effective IT support performance is managed and evaluated by executives and users and directly correlated with results.
3. High-impact results are engineered from the beginning through collaboration among executives, operations management, and enabling technologists.
4. Measurements are meaningful and expressed in ways that tie directly to optimizing enterprise performance.
5. A strategy and detailed plan is needed that includes roles, responsibilities, and accountability. For the government, the plan must be understood by ultimate citizen consumers. For commerce, the plan must be understood by employees, stakeholders, partners, investors, and ultimate consumers.

Today, enterprises are swamped in a sea of data and their vision and goal attainment are impaired by a fog of unwanted and unnecessary costs. In addition, often infrastructure-processing and communication infrastructure cannibalizes operational budgets to the point that more resources are consumed by IT than by the direct mission. This observation is substantiated by remarks by U.S. Army CIO LTG Jeffrey Sorenson:

> While our purpose remains true, we were eclipsed by events. The global economy collapsed and caused us to reprioritize our argument. As a result, our case and appeal to our target audience of senior executives is strengthened. We were on the right track, though may have underestimated the importance of our message.

Enterprises—government and commercial—are confronted by a failed financial system, one that is overburdened by debt, confounded by regulatory deficiencies, and encumbered in production of capital needed to fuel commercial enterprise from which government enterprise raises taxes to operate. Government finds itself in a position of needing to correct deficiencies while stimulating economic growth.

Our priority is improving enterprise management performance through smart data strategy and improving IT support performance.

ORGANIZATION OF THIS BOOK

Chapter 1 present the big picture and introduce terminology and ideas that describe the relationship of data in the context of government and commercial enterprise. Data are of paramount importance in this chapter and a paradigm introduced is called the Service-Oriented Enterprise (SOE) that integrates topics as a basis from which we segregate and segue to focus on data in context with all the other moving parts. Service-Oriented Enterprise implies that enterprises are about the business of delivering services (including products) that satisfy customers' or constituents' needs. Service-Oriented Enterprise is enabled by smart technologies.

IBM recently launched campaigns called, "Change for Smart's Sake," and "Smart Planet." Throughout this book, these themes certainly complements our ideas about smart data, although our intense focus addresses more than the general theme that the world needs to imbed smartness in products and services. Smartness begins with how executives optimize enterprise performance with application of data, methods, and algorithms.

Case studies are used throughout the book to provide evidence and support for our ideas. This manner of presentation is challenging because some of the issues, such as the SOE paradigm, are long and complex. Incorporating cases makes it even more complex, though very rich in content and proof. The boundaries of the case stories are segregated from the main text so that you will not get lost, or if you do get lost, you will find your way again.

Admittedly, some of the cases are quite detailed. This is especially true for the case of demonstration methods and algorithms wrapping data. The cases themselves are technical in nature, which makes for weighty reading; but they are included because ideas with scientific and academic rigor are appreciated by that some readers.

We live in a world of lexicons; a central fact when discussing "data." Every industry has its own vocabulary, as does every profession and vocation. Today, communities of participants introduce "buzzwords" which are important because people use them as a shorthand or notation, with precise meaning for members of the club. So, as members of our club, we are deliberate in defining terminology throughout.

In business life, the challenge remains to accommodate the natural propensity to form clubs and to speak with uniqueness, while seeking to be understood by other clubs. It is fashionable to call these clubs, *Communities of Interest* (COI). Often COI are supported by social media, which are mechanisms for socializing ideas in addition to other meanings to be discussed here. Social media add dimension to the date environment as well as depth to the data itself.

The SOE is not to be confused with the popular Service-Oriented Architecture (SOA), although SOA is a part of SOE. SOA is an architectural design that uses technologies to enable long-term cost savings, adaptation of agile processes, and enterprise integration.

Government customers procure products and systems having useful life spans that often outlast their suppliers. Sustaining these items is dependent upon data whose life cycle often outlasts the information technology (IT) paradigm under which it was created.

For these reasons, smart data strategy advocates embracing openness and interoperability made possible by a combination of smart application of standards and technology that accommodates natural diversity. It discourages data being held hostage by proprietary software.

In what we characterize as the "old paradigm": People (organizations, corporations, enterprises) seek harmony and integration by imposing rigid standards, forcing strict compliance. Compliance was implemented in hard code, computer terminology resulting in brittle, difficult, and costly to maintain interfaces which would enable diverse and disparate applications to communicate.

George spent nearly 10 years facilitating Electronic Data Interchange (EDI) meetings between the Aerospace Industries Association (AIA) and DoD for the purpose of harmonizing standards and implementation conventions. At the very first meeting, an executive from Hughes Aircraft Company said, "What you need is a universal translator." The executive had identified the solution in the very first meeting. The second meeting presented the idea to exchange data employing a common data model, similar that employed for product data management on the B-1 bomber at Rockwell International (now Boeing). Yet, this idea was overpowered by those wanting a rigid standard for data exchange, resulting in thousands of hours and years of time devoted to maintaining brittle standards, when an agile alternative was a better way.

Why do these things happen? The executive was insightful with the right idea, yet no other executives joined him or had the staying power to advance a concept that would save much time, labor, and money.

Part of the resistance was cultural. Aerospace companies at that time were reluctant to share best practices that included processes and technologies. They were unaccustomed to collaborating in approaching electronic commerce—a new initiative in the early 1990s. It look 10 years for aerospace, automotive, and other industries, including healthcare, to realize that industry-wide strategies are desirable.

The smart data paradigm advocates substituting application of new technologies with a strategy that permits organizations to operate with high degree of independence and natural diversity whereby data exchange is accomplished through a neutral exchange utility. This is just one aspect of smart data thinking, which pertains to a means of technical achievement.

SMART DATA IN THREE DIMENSIONS

Our ideas about smart data include three dimensions:

1. Enterprise Performance Optimization Context
2. Interoperability Technology
3. Data-associated Methods and Algorithms

Both government and commercial enterprises generate massive volumes of data in their day-to-day operations, little of which is directly useful for nimble and wise management.

However, if these same organizations adopt the SOA paradigm in concert with smart data, they can enjoy benefits of procedure, standardization, visibility in decision-making, agile adaptation to change, interoperability of legacy systems, reduced costs through elimination of redundant services for redesign of existing business processes.

Corporations serving government customers must address unique customer needs, for which data is a core requirement. High visibility problems, such as the U.S. Air

Force sending nuclear fuses instead of batteries to foreign government customers, very likely have data deficiencies at their root cause. These problems do not stem from a lack of data; they stem from a lack of actionable data. This fact remains in spite of years of advancement in the technology of data. The technology of data is not lacking. What is lacking is an emphasis on data as a corporate and enterprise resource.

Service-Oriented Enterprise is a paradigm in which data-focused strategy can flourish when accompanied by the SOA that has been implemented enterprise-wide. Although, the SOA can be implemented at the department level or within a single application or functional domain, the integration of individual SOAs into an enterprise-wide philosophy leads to the SOE paradigm shift.

Don't confuse talk about information technology infrastructure and software applications as a substitute for straight talk about data. However, it is necessary to understand the relationship among smart data, smart grid, and smart enterprise services as the context for developing a data focus. These relationships can exist in government and industry settings because progress is being made today in both.

Today's progress is often invisible because it is applied in such narrowly applied spheres. For instance, Sarnoff Corporation addressed difficulties in integrating battlefield data into a cohesive picture because the sources of visual surveillance data and facts were varied and nonstandard. Employing a combination of metadata management and application of standards, war-fighters are now able to leverage all the actionable data available to them, resulting in a dramatic reduction of improvised explosive device roadside bomb casualties.

Sarnoff's technical solution was an application of smart data strategy at a local level, providing proof that the technology works. Now, the challenge is for the Army to extrapolate and aggregate the approach to exceedingly higher echelons in the military enterprise. Herein lie a problem and challenge for the Army CIO to connect the dots and to apply a local solution to an enterprise-wide paradigm shift. A higher strategy to guide and pull together individual accomplishments is needed.

New government leadership with new priorities demands higher performance from government enterprises and from the commercial enterprises supporting government:

- Budget pressures continue to demand higher return on information technology investments.
- Defense, Homeland Security, National Security, State and other Government Agencies including Health & Human Services departments each extend their already large enterprises increasing dependence on supply chain participants further into the international community.
- Federal government strategy stresses best commercial practices and commercial off-the-shelf technologies with the intent of achieving best value in its conduct of the people's business.

Therefore, our audiences comprise commercial and government businesses, their executives and enabling technologists.

What is needed is a strategy that traverses the enterprise from top to bottom, with breadth and depth to leverage automation technologies to dramatically optimize enterprise performance. Attention must be paid to smart data strategy as an organizing element of enterprise performance optimization.

A number of popular enterprise performance improvement strategies exist today, for instance, Six Sigma. According to Smith and Blakeslee [7], "Because Six Sigma methodology, at its core, relies on the use of factual data, statistical measurement techniques, and robust feedback mechanisms to drive decision making, it's able to unify top leadership teams behind a common language (and set of data points), making strategic planning and execution more efficient and successful."

Smart data as a corporate priority can revolutionize government or commercial enterprise performance much like "Six Sigma" or "total quality" as organizing paradigms have done in the past. The difference is that focus on data is most direct, and establishing high value at the top is paramount for exacting higher performance from subordinate strategies.

This revolution has not yet taken place because data historically resides in the province of the information resources organization. Solutions that render data smart are articulated in technical terms versus the language of the board room. While books such as *Adaptive Information* by Pollock and Hodgson [8] ably describe the current state of the art, their necessarily technical tone is not conducive to corporate or agency-wide qualitative change.

Pollock and Hodgson promote frictionless information whereby "Autonomic computing concepts will drive strategic technology development in a number of industries and software spaces" [8]. They mention progress by the U.S. intelligence community in examining semantic interoperability architectures, for instance.

Smart data is interoperable by leveraging open attributions about authorized use and meaning. What makes the data "smart" is the combination of interoperable characteristics operating in a "system of data exchange" that is agreed upon by the user community for which the premise is to map various data through a "neutral mechanism for exchange" to any other user that may have different characteristics.

Note the use the term "mechanism" in contrast with the term "standard." This illustrates that the exchange utility is not intended to be an inflexible and brittle exchange standard. Rather, it is to apply information technology that supports adaptive information and flexible interfacing.

BUSINESS RULE

The business rule for smart data would say something like the following: Make data exchangeable enterprise wide by requiring members to map data from their unique environment to a neutral mechanism for exchange. If a member is already employing the neutral exchange mechanism, then it would not be required to do anything. If a member has a proprietary or different application standard, then it would map data one time to the neutral exchange after which exchange would be automatic.

Since members of the community are likely to have some of the same proprietary or other application formats, when one enterprise member performs the mapping, with agreement, others may leverage it. Under a smart data strategy, enterprise participants collaborate to address such diversity from which participants would accelerate resolution and achievement of automated exchange. Efficiency is a logical outcome. The intent is to greatly reduce the amount of labor intensity required to support data exchange, while greatly increasing actionable information for executive use.

The neutral exchange translator affords robust utility for enterprise participants, for which various options regarding deployment exit, ranging from dispersed exchange servers to centralized services and combinations thereof depending on the unique needs of member participants and their respective enterprises. Participation in supply chains has provided the organizing force for similar challenges, and that may serve application of our ideas as well.

The *smart data system* permits a high degree of freedom, agility, and adaptability among participants of the user community independent from their unique environments or legacy systems. Unlike "standardization" that requires dumbing-down data and rigorous compliance to standards, the smart data system allows for natural behavior among free enterprise participants in a global environment. Technology mitigates variable behavior and assures accurate use and interpretation among disparate users. Technology mitigation makes data smart.

Another view of smart data advanced here is that it incorporates attention to methods, heuristics, and algorithms to use data in support of executive needs. This description is in addition to preparing data for exchange as it addresses aligning data to specific executive uses. Smart data is asymmetrical, having dissimilar appearances, and performing in vertical and horizontal dimensions throughout the enterprise. In this regard, smart data is agile.

Smart data provides:

- The ability to make more timely and responsive enterprise decisions.
- Increased agility of data which is more responsive to change, lower IT cost to change, and
- Provision of the foundation for enterprise performance enhancement—end-to-end process optimization, continuous optimization, etc.

For example, an important concept is the optimization of enterprise performance, where enterprise data is managed to optimize capital budgeting problems within the organization. It is important for managers to use smart data and an SOE vision to eliminate and consolidate redundant services to reduce the costs of maintenance and operation. Managers need smart data and a methodology to both phase out legacy systems and upgrade their SOA. As a result, these lower operation and maintenance costs provide funds that can be applied to building new services.

CASE STUDY: IT CAPITAL BUDGETING USING A KNAPSACK PROBLEM

An example of more efficient and cost effective information technology (IT) support performance is demonstrated in the case example *IT Capital Budgeting using a Knapsack Problem* [9]. The knapsack problem addresses combinatorial optimization.

It derives its name from the maximization problem of choosing the best essentials that can fit into one bag to be carried on a trip. Given a set of items, each with a weight and a value, determine the number of each item to include in a collection so that the total weight is less than a given limit and the total value is as large as possible.

Presenting this case utilizes the IDEF lexicon. The activities are proposed heuristic solution procedures. The inputs are the data for the simulation. Controls consist of the performance of the two ranking methods. Outputs are the higher after-tax profits. Finally, the mechanism is the simulation method used to solve the problem.

In SOE, much of the smart data is heuristics, algorithms, and mathematical methods to implement these improvements. For example, the capital budgeting problem is an activity that decision makers face when selecting a set of capital expenditures that satisfy certain financial and resource constraints. In this example an information technology capital budgeting (ITCB) problem is described and it is shown that the ITCB problem can be modeled as a 0–1 knapsack optimization mechanism. Two different simulated annealing (SA) heuristic solution procedures are proposed to act as controls to solve the ITCB problem. Using several simulations, the performance of two statistical analysis (SA) heuristic procedures are empirically compared with the performance of two well-known ranking methods for capital budgeting.

Output results indicate that the IT investments selected using the SA heuristics have higher after-tax profits than the IT investments selected using the two ranking methods. Although the development of the algorithm is itself very important to the decision-making process, it is useless without the integration of smart data into the organizational memory and learning, so that it can be incorporated into the SOE as a distributed module that is the standard used whenever capital budgeting problems are encountered.

Addition of the SA module into the SOE would improve the standardization of decision making, by alerting the decision makers that this methodology exists and gives superior results, under the conditions described in the case.

Smart data also addresses the concept of better enterprise decision making, which is timelier, more responsive, and has improved visibility. Smart data and an SOE philosophy leads to improved visibility for decision makers by enabling rapid integration into the portals of the SOE by providing real-time monitoring that aids business decision making.

As stated previously, the context of smart data is to optimize enterprise performance, provide more efficient and cost effective IT support, engineer high impact results, make meaningful measurements, and develop a strategy and detailed plan that includes roles, responsibilities and accountability. Distributed computing is a mechanism that enables the smart data paradigm and the supporting points listed previously.

A smart data philosophy embraces the concepts of distributed computing properties. One of these properties is openness, where each subsystem is continually able to interact with other systems. While the goal of a distributed system is to connect users and resources in a transparent way, this must be accomplished by overcoming the challenges of *monotonicity* (once it is published, it cannot be taken back), *pluralism* (different subsystems include heterogeneous and possibly conflicting information), and *unbounded nondeterminism* (different subsystems can come online and go offline).

CASE STUDY: BETTER DECISION MAKING: FIELD TESTING, EVALUATION AND VALIDATION OF A WEB-BASED MEDWATCH DECISION SUPPORT SYSTEM (MWDSS) [10]

An example of better decision making through the adoption of SOE principles, smart data and the application of distributed computing principles can be seen in the case titled, "Field Testing, Evaluation and Validation of a Web-Based MedWatch Decision Support System (MWDSS): An Empirical Study of the Application of Mobile Tracking of Epidemiological Disease Trends and Patient Movements in a Military Scenario" [9]. This case shows through the following evidence, how the use of smart data can improve enterprise decision making: The MedWatch decision support system (MWDSS) is a promising mobile medical informatics technology approaching maturity.

As such, the MWDSS has potential to be an intricate component of the DOD SOA and an important module within the SOE. MWDSS was originally commissioned as a Web-based mechanism tool for medical planners to facilitate resource management by medical facilities and to track disease trends and patient movement. It is a system module that can interact with other modules in an open and transparent manner. Given the present military activities and climate, automated medical surveillance capabilities are valuable mechanism tools for medical support of the armed forces. MWDSS promises to enhance the output of medical support accomplishments for the military. The development of MWDSS is consistent with DOD directives, controls, regulations and policies which have identified medical surveillance as important for maintaining force readiness.

It is easy for executives to learn and employ the technique used to understand enterprise performance and to express expectations for improvement with sufficient precision. In this case, top military executives have the ability to compare processes that are essential to 1) changing from current performance to expect improvement by adopting the MWDSS, 2) integrating medical processes either internally or externally, and 3) enabling continuous improvement for medical surveillance.

This case documents the results of the inputs of a subject matter expert (SME) survey, conducted to evaluate the MWDSS software. Thirty-nine SMEs agreed to undergo any necessary input training to use the software and provide feedback on its performance during the simulation.

Outputs of the simulation varied. While some users did not require follow-up training, others required some refresher training to perform certain functions. SMEs for this exercise were able to quickly and successfully train to use MWDSS and were satisfied that it performed most of its functions well. They also stated that it could be a useful tool in the SOE as well as very useful in understanding the common operational picture.

The transparency of the system provided features that would allow users to obtain, evaluate, and present information more efficiently than previous methods. The MWDSS provided concurrency and helped to interrelate with current programs so that they are seen as similar and not distinct subjects. For example, the epidemiological data is viewed in conjunction with the other military information in a database-centric common operational picture.

The MWDSS case contributes information regarding optimizing enterprise performance with a strategy that vastly improves enterprise data resource management reflected in more efficient and cost effective IT support that produces high impact results for executive and operations management users.

Smart data describes data that has been engineered to have certain superior characteristics, such as being flexible and current and is a product of state-of-the-art data engineering discipline, such that it is interoperable and readily exchangeable among qualified members of the military enterprise user community.

Smart data as demonstrated in this case, is an executive strategy and tool for exacting higher performance enterprise-wide. MWDSS operates by leveraging state-of-the-art infrastructure and enterprise services. Overall, participants indicated that MWDSS had significant potential utility for mobile medical informatics, especially by utilizing a shared database. Some of the SME's reluctance to accept new mobile medical informatics may be viewed through a cognitive perspective.

The research drew on personal construct theory (PCT) as a mechanism to argue that SME's reluctance to use new mobile medical informatics are influenced by inability to change established group personal constructs, related to information systems development and delivery activities.

In this scenario, the smart data and SOE philosophy along with distributed computing mechanism allowed for a method of communicating and coordinating work among concurrent processes. The MWDSS used web protocols and a shared, database-centric architecture to deliver a SOE environment that stretches across the globe. The case study also demonstrates outcomes, such as the reluctance of some end-users to embrace the smart data and SOE philosophy due to their inability to change their group personal activities.

As stated before, smart data also provides increased agility—more responsive to change, lower IT cost to change, continuous optimization, etc.

By adapting flexible standards that come with smart data in an SOE setting, implementers can rapidly adapt to changes as their business needs change. Business dashboards have become an important Business Intelligence (BI) mechanism tool available to managers and executives to monitor various aspects of organizational functioning including continuous performance monitoring. In general BI systems are an SOE component and represent smart data driven-decision support systems (DSS). Similar to a car dashboard which shows current trip status and monitors vital car statistics, business dashboards can offer real-time information about the organization and aid in better managerial decision making and agility. SOE application modules are utilized to investigate production, sales, financial, supply chain, and other business data to make output comparisons to other companies' performance.

Business dashboards are a fairly recent phenomenon and, as a result, offer several challenges and opportunities. Our research examines characteristics of dashboards as a tool that should have a unique place when compared to other BI mechanism tools, in the smart data, SOE paradigm. Automated systems have increased the amount of data, but application of SOE methods for utilizing smart data are necessary if this data analysis is to be used for improved long-term decision making. Dashboards are one method for increasing agility so that decision makers can be more responsive to change, lower IT costs, and provide the foundation for continuous optimization.

Issues related to dashboards are discussed in this book and some suggestions and recommendations for their importance in the SOE paradigm are put forward. A post-implementation survey for a dashboard implementation was used, which was conducted in a regional North American bank, to support the issues and recommendations proposed in our research and to provide output results.

This case demonstrates that the SOE philosophy embraces the concept BI and the use of information as a vital resource in decision making and DSS. The BI mechanism component of the SOE is used to gather data into data marts, which are then placed in a data warehouse. SOE modules are developed to extract, analyze and report the output results of this information, in order to make smart decisions, using smart data.

The challenge to make smart data strategy an explicit element of enterprise performance optimization is persistent in part because its meaning has not been made explicit and understandable among layers of owners and user participants. The gap between enterprise executives' understanding and that of enabling technologists must be closed to maximize return on information technology.

ENGINEERING AN UBIQUITOUS STRATEGY FOR CATALYZING ENTERPRISE PERFORMANCE OPTIMIZATION

Computers are incredibly fast, accurate, and stupid: humans are incredibly slow, inaccurate and brilliant; together they are powerful beyond imagination.

—Albert Einstein

Short Fuse and Attention Span

More than 10 years ago, an executive from Hughes Aircraft Company addressed an audience of Aerospace Industries Association EBusiness Working Group members after listening to various descriptions of the problems with harmonizing electronic commerce standards and implementation conventions. He said, "You need to develop a universal translator." With that, he arose and left the meeting never to return again. Either he thought that he had provided the right direction and his mission was accomplished or he anticipated the insurmountable difficulty in doing this and left in great frustration.

Thorough Process and Longer Attention Span

Training and testing for the National Center for Manufacturing Sciences (NCMS) *Achieving Manufacturing Excellence Program* was developed to address the needs of small business to participate in their customers' formal improvement initiatives, such as the Malcolm Baldridge National Quality Award, and proprietary programs from major automotive, aerospace, and defense manufacturers (13 different programs at the time of development). The NCMS team objective was to produce one self-assessment that would satisfy all the requirements for self-assessment from all 13 "top quality" programs. Performing one self assessment is work and investment enough.

To accomplish this, NCMS identified 14 areas of excellence accompanied by 1400 common questions for which evidence (data) was required for compliance. This experience illustrates the power of self-assessment based on the requirement to produce evidence of compliance with stated improvement criteria.

Since then, the supply chain improvement concept has evolved and become more refined, focused, and sophisticated. Anyone addressing performance improvement in a large enterprise, such as the U.S. Defense Enterprise, is humbled by the complexity. How in the world can individual leaders of organizational elements of a complex enterprise make a difference?

By developing and advancing a strategy engineered to produce desired results under constraints; one that is immediately relevant to participating organizations whereby the local benefit is immediate and the overarching benefits aggregate. Elegance and effectiveness come from simplicity.

Critical to the success of enterprise-wide strategy is CEO ownership in private enterprise and Secretary ownership in a government enterprise. An enterprise organization, such as the Department of Defense led by the Secretary of Defense, comprises a vast network of commercial entities that are held together by agreements that consultant Daniel S. Appleton described as *contingent claim contracts*. The first level below the top is a hierarchy of relationships whereby prime contractors, subcontractors, and suppliers work together, and where each represent their own unique enterprise.

Leaders in the top positions of any organization command, lead, and integrate with the following elements:

- Power
- Time Constraints
- Focus
- Attributes
- Outcomes
- Processes

Power translates into value-setting, business rule making, resource allocating, and balancing consequences, that is making certain organizational and people participants clearly realize the difference between positive behavior and deficient performance.

Time constraints force the identification of milestones, metrics, and outcomes in a meaningful period with associated metrics such as rate and frequency.

Focus describes the domain in which change and improvements are to be realized, lending to the formation of themes.

Attributes are the detailed characteristics that describe desired process and associated performance metrics, such as cost and quality, in addition to previously mentioned budget (resource allocation) and time.

Outcomes are the end results that include everything from physical products, assets including information assets, and performance results.

Processes for management are different from other enterprise processes in that management processes are what executives use to carry out their responsibilities. Management processes can be a source of proprietary difference, though there is a baseline common among all enterprises and shared by all management. Management processes are driven by data and produce data some of which is company private, though much of which is the source of transparency.

Now, let us apply this general consideration to developing a strategy that will address the following:

- It is generally accepted that "integrating" the elements of an enterprise is a good thing. The question is what do you mean by *integrate*? What are the attributes of integration that produce corresponding benefits?

- It is generally recognized that "information" (information = data facts + meaning, and information + context = knowledge) is an asset that can sometimes become noise.

- It is generally accepted that information technology (infrastructure and software) has evolved into something that is useful and essential while costly, unwieldy, and unmanageable.

- It is generally accepted that "interoperability" among data, processes, interfaces, applications, taxonomies, policies, and social networks is highly desirable, including the latest, semantic interoperability.

The highest order activity should be shared by all enterprises independent of organization type with a role to "optimize performance." Enterprise data is a most treasured asset and that strategy focused on improving enterprise data is of equally high importance and aligned with enterprise performance optimization.

A smart data strategy is intended to act as a primary catalyst for enterprise-wide change and improvement.

It is a well-known axiom that "knowledge is power". As it relates to commercial or government enterprise operations, knowledge, in the form of accurate, complete and timely information, appropriately applied to the operational context at hand, can greatly enhance enterprise operations enabling, for example, optimization of enterprise operations or increased enterprise agility.

In spite of this axiomatic and demonstrated fact, much of the information generated, managed, or used in the progress of enterprise operations is not yet treated as enterprise (corporate) knowledge. For the most part, it remains in the practical possession of the operational subunits of the enterprise, rendering it difficult to extract for enterprise optimization purposes. Worse, the advent of extensive enterprise IT assets, such as enterprise resource planning (ERP), has further locked down corporate knowledge in the domain of the IT department.

Whether corporate or government, most enterprises employ extensive cadres of analysts. The majority of analyst time is spent on extracting and interpreting enterprise data rather than using it to advance enterprise performance modeling. Whether a symptom or a cause, it is worth noting regarding data as a cost item rather than profit-generating center inhibits change to the environment just described. The bottom line is that good data, properly understood and applied, can greatly enhance enterprise operations; but most enterprises lack the culture of transacting in good data.

The issues of enterprise cultural change in which enterprise data is readily available and useful for enterprise optimization need to be explored. To be effective, this cultural change must take place outside of the IT department. Once data is relegated to IT, then suffers much of the limitations it has in the past, such as lack of ownership by operational units that generate data, resulting in issues of quality, interpretation, and timeliness. The culture must start at the operational side, with its "owners" and migrate as requirements to IT. Although the culture of "smart data" can't be an IT transformation issue, it will strongly influence IT investments in the future.

WHAT *SMART DATA* PROVIDES

To summarize, smart data provides:

- Enterprise performance enhancement: end-to-end process optimization, continuous optimization, etc.
- Better enterprise decision making: more timely, more responsive, etc.,
- Increased agility: more responsive to change, lower IT cost to change, continuous optimization, etc.

The end product from this effort is Chapter 5, the *CEO's Smart Data Strategy Handbook*, which handbook spells out what the CEO and Operations Management must do to lead their IT Professionals to adopting and implementing smart data strategy.

REFERENCES

1. Joshua Cooper Ramo, *The Age of the Unthinkable*, Hachette Book Group, New York, 2009.
2. en.wikipedia.org/wiki/Resilience_(network)
3. *The American Reinvestment and Recovery Plan – By the Numbers,* http://www. whitehouse.gov/assets/documents/recovery_plan_metrics_report_508.pdf.
4. Lori Montgomery,"U.S. Federal Deficit Soars Past Previous Estimates," *Washington Post*, March 20, 2009.

5. George W. Bush, *The President's Management Agenda*, FY 2002, Government Printing Office, Washington, D.C.

6. Barack Obama on Technology and Innovation, December 8, 2008.

7. Dick Smith and Jerry Blakeslee, *T + D*, September, 2002.

8. Jeffrey T. Pollock and Ralph Hodgson, *Adaptive Information*, Wiley, Hoboken, NJ, pp. 39–65, 2004.

9. P. C. Pendharkar and J. A. Rodger, Information Technology Capital Budgeting Using a Knapsack Problem, *International Transactions in Operational Research*, **13**, 333–351, 2006.

10. J. A. Rodger, P. C. Pendharkar, "Field Testing, Evaluation and Validation of a Web-Based MedWatch Decision Support System (MWDSS): An Empirical Study of the Application of Mobile Tracking of Epidemiological Disease Trends and Patient," *Eighth World Congress on the Management of eBusiness*, IEEE Press, Piscataway, NJ, 2007.

Chapter 1

Context: The Case and Place for Smart Data Strategy

Experience is the name everyone gives to their mistakes.
— Oscar Wilde

1.1 VALUE OF DATA TO THE ENTERPRISE

The context in which executives address performance is truly selective in that one can choose to consider performance of a function or department, of a product or asset, or even of an individual. When we talk about performance optimization it is in the enterprise context that is discussed at some length. Only the CEO and executive staff sit in a position of neutral view, the all encompassing enterprise view of performance. Performance metrics from all the pieces or elements of the organization aggregate to comprise a composite view.

Performance management expert Frank Ostroff, speaking to the Association for Enterprise Information (AFEI) on the subject of "Designing and Implementing Horizontal, Collaborative Organizations to Improve Performance," emphasized the criticality of having executives spend at least 20% of their time addressing managing change or he said "it isn't worth it."

Well, we are not talking to people who somehow make change happen through consultation and cajoling from the outside; we are talking to *you*, the executive, about what *you* can do personally to make change happen. We are not talking about a special program; we're talking about executives asking the right questions and demanding the right data to answer them. That is the strategy, plain and simple.

Smart Data: Enterprise Performance Optimization Strategy, by James A. George and James A. Rodger
Copyright © 2010 John Wiley & Sons, Inc.

The easier and faster it is for you to answer questions and make decisions, the greater your capacity to optimize enterprise performance. That is a critical element of the smart data focused strategy.

However, to optimally manage performance, the executive must be able to identify deviations and improvement opportunities by interpreting indications and drilling down to root causes. That is particularly the case for mature requirements and demands. To accomplish this you need a framework, methods, and algorithms upon which you can depend with confidence, and which we advocate be defined through the smart data strategy.

Responsibility to lead and to see ahead, to accurately anticipate the future, and to formulate vision rests with senior leadership. In both instances, managing the operation and envisioning the future require actionable data and analytical support. While ultimate decisions rest at the top, end user customers and employees are major contributors. Moving data from them in an actionable state to that of the executive is a deliberate engineering challenge. MIXX.com CEO Chris McGill says that "sifters" are needed to help cull the data, referring to application of his social media service to help accomplish this.

On April 19, 2009, the *Washington Post* headline [1] is "Obama Picks Technology and Performance Officers."

Aneesh Chopra is the nation's first Chief Technology Officer (CTO). Jeffrey Zients is the Chief Performance Officer (CPO). The Chief Information Officer is Vivek Kundra. Significant is the fact that there are three positions—Technology, Performance and Information—and that they are expected to collaborate, in our words to help optimize enterprise performance. They do this not as operations executives managing complex bureaucracies, but by providing leadership and expert direction and input to the president and to the cabinet secretaries.

Can we assume that the CPO will be a performance analyst, examining government operations to seek out improvement opportunities? Can we assume that the CTO will be evaluating and improving the technology underpinning that spans the spectrum of research beginning in a discovery mode to identify strategic technologies that will not only optimize government enterprise but will energize the economy? Can we assume that and the CIO will focus on hardware and software systems that enable all aspects of government to optimize performance, by leveraging the best information technologies? Those *are* our assumptions.

According to the Department of Labor is *Occupational Outlook Handbook 2008–2009*, "chief technology officers (CTOs), for example, evaluate the newest and most innovative technologies and determine how these can help their organizations. The chief technology officer often reports to the organization's chief information officer, manages and plans technical standards, and tends to the daily information technology issues of the firm. (Chief information officers are covered in a separate *Handbook* statement on top executives.) Because of the rapid pace of technological change, chief technology officers must constantly be on the lookout for developments that could benefit their organizations. Once a useful tool has been identified, the CTO must determine an implementation strategy and sell that strategy to management."

The last sentence is a problem because it implies all technology is a tool, and that is too narrow a definition and context. Broadly, technology is the application of science. The National Institutes of Health (NIH) defines it as "a body of knowledge used to create tools, develop skills, and extract or collect materials; the application of science (the combination of the scientific method and material) to meet an objective or solve a problem" [2].

The Department of Labor (DoL) defines the duties of chief information officers as follows. "CIOs are responsible for the overall technological direction of their organizations. They are increasingly involved in the strategic business plan of a firm as part of the executive team. To perform effectively, they also need knowledge of administrative procedures, such as budgeting, hiring, and supervision. These managers propose budgets for projects and programs and make decisions on staff training and equipment purchases. They hire and assign computer specialists, information technology workers, and support personnel to carry out specific parts of the projects. They supervise the work of these employees, review their output, and establish administrative procedures and policies. Chief information officers also provide organizations with the vision to master information technology as a competitive tool."

General descriptions such as these are helpful in understanding normal expectations, although these are extraordinary times and keen attention must be given to how these positions fit into the enterprise schema. In fact, the enterprise must be engineered to perform in pursuit of aims that are defined by the chief executive officer (CEO) and supported by the executive team. The CEO tweaks and guides staff leadership positions to accomplish priority requirements.

First and foremost, the CEO needs and should want greatly improved data and data analytical support in order to optimize enterprise performance.

Comparing our brief assumptions with the DoL descriptions, we believe that the CTO position as staffed by the government and by many commercial organizations should not be so tightly coupled with the CIO because much technology is not information technology per se. New technology can shake the paradigm completely, such as synthetic fuels that can make all current engine technology obsolete overnight.

The chief performance officer (CPO) position is a green field notion that a dedicated professional needs to be able to objectively assess departments' performance and to make recommendations about their metrics. We embrace this idea as it emphasizes objective performance analysis that is most certainly data centric.

Commercial enterprises too are staffing these positions with parallel expectations in the commercial context. Perhaps the most significant assumption, one that is shared, is that superior products and services are the expected outcomes. Nothing less than superiority wins in a global economic environment and that means optimizing the interplay between government and commerce.

While regulation is receiving a resurgence of attention, complementary and balanced attention must be focused on the government providing incentives for innovation in the pursuit of superior commercial performance such that the capital engine is reinstated and made healthy again. Therefore the aim must be to achieve higher performance and higher standards than ever before.

These pursuits are made real with associated transparency and accountability with a data focus and with a smart data strategy such as we recommend.

For the *Washington Post*, referring to the CPO position, Max Stier wrote [3]: "When it comes to government performance, one of the best ways to improve it will be to improve the way we measure it." He added: "Right now, our federal government relies on lagging indicators to let us know how our government is doing. In other words, we need leading indicators, not lagging ones."

The CPO is a new position and after a false start at filling the post, the challenge is to grasp what President Obama wants to do with it. We have some ideas as discussed herein. Needed is a management approach to optimizing enterprise performance that is shared from top to bottom, and that is known and understood by everyone—a *transparent* management approach.

It is a nontrivial activity to change the way our enterprises work, from being backward looking to forward looking.

While we are surely data focused, we want to clarify that it takes more acumen than numbers to run an enterprise. A story by Al Ries says it all in the title of an article by him: "Metric Madness: The Answer to Mathematical Failure Seems to Be More Math, If You Run a Company by Numbers Alone, You'll Run It into the Ground" [4]. The article describes the need for people with marketing and creative skills, to be able to generate new products and services that people need and want. Smart data is needed as much by the creative professionals as it is by accountants.

The contemporary performance environment includes an internal view and an external one. The internal view is something that we will discuss in this and later chapters, which describe technology that you need to have on board, or access to, to support smart data. Having access to technology accommodates the notion of "cloud computing," that is, where your enterprise subscribes to computing and communication services on a pay-as-you-go basis versus having in-house infrastructure. Cloud computing is advancing and emerging rapidly and will surely enable smart data strategy as an extension to or as a replacement for how you manage computer and communications technology enablement.

On April 23, 2009, Reuters carried an announcement that "International Business Machines Corp. plans to launch cloud computing services this year, taking on companies such as Amazon.com Inc., Microsoft Corp., and Google Inc." [5].

You also hear terminology describing various levels of Web computing: Web 1.0–4.0. These generations of Web development are defined as follows:

- Web 1.0—First generation and we are all there: interlinking and hypertext documents that are web accessible via browsers, web pages containing text, graphics, pictures, and so on.
- Web 2.0—dmiessler.com distinguishes Web 2.0 as being a difference in how applications are implemented versus being a change in technology. Differentiators include web-based communities, hosted services (i.e., social networking sites), wikis, and collaboration.

- Web 3.0—This is Tim Berners-Lee's dream come true: the semantic web featuring web content expressed in natural language and in a form that can be read by software agents. Herein is one major characteristic of smart data.

Can you have smart data without semantic interoperability or without the World Wide Web Consortium (W3C) completing a standard to tag data to make it smart? Well, our answer is that you can have smarter data by attending to the three domains of our definition. You can make considerable progress until the tagging schema pushes us to the pinnacle of this round of achievement.

- Web 4.0—Marcus Cake (www.marcuscake.com) provides a graphic (Figure 1.1) depicting the transformation that concludes with Web 4.0 being called *transformational*, including the notion of global transparency, with community sovereignty over channels of information.

Our view is that we already have a toe into Web 4.0 as social media pushed us there as did the global economic calamity. So, for our executives, you need to be aware of your changing environment and know that a part of this is relevant to achieving our

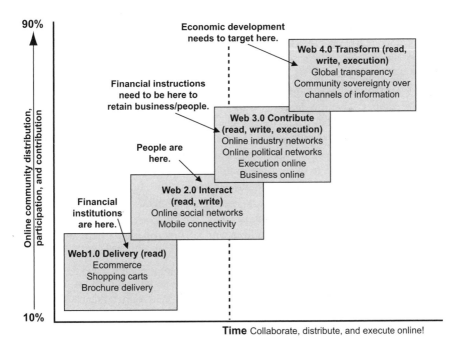

Figure 1.1 Economic development: Web 1.0 (distribute), Web 2.0 (participate), Web 3.0 (contribute), and Web 4.0 (transform). (used with permission from Marcus Cake, http://www.marcuscake.com/key-concepts/internet-evolution [6].)

highest expectations for smart data. However, there is much work to be done by you right now before being concerned about all of the technical details.

Social media are technologies and practices leveraging the Internet that enable people to share opinions, insights, experiences, and perspectives, including user-created media such as a blog, wiki, or hosting site. We propose another definition for executives: social media accelerate communications among communities of practice and interest.

How does this relate to smart data? Smart data moves on the Web and it moves among your enterprise community, sometimes leveraging applications of social media based on your strategy and active participation. The process and the medium become a generating source of data as well as a user of data.

Distributed computing has evolved from the days of remote method invocation. Today, cloud computing, the semantic web, SOA, and SOE have transformed what was once simply referred to as a three-tier client–server architecture. In our discussion of smart data, we have portrayed it in the context of the enterprise view, interoperability of technology, and a view of the data that addresses appropriate methods and algorithms.

Companies such as Google and Amazon have embraced this full-service cloud and definitely have moved the ball up the hill. However, today's executives who are operating in the shadow of a global crash must keep their eyes on the target. They are aware of their enterprise needs and are surrounded by technical support and a lot of data that may be used to optimize their systems. The answer is not to buy more legacy systems or to recapitulate the phylogeny of data from RDF and OWL. The answer is to produce business rules that direct the executive toward the right answers and outcomes.

President Obama has decided to discuss privacy issues at www.data.gov and he has empowered his CIO and federal CIOs to make government data transparent.

In a similar vein, Health and Human Services (HHS) has always operated with a socialization approach to policy development and implementation that can greatly benefit from applying social media such as www.mixx.com to accomplish this mission. Mixx.com is one example of a company that has a public face to a user community while offering customization services and branded facing through government and commercial websites. For organizations that are communications intensive in developing and advancing policy, this can have significant advantage by (1) reducing conference and meeting time while (2) increasing participation and (3) speeding the communications process.

Much of the transfer of this information will take place on what is currently termed Web 2.0. The term Web 2.0 describes the changing trends in the use of World Wide Web technology and web design that aim to enhance creativity, communications, secure information sharing, collaboration, and functionality of the Web.

The concept of cloud computing is based in distributed computing concepts such as grid computing and incorporates components of the grid such as infrastructure as a service (IaaS), platform as a service (PaaS), and software as a service (SaaS) as well as Web 2.0 and other recent technology trends that have the common theme of reliance on the Internet for satisfying the computing needs of the users. Some examples of Web 2.0 include social-network or video-sharing sites, blogs, peer to peer file sharing, and WebTV.

WebTV can be software-based, in which you access the online TV by downloading a specific program or software. WebTV can also be webpage-based, in which you can access the online TV by entering a specific website and run ActiveX. Software-based examples include Live TV and PPLive. Webpage-based TV includes www.hulu.com and www.cnn.com/live. There are advantages and disadvantages to WebTV.

Pros

- Most are free
- No location restriction
- International TV
- No hassle schedule
- Commercial-free
- More choices
- Watch movies at home

Cons

- Requires a high speed connection
- Sometimes low audio or video quality
- Risk of programs bundled with virus and malware
- Lost productivity at workplace
- Piracy
- Not a very large selection of things to watch
- Copyright and illegal links

So where is the payoff for President Obama and his CIO in making government data transparent and how does this fit our concept of smart data? Executives such as President Obama must realize that today's Internet users are deserting the mass media. Whereas previous executives used fireside chats and television, today's leaders must realize that the smart data resides on the Web. There is an old query that asks if a tree falls and no one is around to see it fall, did it really fall? Today's leaders ask if they have communicated.

Also at an AFEI luncheon, we listened to president Obama's former campaign strategist, David Plouffe, talk about the Obama campaign strategy. He made a number of points worth quoting.

Referring to the Obama campaign, Plouffe said: "We were not a democracy." By that he meant that once the executive had chosen the strategy and message, it was time for everyone to execute in synchronization and to be consistent and stable, such that grass roots could absorb, understand, and reinforce. He said: "We had to model the electorate."

Plouffe went on to say that "when we had problems it was because we did not step back to test our assumptions. We had to ask, what's the strategy and do our assumptions align properly? Decisions are based on strategy."

Optimizing enterprise performance requires knowing the strategy; floating the strategy to subordinate organizations such that they are contributing properly; and measuring contributions against the strategy.

How frequently do you change strategy? Our answer is: as often as the data indicates that you should.

Bob Bragdon, Chief Security Officer (CSO.com), commented: "Assigning monetary value to data is a concept that few businesses seem to be embracing, although, even from a cursory examination, it would seem to make a great deal of sense to do. We treat everything else in our organizations as assets with measurable financial value: inventory, people, property, etc. But when was the last time you sat down and calculated a dollar value for your customer database? My guess is that it has never been done" [7].

Our view is that the situation is worse than Bragdon describes. Executives do not associate assets with data and data with costs and value. Accounting systems do not work that way, unless you move to activity-based cost accounting, for example. In fact, the effort to introduce these practices needs resurgence and reinforcement. Accounting rules and methods constrain the ability of companies to plan ahead and require addition and change to support capacity to forecast and predict.

In 2003, one of the authors (George) was at a conference at which Michael C. Daconta talked about designing the smart data enterprise [8]. Daconta was talking about how semantic computing would impact enterprise IT. We really didn't remember this until Dr. Brand Neiemann, at the CIO's office at the Environmental Protection Agency, brought it to our attention. So, with this prompting we went back in time and found the reference. If Daconta was talking on this subject in 2003, why is it not in wide practice today? In fact, George was speaking about semantic interoperability at the same conference and is astonished at the lack of progress.

Daconta went on to become a metadata guru for the Department of Homeland Security, where we are certain he made significant contributions. He laid the foundation in thought that should have advanced much further today. Because IT messages do not often resonate with the executives and because executive leadership was in constant flux, progress is less than expected.

Daconta went on to write a book, The *Semantic Web* [9], published by Wiley, that surely made an impact. Yet, how can we accelerate the progress that the concept deserves? The answer is to engage executives in the subject in ways that matter to them most, as it should.

If we made a mistake, it is from not aligning the subject with the executive audience that most needs the solution.

A demonstrable example of how technology can take off when aligned with executives' needs and interests is social media. President Obama put the technology to work and now it has sprung an industry. Of course, we may not imply that President Obama is responsible for social media, though we can say that he and David Plouffe demonstrated how to use it, as it put the community organizer on steroids, so to speak.

One could assume that data is valuable or we would not be all consumed by the process of creating, storing, publishing, and otherwise transacting it in every aspect of our daily lives. However, how much do we respect data? For many executives and their enterprises, respect does not manifest in a strategy. Data does not appear as the first words in the annual report, though it very well should.

The following excerpt is from the 2006 Annual Report from AIG. "Dear Fellow Shareholders: 2006 was an excellent year, starting with the resolution of our significant regulatory challenges and ending with record financial results" [10].

To be fair the historical data was there, it just didn't say anything about the future. The place to begin is to make certain that we have a common agreement about critical terms. What do we mean by "data"? What do we mean by "smart data"? We can address this need straightforwardly, though the context would be missing. Which should come first? We'll try it two ways.

We don't like trite sayings, one of which is "smart data is that which knows what to do with itself." Yet there is merit for this and that is where we are headed. Data is engineered and with application of today's technologies it can be attributed such that it operates in a system enabling it to appear to qualified users in an actionable state and in a timely, accurate, and complete manner. Internet inventor Tim Berners-Lee proposes accomplishing this by tagging the data, although the W3C has not yet completed the standards for doing this as part of the semantic web initiative, where he has renewed commitment: "W3C refers to the World Wide Web Consortium created in October 1994 to enable the WWW to realize its full potential by developing common protocols to promote its evolution and ensure its interoperability. Having developed more than 20 technical specifications for the Web's infrastructure, including XML, DOM, and SOX, the W3C is now laying the foundations to meet the needs of the next generation of the Web as computers, telecommunications, and multimedia technologies converge" [11]. We advocate that executives commit to accelerating this work.

From Michael Daconta's posting, we summarize what he describes as the historical evolution of data: "Historically, data began as a second-class citizen locked away in proprietary applications.... Data was seen as secondary to the programs processing it. This incorrect attitude gave rise to the expression garbage in, garbage out or GIGO. GIGO basically reveals the flaw in the original argument by establishing the dependency between processing and data. In other words, useful software is wholly dependent on good data" [11].

Computing professionals introduced the notion of data quality and that data are assets.

Daconta reported that object-oriented facilities developed that made data more important, although internal to applications, so that vendors could keep data proprietary to their applications for competitive reasons. Daconta commented that data proliferated in volume and number of formats in the era of personal computers. "The most popular formats were the office document formats like Microsoft Word. These data formats and databases had proprietary schemas understood only by the applications processing the data" [11].

"The growth of the Internet in the 1980s and the World Wide Web in the 1990s began the shift away from proprietary formats." Daconta wrote that in 2003 [11] and we observe that many organizations remain trapped by continuing investments in old paradigm software.

He observed that the evolution of data has sped up in recent years, with major shifts occurring more rapidly. This speed is an effect of a more mature understanding of how to model data. "We are now moving into a new phase of the data evolution— the age of semantic models—in which the standardization of the World Wide Web Consortium (W3C)'s Ontology Web Language (OWL) will be the catalyst" [11].

It is important to understand that data evolution will not end with this phase; there are more fine-grained shifts and more follow-on phases to come. Daconta's treatise pretty well brings us up to today's fine-grained refinements.

Executives should be aware that their organizations need to support and participate in W3C standards initiatives.

We will present a case example in which we demonstrate model-based data exchange employing an open international standard, ISO10303 Product Life Cycle Support (PLCS), for instance.

At the time of this writing, the federal CIO Council is grappling with how to implement "data.gov," which is intended to make a broad array of government data available to everyone. From this, we expect improved policies regarding individual privacy protection and overall data security. The subjects related to smart data are a moving target.

House Subcommittee on Telecommunications and the Internet

Privacy, secrecy, and security are separate and related topics that we cannot ignore as part of smart data strategy, although the subjects and disciplines are a specialty that calls for independent treatment.

The follow excerpt is from the testimony of Larry Clinton, President, Internet Security Alliance, to the House Subcommittee on Telecommunications and the Internet, May 1, 2009:

> At her confirmation hearings two months ago, Secretary of State Hillary Clinton said that the single biggest threat to our country was the proliferation of weapons of mass destruction, and she identified four categories of these weapons: nuclear, biological, chemical and cyber.
>
> The former Director of National Intelligence Advisor to President Bush, Mike McConnell, has argued that "the ability to threaten the U.S. money supply through a cyber attack is [the] equivalent of today's nuclear weapon."
>
> Just 10 days ago, Melissa E. Hathaway, Acting Senior Director for Cyberspace for the National Security and Homeland Security Councils, previewed the report on cyber security she has provided to President Obama by saying: "The Internet is neither secure enough nor resilient enough for what we use it for today and will need into the future.

This poses one of the most serious economic and national security challenges of the 21st century."

The cyber security threat is much more serious than the well publicized massive losses of personal data. There are now recorded instances of our government and industry's electronic infrastructure being compromised by criminals, nation states, and even potential terrorists. The result of such attacks could be disastrous ranging from the possible shutting down of our electrical grid to having our own military weapons turned against us.

On a purely economic basis, if a single large American bank were successfully attacked, it would have an order of magnitude representing a greater financial impact on the global economy than September 11. But the threat is not just speculative. Today, cyber security injuries are already substantial: some estimates now place the economic loss from known cyber thefts at more than $300 million per day. [12]

All of that is sobering to say the least. When accomplished to the level that we envision, smart data and associated tagging schemas combine with credentialing and privileging procedures that should greatly increase security and reduce vulnerability, although work must continue.

Definition of Smart Data as Envisioned

Generally, smart data is the product of engineering, such that it presents itself to the right users at the right place and time in support of planning, problem solving, decision making, sense making, and predicting. It comes with methodology and algorithms that support intended uses. It is interoperable and supports qualified users with diverse needs in an ubiquitous environment. It is the product of advanced data engineering technologies, including modeling and metadata management, and smart application of open standards. It accounts for credentialing and privileging as a dimension of security.

This definition will continue to expand as you read further. To be certain, smart data is our invention, as we want to make smart data and smart data strategies more explicit to people that depend upon it.

The U.S. government, on one hand, says that it wants to invest in technology that is readily available; on the other hand, it wants to make investments that are strategic. Investing in smart data capability is a strategic investment that includes (1) maximizing use of available technology while (2) pushing forward initiatives like the semantic web that will yield substantial enablement of smarter data.

General Discussion About Data from the Information Technologist's View

As an executive, you may hear the following perspective from your IT professionals. Our discussion here will help you understand what they are saying.

When using the term "data," it can be confusing because nearly everyone has an idea about what it means to them. IT guru Daniel S. Appleton instructed that "information = data facts + meaning." Data comprise entities that have attributes. George moved this definition along and said that knowledge = information in context. Wisdom = applied knowledge = knowledge + experience.

That sounded pretty good until we thought about it some more. In a layperson's terms, data describes a person, place, object, event, or concept in the user environment about which the organization wishes to maintain information. Surely, IT professionals will describe data as comprising entities and attributes. Open Source defines entity as "a self-contained piece of data that can be referenced as a unit; you can refer to an entity by a symbolic name in the Document Type Definition (DTD) or the document. An entity can be a string of characters, a symbol character (unavailable on a standard keyboard), a separate text file, or a separate graphic file" [13].

Attributes provide more information about data elements. In computing, attributes are entities that define properties of objects, elements, or files. Attributes usually consist of a name (or key) and value.

Is such a definition compelling to a CEO or cabinet secretary? It is probably not captivating because it sounds so abstract, however technically essential.

Some say that "facts" are things that can be shown to be true, to exist, or to have happened. Therefore data can be a representation of facts such that humans and machines can process it. The data facts require additional attributes that provide meaning and the attributes are themselves data. Data are not always factual. Data describing data is called metadata.

While writing this section we read a newspaper headline in the *Washington Post*: "Math Error to Cost Maryland $31 million." Apparently the state overpaid 17 schools for which it really didn't have the funds. The mistake was said to be a computational error, made by someone using a calculator in the Department of Taxation and Assessment. This is an example of dumb data, and surely not the product of smart data strategy.

The source erroneous: 17 school systems accepted money and intended to use it, even though the amount was more than they had budgeted.

Data are the equivalent of the atomic elements of information. Data is the plural of datum. Data are used to describe virtually anything including their representation in any medium. Information technologists need to differentiate data and programs such that programs are instructions for manipulating data: they say data on its own is without meaning.

Discussion About Data with an Enterprise CEO and Department Secretary

Euclid addressed the subject of data, associated with his work called *Elements*. It is a complex work and King Ptolemy requested a more elementary text on geometry. Euclid replied, "There is no royal road to geometry." We might give a similar reply to an enquiring executive about data, except to say that a data strategy map will make it easier to chart the course to higher performance.

How might the data strategy map appear? It most surely would include a comprehensive enterprise data model depicting the entities and attributes of all the data needed to manage and optimize performance in the enterprise. It most surely would include process models describing how work gets done that are attributed with data inputs and outputs, controls, and mechanisms. It might even depict data flow, and who uses it.

As you might imagine, a data strategy map would be complex, stored in a computer, supported by software, and served up for executives, users, and technologists to analyze and to act upon.

Data is the atomic matter behind government and commercial enterprise performance. It is the essential grist for the mill, although this metaphor stops far short of complex meaning.

Some people ask "Is not your book just as valuable to commercial enterprise as it is to government enterprise?" The short answer is yes. Others ask, "What is the relevance of government to commercial enterprise?" A longer reply is needed for this question. Although current history certainly illustrates the close relationship, albeit too close for some.

A prevailing notion for the past 12 years is that best commercial practices and technologies are best for government. While there is merit to this idea, application is not always direct and universal. We observe that government is in a perpetual catch-up situation, lagging commercial enterprise by 4 to 20 + years. There are structural, bureaucratic, political, economic, and cultural reasons for the lag, and we are certain that commentators like CNN's Lou Dobbs will find no good excuses.

Our approach aims at cutting through excuses. It also introduces another set of views about the relationship between commercial enterprise and government.

Global asymmetrical threats to national security demand that the U.S. government performs with superiority in the comprehensive realm of enterprise management that is enabled by superior technology. Since the government is in the lead of enterprise management, the institution cannot afford to wait for commercial enterprise or to wait to catch up—it must lead in the role of ultimate customer representing citizens. Note that this would be a strategic policy shift to demand that government not default to commercial industry in pursuit of technology leadership.

Commercial enterprises that support the government as prime contractors are often not servicing commercial customers. They are locked into servicing government and therefore are not in the position to be the sources of "best commercial practices." (There is a much larger question related to the military–industrial complex, although that is out of scope.)

Business algorithms for commercial enterprise are often the antithesis of business algorithms for government enterprise. Therefore one-to-one applicability must be challenged.

The common thread among government and private enterprise is data.

There is a pecking order in this book. Government executives, called department secretaries, are the equivalent of CEOs. Commercial executives in this book are considered in the context that their organizations are participants in the government-led enterprise as prime contractors, subcontractors, and suppliers. They are stewards of their shareholders and employees, and their being served is contingent on serving customers in a superior manner.

Citizens are the recipients of services provided by government enterprise and they are the end-user customers. As stewards of citizens, government executives are customers who are served by participants in the supply chain. All participants in this scenario have symbiotic relationships. Success is measured by the following aggregation:

- Best for citizens
- Best for government
- Best for government trading partners
- Best for allies

A nation's capacity for "best" is constrained by capital, infrastructure, resources, institutional design, and intellect, among other things. A factor critical to overall success is the ability of the elements of enterprise to collaborate in achieving ultimate outcomes.

Collaboration is a characteristic describing how enabling mechanisms interact to produce outcomes from processes that are shared under complex rules by members of the supply chain and by customers themselves. Data are provided as inputs and created as outputs throughout the process, and agreements determine to whom the assets belong.

There are only a few executives who are performing on behalf of citizens. They are supported by a large number of organizations, operations management, and bureaucracy. They are dependent on a large number of commercial contractors and suppliers. The relationships are governed by laws, regulations, business rules, and contracts.

The term "enterprise" is employed to describe the aggregation of organizations—public and private—that work together to satisfy missions, goals, and objectives on behalf of citizens aka constituents aka communities. Enterprises have certain building blocks:

- Control architecture (more contemporarily called leadership and integration architecture)
- Inputs (capital, material, data)
- Processes that are prescriptions for how work gets done (activities that define how work gets done to produce required and desired outputs)
- Outputs (outcomes, products, services, assets, data, and results)
- Enabling mechanisms that may be considered the technical architecture (people, organizations, and technologies)

Note again that data are both an input and an output. Data can be an asset and can also be noise, just as outcomes can be positive or negative (i.e., value or costs, revenue or expenses). This is the construct by which Dennis Wisnosky, Chief Technology

Officer at DoD, led the Defense Enterprise Modeling effort. The artifacts from this effort will become assets and input for future administrations.

To optimize enterprise performance, CEOs and secretaries must be able to track data through processes and to realize data as assets, or to otherwise assess it as being costly noise. The degree to which executives and their operations management are competent at accounting for data assets can be realized as competitive advantage or operational superiority.

From a citizen's perspective, government should be performing optimally, maximizing service while exploiting scarce resources to best use and advantage. From a shareholder's perspective, commercial organizations are expected to provide the best value to customers while returning optimal profitability that is competitively best. Both scenarios are dependent on best uses of data and associated accountability.

The Information Clearing House (ICH) definition of enterprise is "a system of business endeavor within a particular business environment. Enterprise architecture (EA) is a design for the arrangement and interoperation of business components (e.g., policies, operations, infrastructure, and information) that together make up the enterprise's means of operation" [14].

In the U.S. government there have been and continue to be "enterprise architecture" initiatives intended to engineer performance as expressed in models. Making enterprise performance explicit is a good first step. It is also a continuing necessity. It is the modern day translation of laws and regulations into automated operations and automated contracting environment.

Unfortunately, government leaders are rarely around long enough to see the effort completed. Worse still is that leadership often does not know how to use the models because the IT people have not made them operationally useful.

The reality is that government leaders are elected to manage and make a difference *today*, and not so much in the years after they are gone from office. It is a rare occurrence when architectural assets trickle down to operational management and bureaucracies with any positive effect as they are trapped in the day-to-day necessities, by the legacy as it were.

Something gets done anyway. For instance, wars get started and waged. Departments' consolidation gets initiated on massive scale. Executives get in trouble for short-circuiting the system in ways that appear to be abusive of the system.

Government auditors declare as foul the discovered waste, fraud, and abuse, while they too are on a short term and subject to resource manipulation that undermines their effectiveness. This is not an acceptable picture and at the bequest of American constituents, new leadership is pressured to find better solutions to which this book contributes by addressing data, the atomic matter behind government enterprise performance.

Automated real-time auditing enabled by smart data is an area worth exploring and improving in this regard. Such would provide management with actionable controls as it would simultaneously provide independent flagging of potential problems.

Further complicating government enterprise performance is the fact that government leadership performs in short terms while addressing problems and needs of great scale and magnitude. As a nation, we have trouble reconciling this circumstance.

The way to stabilize the situation and to assure continuity in focus and progress is by providing a legislative framework that is supported by operational management policies. While government laws and regulations may flow to participants in the government supply chain, private enterprise may address its own continuity issues via accounting and auditing practices.

Assets

All assets are defined by data; however, not all data are assets.

Definitions of "asset" include all real or intellectual property owned by the enterprise that has a positive financial value; resources and/or property in the possession of an individual or business entity; everything a corporation (or government entity) owns or that is due to it—cash, investments, money due it, materials, and inventories, which are called current assets; buildings and machinery, which are known as fixed assets; and patents and goodwill, called intangible assets.

One cannot just substitute government for corporation and apply the definition because there are definite nuances in how government accounts for assets versus corporations. However, for our purposes the first definition is probably best.

Government may make investment in assets at various stages in the process of transforming from raw materials to end-use products. Government may contract with private enterprise to care for the assets, and while in custody by private enterprise where legal title may reside subject to contingencies. The assets may be destined for ultimate possession and use by government customers. Government may do this for the purpose of managing investments and limiting risk and liability, and to preserve flexibility in reassigning assets to alternative uses and purposes.

We observed this specifically in the case of material stockpile management, where investments are made in staging metal products in certain quantities at various stages in the production process, for instance. This is done to have surge capacity.

The Health and Human Services Department maintains surge capacity in medical suppliers. The Department of Homeland Security maintains surge capacity in logistics capability for use in nationally declared emergencies. All such departments must track materials for these purposes as assets for which they must provide visibility to planners and prospective users, as well as to property managers.

To keep track of assets, one needs to know the contractual business rules as well as the associated processes and how information feeds processes, and how information is transformed and transferred as a result.

Information transformation describes how data is changed. Information transference describes the status of the asset as it moves through various processes and by various custodians. These are separate, though associated, ideas. Complicating this further is the fact that some assets are perishable and have limited life cycles.

The authors recall an instance where the DoD procured a new jet fighter under a performance-based logistics contract. In the process of developing the autonomic logistics component of the contract, disparity between what the government thought it was buying and what the contractor thought it owned was revealed in discussions about protecting weapon system data. Such revelations are common as tracking layers of data and elements is dynamic and subject to interpretation and legal determination.

Sometimes multiple government departments and agencies contract with the same supply chain for materials needed for surge capacity, for instance. The chances are that all departments will call on the surge inventory at the same time. Will the suppliers be able to respond to these multiple requests? We suspect not, because from our direct experience we believe the process for accounting for materials in such events is flawed as a result of the absence of smart data and smart data strategy.

Data Assets

All assets are ultimately expressed, represented, and defined as data. Therefore it is imperative to be able to track and manage the disposition of data assets. It is equally important to keep the system clear of extraneous data that is simply noise in the system.

All data that is in a system consumes resources to be processed, stored, and maintained. Delinquent data, that which is not maintained, can dilute the system's scarce resources. That is why Sid Adelman, Larissa Moss, and Majid Abai, authors of another book, *Data Strategy*, are consumed by the idea to rid the system of redundant data, for instance [15].

While eliminating unnecessary redundancy may be a worthwhile pursuit, what are the other options and priorities that represent strategic focus? It seems to us that a more direct strategy is necessary, such as knowing your data, tracking and protecting your data assets, and optimizing data transacting with your enterprise trading partners.

What Is the Work of Management?

From an executive perspective, managing the enterprise is all about the data.

Smart data will improve the enterprise via enterprise level data resource management with improved operational management reflected by more accurate and timely information, situational awareness, and decision making at all levels.

Enterprise management, executives, and operational management own the data. Information technology and information system specialists provide enabling support.

Using the three-schema architectural view of data, there is (1) an external view and that is as seen by the end user and their applications and in our case that is the executive; (2) a conceptual view and that is from the system designer's view and independent of storage mechanisms; and (3) internal schema that describes how data is organized, stored, and manipulated independent of the application. In our discussion about smart data, we traverse these different views and schemas but always with the executive view and the enterprise view assuming paramount importance.

For executives to receive the level of support they need from information technologists servicing a smart data strategy, they must know what they require, and this involves expecting data to be engineered with certain characteristics that make it smart, as well as expecting the infrastructure and services to be engineered to support smart data strategy. Because you are pioneering a developing new idea, albeit with existing and emerging technology, it is an iterative process requiring close collaboration between executive customers and IT solution providers. You will be producing a capability that is best for your enterprise and that will produce superior results.

It is important to say this because we believe that, too often, the data ownership and responsibility is passed to IT or slips away from management control and visibility, getting lost in a sea of information technology. We want to make explicit how to channel data, and how to optimize enterprise performance with complete visibility.

For instance, the DoD embarked on a strategy to employ enterprise resource planning software as a panacea for escaping a myriad of legacy systems supporting defense logistics. The notion is that by having everyone on the same software, all organizations will be able to integrate operations. Trouble began when organizations were asked to throw out their business rules and to adopt those embedded in the software.

First, the ERP software, for instance, was originally developed for private enterprise that is motivated by a completely different set of metrics than government enterprise, notably profit motive versus constituent service utility motive. Second, the solution of choice is often from a foreign vendor, making U.S. defense systems and operations dependent on foreign software.

These large issues loom among many others and the circumstance has moved responsibility for enterprise data away from management and into the hands of supporting information technologists. How did this happen?

Some key government executives and former military flag officers ended up working for technology vendors as a part of the revolving door practice that compromises objectivity.

Alright, we know that American defense is dependent on foreign trading partners, customers, and allies. It is not necessarily bad to engage foreign suppliers in U.S. government programs. However, one must ask what capabilities and assets should a nation own or be able to deliver domestically in order to remain secure? These are executive questions that need constant attention as they affect policy and practice and, most important, data.

Accept the Premise

Concluding the discussion about the value of data to the enterprise, accept the premise that all assets are ultimately expressed, described, and accounted for as data. To optimize enterprise performance, the process is to take capital and material and, through enterprise processes, produce outcomes that are higher yield products and results, whereby the processes operate under constraints and the work is performed by people and technologies performing in concert. The people are aligned with organizations, and the relationships among organizations are contractually bound.

From the viewpoint of the enterprise head, the enterprise creates certain data that is critical to its performance. The enterprise receives data from external sources and transforms this data for internal purposes that translate into higher yield outputs. The outputs are sent on to other users in the community—some of which are end users while others will employ the data as input into their own processes.

The purchase, lease, and use of data falls under constraints and agreements whereby there is consideration—monetary value exchanged for use either explicitly or implicitly.

As an enterprise executive, you must know your data assets, as that is what you must protect, and it is the collateral for which you receive revenue or benefit. You must know the data on which your enterprise performance is dependent. You must know about the sources of data. You must understand that you may influence the sources of data such that the data you receive is optimally engineered for your use.

Likewise, you must know where your data outputs are headed. You must be concerned about how well your outputs satisfy the needs of the user community.

Of parallel importance is to know where and how data are applied to support

- Sense making and predicting
- Planning
- Problem solving
- Decision making

A prerequisite is to know enterprise processes and how data feed processes, and how processes produce data. A part of describing or modeling processes is defining and accounting for business rules as constraints. Conversely, executives need to identify what data is needed to optimize enterprise performance. As Gertrude Stein might have put it, "What is the answer? What is the question?"*

Understanding data is dependent on semantics (meaning) and syntax (format) that manifest in lexicons (vocabulary), data dictionaries (reference books), and ontologies (organization of knowledge).

In a complex enterprise, associated with data are functional profiles. That is, your data has certain attributes that affect associated costs of access, storage, processing, publishing, and usage. The data attribute profiles have direct bearing on your own internal costs as well as having impact on the aggregate enterprise performance where you are a part of something larger.

There remains a clear field for accountants to think about how to ensure that aggregate benefits are distributed among the community, and how those that introduce excess costs pay for their deficiencies as the expense is born by the source and by users.

Exiting the disaster of 2008 and the Bush era, U.S. business must reestablish credibility, trust, and integrity, as must the government. Stimulus from government spending is intended to buy time for commercial enterprise to regain footing, although such cannot begin until the banking, finance, and lending system is reengineered to a suitable international standard that is under development simultaneously with the fix.

---

*In one account by Toklas, when Stein was being wheeled into the operating room for surgery on her stomach, she asked Toklas, "What is the answer?" When Toklas did not answer, Stein said, "In that case, what is the question?"

The management approach proposed in *Smart Data* can surely increase the probability of success while reducing associated risks as it promotes a scientific and quantitative process to enterprise performance management.

1.2 ENTERPRISE PERFORMANCE VERSUS ENTERPRISE INTEGRATION

CEOs and department secretaries talk about enterprise performance. When speaking with IT professionals, they talk enterprise integration. What's the difference? Answering this question takes considerable explanation, although in so doing, we can introduce some perspectives that will ultimately permit our readers to appreciate the focus on smart data to which we are headed.

Enterprise Integration

There are two principal definitions on the Web, for instance. One says that enterprise integration (EI) "is the alignment of strategies, business processes, and information systems, technologies, and data across organizational boundaries." We discussed this with electronic magazine publisher guru Bob Thomas, who had a magazine called *Business Integration Journal*. He changed the name to *Align Journal*, emphasizing that alignment of IT strategy and enterprise strategy is essential to optimizing performance. We subscribe to this idea.

Another definition of enterprise integration "refers to internal coordination processes among different core activities." The first definition expresses a more external and global view, while the second definition is more inwardly focused. The trouble with inwardly focused enterprise integration is that the result is often stovepiping.

What is wrong with stovepiping? It adds extra costs and often results in islands of automation that require extra effort, ranging from manual intervention to extra programming and maintenance to share and process data.

A popular notion the past few years is transformation. Applied to the enterprise, it means qualitative change. This was the primary topic in the 2008 political campaign and is a central theme of the new presidency.

According to some, transformation applied to defense is the term used for new methods in warfare integrating communication and technology. We think that it is much more and note the absence of the key term, data, from this definition.

We advised the DoD Business Transformation Office that transformation requires an enterprise view. With that we developed what we called the service-oriented enterprise (SOE) management paradigm. We draw from this experience to explain our views about enterprise integration. SOE is a management paradigm and strategy directed at improving performance from enterprise integration and from engineering information for interoperability. The Electronic Logistics Information Trading Exchange (ELITE) was a specific program managed by the Office of Secretary of Defense, Unique Identifier (UID) Program and in concert with the Logistics Enterprise Service Program Office at the Defense Logistics Agency (DLA).

The SOE paradigm was advanced as a best-practice approach to achieving results from enterprise integration placing enterprise context at the forefront and endpoint of all enterprise integration initiatives. SOE is presented as a replacement for traditional systems integration approaches that are too often narrowly focused, and from which results are functionally siloed, rarely producing the magnitude of improvement planned and costing far more than expected.

The SOE initiative addresses the causes and the remedies and identifies what actions and elements can improve optimizing enterprise performance with computer automation.

SOE was developed in concert with a subset of U.S. Department of Defense (DoD) customers who manage the most complex enterprise in the world. Because optimizing performance in defense enterprise is dependent on supply chains and complex relationships between government customers, contractors, subcontractors, and suppliers, this may be the source of best practices in addressing associated problems and needs. Needed is leadership that is aware of this: perhaps retaining Secretary of Defense Gates will increase the possibility that a strategy may take hold.

Transforming the U.S. government, including the DoD, is a long-term challenge for which economic pressure demands improvement from information technology investments and higher results from performance improvement initiatives. Continuing to keep U.S. commercial enterprise competitive in an increasingly global market also demands superior application of advancing technology, and the SOE initiative demonstrates what leaders of enterprises must do to achieve and sustain leadership positions through this means.

At the outset, it is imperative to have executive sponsorship. It is also imperative to staff the initiative with those who have domain expertise, as this is essential in garnering support from those who are responsible for the area of change and improvement. Improving operational performance with advanced technologies will result in changes in work design and resource deployment over the life cycle of the initiative.

A part of the SOE argument is that upfront investments in engineering information for interoperability will have significant downstream payback. From past experience, we know that organizations must be prepared for change such that they anticipate improvements from doing things differently.

In this section we outline topics that are prerequisites for team participants in helping adopt an enterprise viewpoint for developing solutions that contribute to optimizing performance.

A Comprehensive Review of the Service-Oriented Enterprise

SOE is the wide angle view of the enterprise and presents a context in which to address smart data. A film maker might open a movie with a wide angle view of the landscape before zooming into the cowboy on a horse, for example. SOE is our landscape.

Is data the horse or the cowboy in this metaphor? Neither, as data is more like the air we breathe in wide open spaces to keep living.

The goal of enterprise integration is to optimize performance through seamless automation of business and technical operations. Accounting for this, consider efficiencies realized from better work design and operational performance required to support the enabling information technology. Also, consider the impact of resulting automation realized as benefits among the community of participants: users, customers, members of the supply chain, and other stakeholders.

Accounting for benefits is addressed by answering what they are, who realize them, how they are measured, and how the enterprise rewards their attainment and discourages deficiencies.

Up to now there have been different strategies:

1. *Standardization-oriented strategies* emphasize getting members of the enterprise to adopt standards and implementation conventions and rigorously comply with them.

2. *Enterprise applications integration(EAI) strategies* depend on middleware and the development of interfaces that translate and convert information from one form to another for automated processing.

3. *Enterprise resources planning(ERP) strategies* depend on wholesale adoption of proprietary enterprise software to effect complete change to a common platform.

4. *Enterprise information integration and interoperability (EII) strategies* leap beyond standardization, EAI, and ERP and create a new generation of capability based on preparing information for integration in a manner that is open, nonintrusive, and loosely coupled.

It is in the latter category that we are developing and promoting smart data, although we extend the idea further with consideration of semantic web and data tagging techniques.

Introducing the SOE framework or paradigm reminds me of how Steve Martin introduced the notion in a play called *Picasso at the Lapin Agile* (the agile rabbit) [16]. EINSTEIN asks a question about a Matisse painting.

Sagot: I'll tell you what makes it great [taking the painting from the frame].

Gaston: The frame?

Sagot: The boundaries. The edge. Otherwise, anything goes. You want to see a soccer game where the players can run up into the stands with the ball and order a beer? No. They've got to stay within the boundaries to make it interesting.

Einstein: That frame is about the size of my book.

Sagot: Well, I hope that you chose your words carefully. Ideas are like children: you have to watch over them, or they might go wrong.

Preparing the canvas involves frameworks, architectures, reference models, and concepts of operations. The process is telescopic, microscopic, and otherwise kaleidoscopic.

Service-Oriented Enterprise (SOE) Topics Overview

- SOE Main Idea and Key Terms
- SOE Demand: Problems and Opportunities
- SOE Description: Framework
- SOE Engineering Disciplines: Practices
- SOE Business Case: Customized to the Situation
- SOE Enabling Technology: Methods and Tools
- SOE Implementation Plan: Strategy and Process

The information presented here is a review of a briefing that is intended to prepare participants embarking on enterprise performance improvement employing the SOE approach that is a framework in which smart data strategy may flourish.

SOE Main Idea and Key Terms Understanding the demand for change and improvement establishes context. "Enterprise" is the context for application.

Today's demand for change and improvement begins with the requirement for government to right the U.S. banking and financial system so that capital is available to drive the commercial enterprise engine. In all cases, resources are scarce and there will be increasing demand on government for higher performance from a much smaller footprint. Therefore automated support systems must be smarter and reliant on smart data.

Defining terms is essential to improving performance in an enterprise. Understanding the meaning of terms in context with their use is called *semantics*. Semantics is crucial to applying today's most advanced technologies that produce seamless automation among organizations, corporations, and individuals sharing information and completing functional transactions among them.

You may have heard the terms semantic web, semantic mediation, and semantic interoperability. These terms are used to describe the newly emerged and emerging state of modern computing.

They describe the circumstance whereby entities, organizations and people, can communicate automatically (self-governing) from their unique viewpoints while retaining their own lexicons of words, and are able to be understood accurately and completely by others with different viewpoints and terms. This describes our current state of pursuit, although information must be engineered to accomplish the result.

By contrast, heretofore, information technologists were more preoccupied with *syntax* the rules for formatting or structuring of information for automated processing. Semantics and syntax are subjects essential to understanding new strategies for producing higher enterprise performance through better use of information technology.

Wrestling with syntax and semantics, aerospace and defense prime contractors worked for over 10 years to harmonize the basic terms used to execute sales to the government that centered to a large extent on business rules and the terminology for conducting business using electronic data interchange (EDI) standards and implementation conventions.

At the start of the process, EDI was already 10 years mature but had penetrated only a small percentage of the supply chain. Ten years later, harmonized EDI was accomplished among a shrinking number of remaining aerospace prime contractors and about 20% of the suppliers. EDI was too hard, too brittle, and too inflexible to afford full potential from electronic business automation.

EDI is an example of a standard that is by design inflexible. It results in users having to design brittle interfaces to support changes and improvements, which require a high maintenance effort and expense. Selecting this standard was not smart and therefore not reflective of smart data strategy. The goal was elusive, and the introduction of XML and Web-based service oriented architectures (SOAs) destabilized the effort. XML has more desirable characteristics than EDI and affords greater flexibility, though no less rigor among user participants to maintain.

On the one hand, great investment was committed to "standardizing" with the use of EDI standards. On the other hand, introduction of new Web-friendly standards opened the window for greater flexibility.

Industry faced a dilemma: (1) continue to wrestle with a rigid standard or (2) expect potential for improvements from a new standard. A third alternative was to consider the application of new data exchange technologies possible in SOA that would increase flexibility in applying standards that are inherently rigid.

Using advanced methods and technologies, enterprises can greatly reduce the effort required to maintain standards-based interfaces by engineering information for interoperability upfront.

The ultimate goal is to provide benefits from automation to all enterprise participants, increasing information sharing and reducing operational costs.

Returning to the term "enterprise," CEOs and senior executives and managers have active roles and responsibilities for managing technology-oriented improvement strategies. While CIOs and information technology professionals are partners in the process, superior results are produced through optimal working relationships throughout the enterprise.

Generally, "enterprise" refers to the undertaking necessary to accomplish something. An undertaking needed to produce an automobile that solves consumers' transportation needs and the need for fuel economy must certainly be considered a complex enterprise. An undertaking to develop a metrology machine used to measure an automobile piston or to measure the precision of jet engine turbine blades may also require a complex enterprise.

The enterprise needed to produce national security for the United States of America is enormously more complex than any of the previous examples. Integrating 22 disparate departments and agencies into an effective homogeneous system for delivering security to the homeland is a task of similar complexity. On the premise that these initiatives began with a flawed management approach, it is unlikely that the condition today is as it should be.

The National Institutes of Health describes enterprise by the following definition in its handbook of "net" terms: "In the computer industry, an enterprise is an organization that uses computers. In practice, the term is applied much more often to larger organizations than smaller ones." we don't find this definition particularly useful.

Another definition is more suitable; this one is from the Interoperability Clearing House (ICH): "Systems of business endeavor within a particular business environment. Enterprise architecture is a design for the arrangement and interoperation of business components (e.g., policies, operations, infrastructure, and information) that together make up the enterprise's means of operation" [14].

Here is an enterprise definition that we prefer from an information technologist's view. "An enterprise is an organization with partially overlapping objectives working together for some period of time in order to attain their objectives. The actors utilize technology, competence, information and other resources in order to transform input to products that satisfy the needs of customers" [17].

So you see that one can shop for terms and meanings that best apply to a certain viewpoint and situation. That is what we all do as individuals, some more precisely than others.

When the number of constituents for your messages is large and complex, there are different strategies for producing timely, accurate, and complete communications that may be used to consummate business transactions.

All of these definitions of enterprise are applicable to our discussion. We recall a definition of enterprise used by Daniel S. Appleton, IT guru. He said that an enterprise comprises organizations that may include commercial partners and government customers that are linked together by contingent claim contracts. He emphasized the importance of understanding the dynamics of business rules and how they operate throughout the enterprise.

As a commercial organization conducting business with the government, or as a government entity operating as a part of a larger organization, both examples are enterprises with their own internal characteristics that link with other entities to form a greater whole that is itself a larger enterprise.

The degree to which the linkages are seamless or efficiently integrated to produce the desired output is what we describe as the enterprise performance domain. If an enterprise can be described by its essential operating components with a degree of consistency in method and technique, we can more clearly understand how to optimize the linkages.

Integrating linkages is not the end game. The end game is delivering the needed services or products in a timely, accurate, and complete manner at the best cost and with the best quality. This is what is meant by "service" in the service-oriented enterprise.

Integrated is a term used throughout this discussion. For software developers it may describe a group of application programs designed to share data. For an enterprise executive, integrated may describe the linking of controls (business rules), processes, and information among different members of the enterprise (i.e., trading partners).

What is common in these two ideas is shared data or information. Data = facts meaning. Information = data + understanding, or data in context. Some consider the term "integrated" to be associated with tightly coupled linkages that are accomplished by application of standards and development of enterprise application interfaces. By contrast, some use the term *interoperability* to describe linkages.

"Interoperability" implies the ability of a system or product to work with other systems or products without special effort on the part of the customer or user. Words such as loose coupling, adaptive, and noninvasive are used to differentiate interoperable linkages from integrated linkages. These ideas are important in understanding the attributes and characteristics of enterprise performance-improving strategies and solutions.

They help the people who must plan, budget, and pay for solutions to better understand the trade-offs from the technologies enabling results. Technologies affect how work gets done and how results are produced with associated attributes and metrics.

When there is a surplus of labor or people needing work, some leaders may be more inclined to use people versus technology to perform work. When there are shortages, outsourcing and worker immigration are alternatives. The sociogeopolitical trade-offs are real for government leaders and for American commercial enterprises alike.

Metaphors are used to help explain complex ideas. Supply chain and value chain are used to describe the relationship among trading partners that comprise an enterprise. *Chains* describe dependent relationships, yet the linkages we are developing with application of new technology may be better described as neurons and synapses. (Neurons are nerve cells that can receive and send information by way of synaptic connections.) Chains are rigid and hard to break. Neurons and synapses work as more agile and adaptable means of linking.

Those performing enterprise integration need to adopt a different view from systems integration. To illustrate this, compare systems integration to that of a castle and enterprise integration to that of the open marketplace. There is only one way into the castle and that is to take the bridge over the mote that is opened and closed by request to the gatekeeper. By contrast, the open market may be accessed by anyone entering from all sides. The difference is context. A commercial enterprise participating in the government market must cross many motes and gatekeepers, though the goal is to make the enterprise more open like the marketplace.

Systems integration is more concerned about performance inside organizational or functional boundaries that are often called *silos*. Enterprise integration must consider how software solutions perform for all of the entities that share information. As much value is given to how systems perform in the enterprise context as is given to the local context.

A more advanced topic has to do with consideration of self-organization versus entropy and striking a balance between freedom and constraints. From the viewpoint of a supplier or subcontractor that is a member of the larger enterprise, and from the viewpoint of prime contractors and ultimate customers, achieving the optimum degree of freedom while ensuring enterprise-wide interoperability is a goal.

SOE Demand This section describes why a new management paradigm and strategy is needed to improve results from enterprise integration initiatives. The concept of service-oriented enterprise (SOE) is rich with substance applicable beyond enterprise integration in that the "service-oriented" values and optimizing enterprise performance are considered first.

When we discussed SOE with the Technical Operations Vice President of the Aerospace Industries Association (AIA), he asked, "Is SOE a business or technical-oriented paradigm?" That is an interesting question because managers and technical professionals often ask this question to determine if and how it fits their viewpoint. SOE addresses both business and technical functional requirements in a comprehensive and integrated manner.

In an article published in *Business Integration Journal* [18]. Hall Bailly, and George outlined the following takeaways.

1. ***Business*** "A stable framework is needed to help enterprise management configure a collaborative enterprise from independently migrated processes and applications. The ability to describe application architecture for EI is essential to the decision process. SOE will help in understanding the immediate costs of directed investment in EI and its benefits in reducing costs for ongoing interface maintenance."

2. ***Technical*** "Without an overarching strategy and associated architecture, many EI activities continue to be ad hoc and, as such, non-predictable and non-repeatable. As a strategic approach to EI, SOE blends structural elements in harmony with three software engineering disciplines. The objective is to achieve the optimum balance between attention to the elements and application of SOE disciplines."

We wanted to present evidence demonstrating the need for a new paradigm and strategy. Software vendors such as Metamatrix offered a commercial example, but we needed an objective third-party source. From a government perspective, the General Accounting Office (GAO) reported that program and project managers need more disciplined processes to reduce risks associated with enterprise integration and systems engineering. There is too much rework needed to stabilize applications after deployment.

Pressure to get systems implemented quickly has shortchanged investment in information engineering to get them right. Our view is that more investment upfront will deliver high reward downstream. This begins with smart data strategy and adoption of supporting values and principles.

Another aspect of the demand for SOE is from continuing to support the value for information sharing. Former Defense Deputy Under Secretary of Materiel Readiness, Dave Pauling, advanced the idea that "improving operational performance is realized through improved planning, sense making, and decision making that is currently too disjointed and costly to maintain." Information interoperability deficiencies are the cause. Correcting the cause eliminates proliferation of deficiencies while maximizing the distribution of benefits.

1.3 CURRENT PROBLEMS AND DEFICIENCIES FROM POOR DATA STRATEGY

Current problems and deficiencies from poor data strategy must be addressed in the context of enterprise resource management and the SOE paradigm. Enterprise resources are constrained and demand for change and improvement often outstrips available resources. Here are some considerations that we imposed in constructing the SOE paradigm:

- Enterprise scope and context
- Acknowledging that resources are insufficient to replace the magnitude of legacy systems
- Necessity to keep pace with advancing technology
- Necessity to deliver significant improvement from constrained investment
- Make application least invasive while delivering immediate tangible benefit

Therefore we concluded that a new strategy is needed based on addressing the need for information interoperability.

All enterprise environments operate under resource constraints that manifest as shortages of capital, and shortages of scarce talent and materials, for instance. Demand for improvement and change nearly always outstrips an enterprise's capacity for making change and improvement.

Different things, often working in combination, create demand for improvement. Demand for improvement and change is different for government and commercial enterprise. While some things are the same, others are the opposite. Government enterprise is about the business of maximizing citizen services. Commercial enterprise is about the business of maximizing return to shareholders. When the government is the customer of commercial enterprise, government negotiates for the best value, while business negotiates for the best margin. Buyers and sellers need one another and in a competitive marketplace, agreeable parties will connect. Correct classifications are essential to this process.

In an environment in which resource constraints manifest themselves as capital shortages, cost containment is a major concern to the SOE. Not only are correct classifications of costs important in a smart data–SOE scenario, but understanding misclassifications merit a closer inspection as well.

Smart data can be utilized, in a combination with artificial intelligence methods, to shed light on cost classification within the SOE. For example, a misclassification cost matrix can be incorporated into nonlinear neural-evolutionary and genetic programming-based classification systems for bankruptcy prediction. This smart data–SOE approach, properly implemented, could have helped to prevent and predict the 2008 Wall Street collapse, if implemented in a smart data paradigm.

Remember that smart data provides the following benefits: enterprise performance enhancement, end-to-end process optimization, and continuous optimization. Smart data also leads to better enterprise decision making—timelier, more responsive, and increased agility—that is more responsive to change, has lower IT cost to change, and achieves continuous optimization.

In statistics, the terms Type I error (α error, or false positive) and Type II error (β error, or a false negative) are used to describe possible errors made in a statistical decision process. When an observer makes a Type I error in evaluating a sample against its parent population, he/she is mistakenly thinking that a statistical difference exists when in truth there is no statistical difference. For example, imagine that a pregnancy test has produced a "positive" result (indicating that the woman taking the test is pregnant); if the woman is actually not pregnant, however, we say the test produced a "false positive." A Type II error, or a "false negative," is the error of failing to reject a null hypothesis when the alternative hypothesis is the true state of nature. For example, a Type II error occurs if a pregnancy test reports "negative" when the woman is, in fact, pregnant[*].

Most classification systems for predicting bankruptcy have attempted to minimize the misclassifications. The minimizing misclassification approach assumes that Type I and Type II error costs of misclassification are equal. This assumption may be a fallacy. Remember that wisdom = applied knowledge = knowledge + experience.

There is evidence that these costs are not equal and incorporating these costs into the classification systems can lead to superior classification systems. The principles of evolution can be used to develop and test genetic algorithm (GA)-based neural and genetic programming (GP)-based classification approaches that incorporate the asymmetric Type I and Type II error costs. By applying these artificial intelligence methods and mechanisms to decision making, it is possible to move toward the smart data "wisdom" paradigm of knowledge + experience, in which data comprise entities that also have attributes.

Using simulated and real-life bankruptcy data, we compared the results of the proposed approaches with statistical linear discriminant analysis (LDA), back-propagation artificial neural network (ANN), and a GP-based classification approach that does not incorporate the asymmetric misclassification cost. In essence, we are providing a data strategy map in order to make it easier to chart the course to higher performance. We are moving toward "wisdom" because we are investigating both a linear and nonlinear approach and comparing the two outcomes.

The results indicate that the proposed approaches, incorporating Type I and Type II error cost asymmetries, result in lower Type I misclassifications when compared to LDA, ANN, and GP approaches that do not incorporate misclassification costs. These smart data truths should be self-evident and useful as a common thread among both government and private enterprise. The smart data "wisdom" here is to investigate both the linear and nonlinear approaches and to compare the outcomes for the one that demonstrates the most cost containment.

We must remember that smart data is the product of engineering, such that it presents itself to the right users at the right place and time in support of planning, problem solving, decision making, sense making, and predicting. In this case, there is

---

[*] Wikipedia definition as of 2008.

evidence that costs are not equal and incorporating costs into classification systems can lead to superior results.

Our smart data approach to Type I and Type II errors comes with methodology and algorithms that support intended uses. Our approach is interoperable and supports qualified users with diverse needs in an ubiquitous environment. It is the product of advanced data engineering technologies including modeling and metadata management and smart application of known standards.

Bankruptcy prediction is an active area of research in finance. Several analytical approaches have been proposed beginning in the late 1960s. Among the popular approaches for bankruptcy prediction are the use of statistical discriminant analysis, artificial neural networks, decision trees, genetic algorithms, and probabilistic approaches, such as logit and probit. Researchers use analytical techniques for prediction of bankruptcy. Using statistical discriminant analysis and data on a set of companies that went bankrupt during the period 1946–1965, Altman [19] showed that statistical discriminant analysis is a viable tool for prediction of bankruptcy.

Following Altman's study [19], researchers investigated the use of probabilistic approaches to predict bankruptcy and some used a maximum likelihood estimation of the conditional logit model with the objective of making probabilistic estimates of insolvency. Logit is a linear technique that does not require any assumptions about the prior probabilities of bankruptcy or the distribution of predictor variables.

Unlike linear discriminant analysis (LDA), logit does not specify a cutoff point delineating bankrupt firms from nonbankrupt firms. The model assigns each firm a probability of bankruptcy. The decision makers can then choose a level that they are willing to tolerate. The trade-off is between choosing a higher Type I or Type II error.

Several machine learning techniques were also used for prediction of bankruptcy. Among the machine learning techniques used for prediction of bankruptcy are artificial neural networks (ANNs) and genetic algorithms (GAs). Since these techniques do not rely on any distributional assumptions about the variables, they avoid problems associated with LDA and logit.

It has been found that although LDA provides superior results, GAs are effective tools for insolvency diagnosis, since GA results are obtained in less time and with fewer data requirements. ANNs have been used for bankruptcy prediction and several researchers have reported a better performance of ANNs against statistical LDA and logistic approaches.

The foregoing discussion, about the value of correctly classified smart data to the enterprise, accepts the premise that all assets are ultimately expressed, described, and accounted for as data. Examining this data can lead to cost savings and optimizing enterprise performance. In this case, we showed how to improve the process of minimizing costs.

We examined capital and material and, through enterprise processes, produced outcomes that are higher yield products and results, whereby the processes operate under constraints and the work is performed by people and technologies performing in concert. From the viewpoint of the enterprise head, the enterprise creates certain data that is critical to its performance.

In bankruptcy prediction, and in other classification problems, the costs of Type I and Type II errors are important considerations for a decision maker. There is evidence of an asymmetric cost structure, with an estimate of Type I error cost that is higher than Type II error cost. The cost of Type I error was estimated from the loan loss experience of banks, and the cost of Type II error was the opportunity cost of not lending to a nonbankrupt firm because it was predicted to become bankrupt.

Past studies employing LDA, ANNs, and GAs do not allow a user to incorporate asymmetric costs of misclassification. As a result, the Type I and Type II error costs are considered equal in most past studies. A probabilistic approach such as logit, however, allows a decision maker to trade off between Type I and Type II error by setting the cutoff probability. The technique, however, does not provide any guidelines for deciding on the cutoff probabilities, which makes it difficult to use. Recently, GAs have been used to learn connection weights for an ANN.

Among the advantages of using GA to train ANN connection weights are flexibility of design of the fitness function and global search. The flexibility of the design of the fitness function makes it possible to use GAs to learn connection weights of an ANN so that Type I error is minimized. Furthermore, the discriminant function learnted by using GA-based ANN is a nonlinear function and is likely to perform better than linear discriminant functions developed by LDA and GAs.

The preceding discussion provides the basic underpinnings of the bankruptcy prediction problem, that is, information = data facts + meaning. However, it is not until a comparison of the various linear and nonlinear techniques are made and the evidence that Type I and Type II errors are not equal, that actual knowledge or "information in context" has been supplied to move the decision maker toward the smart data "wisdom" paradigm shift. The knowledge + experience context is to recognize that most classification systems for predicting bankruptcy have attempted to minimize the misclassifications.

The minimizing misclassification approach assumes that Type I and Type II error costs of misclassification are equal. This long-held assumption may be a fallacy. Remember that wisdom = applied knowledge = knowledge + experience. The lesson to be learned is that there is evidence that these Type I and Type II costs are not equal and incorporating these costs into the classification systems can lead to superior classification systems. The common thread among government and private enterprise is data.

In the following case, we demonstrate what we stated on page 45 how the enterprise receives data from external sources and transforms this data for internal purposes that translate into higher yield outputs and helps to prevent accounting fraud in the organization. The outputs are sent on to other users in the community—some of which are end users while others will employ the data as input in their own processes in order to prevent fraudulent activities in their enterprises.

The purchase, lease, and use of data falls under constraints and agreements, such as generally accepted accounting standards, whereby there is consideration—monetary value exchanged for use either explicitly or implicitly. Material misuse of these assets can be extremely costly to the organization. It is the job of the enterprise executive to know the data assets and protect them, and these assets are the collateral for which the enterprise receives revenue or benefits.

Executives must know the data on which the enterprise performance is dependent. They must know about the sources of data and understand how they may influence the sources of data such that the data received is optimally engineered for their use. Likewise, top executives must know where their data outputs are headed and they must be concerned about how well their outputs satisfy the needs of the user community.

The integration of smart data, application of artificial intelligence techniques, and adoption of a smart data—SOE paradigm can be used to counter accounting fraud fiascos, such as Enron and World Com, and are set forth in the case "Utilization of Data Mining Techniques to Detect and Predict Accounting Fraud: A Comparison of Neural Networks and Discriminant Analysis" [20]. Accounting information systems enable the process of internal control and external auditing to provide a first-line defense in detecting fraud.

There are few valid indicators at either the individual or the organizational level which are reliable indicators of fraud prevention. Recent studies have shown that it is nearly impossible to predict fraud. In fact, many of the characteristics associated with white-collar criminals are precisely the traits that organizations look for when hiring employees. This case proposes the use of information systems and smart data and utilization of the SOE model to deal with fraud through proactive information collection, data mining, and decision support activities.

Here is an example of citizens being recipients of services provided by government enterprise. All participants in detecting accounting fraud have a symbiotic relationship. Success in detecting and deterring accounting fraud is measured in an aggregation that is best for citizens, best for government, best for government trading partners, and best for allies.

Results show that while traditional methods, such as discriminant analysis, yielded 50.4% of original grouped cases correctly classified, no significant relationship was found (0.149) between attitude, morale, internal controls, increases in expenditures, and whether or not fraud was actually committed. Cronbach's alpha of reliability was 0.6626 and offered somewhat reliable results in this exploratory research. Neural networks did a much better job of predicting fraud (75.9%) than discriminant analysis (50.4%). Neural networks were able to find patterns in the training set and then correctly identify more than three-fourths of similar patterns in the testing set.

Therefore it can be concluded that neural networks outperform discriminant analysis by 25.5% in this data set. It is not until a comparison of the various linear and nonlinear techniques are made that actual knowledge or "information in context" has been supplied to move the decision maker toward the smart data "wisdom" paradigm shift. The knowledge + experience context is to recognize that the application of artificial intelligence mechanisms for predicting fraud involves not only the context of a symbiotic relationship between the stakeholders, but also a realization that, in a smart data sense, we do not necessarily live in a parametric, bell-shaped world. Nonlinear relationships must be taken into account as well if we are to find true "wisdom."

Problem and Opportunity Types

Possible Problem Types

- Deficiency in leadership and integration
- Deficiency in mission, values, policy, regulations, rules
- Deficiency in strategy
- Deficiency in framework
- Deficiency in planning
- Deficiency in sense making
- Deficiency in decision making
- Deficiency in systems
- Deficiencies of knowledge
- Deficiencies of skill
- Deficiencies of proficiency
- Deficiencies of execution
- Deficiencies in the balance of consequences
- Deficiencies in tools and equipment
- Deficiencies in methods
- Deficiencies in processes
- Deficiencies in infrastructure
- Deficiencies in organization

Possible Opportunity Types

- Better use of capital
- Better use of people
- Better use of technology
- Better use of materials
- Better competitive advantage

In the past six years, we have seen the introduction of "enterprise portfolio management" to evaluate information technology initiatives and their effectiveness. We have learned that it is difficult to collect all of the information about IT initiatives, programs, and projects to perform objective evaluation. A reason for this is executives' and CIOs' failure to manage performance improvement initiatives as they might other business decisions and operations.

Managing today's highly automated enterprises places increased demand on executives and their management teams to have higher competence in the discipline of management science that is interlocked with the disciplines of computer science and engineering. It is the nature of the enterprise that forces our rethinking about the required disciplines.

Change and change management are actually very mature disciplines; however, application is now integral or embedded among those who are planning, designing, and developing enterprise systems. While the work of employees is changing in modern enterprise, so is the work of their leaders.

SOE Framework

We present the SOE framework in different ways to help understanding. At the core are structural elements of enterprise integration: smart data, smart grid, and smart services. Required software engineering disciplines include data engineering, process engineering, and grid engineering, each of which set requirements for integration into the enterprise.

Admittedly, there are questions about why we chose these labels, and questions about relationships. If this causes you to think about it and to develop a better understanding, we will claim that as our intention. Confusion without resolution is not our intention.

The highest order activity for enterprise management is to optimize performance. Figure 1.2 illustrates the key relationships needed to optimize performance in response to the need for supply chain transformation, for instance.

Inputs are arrows entering the optimization activity and include problems and opportunities, structural elements, and information. The purpose of the activity is to produce solutions with desirable metrics made possible by producing harmonious structures and interoperable information. SOE disciplines and technologies are the enablers, shown as arrows entering the bottom of the activity. Enterprise integration guidance and control is provided by the SOE paradigm and strategy, accompanied, of

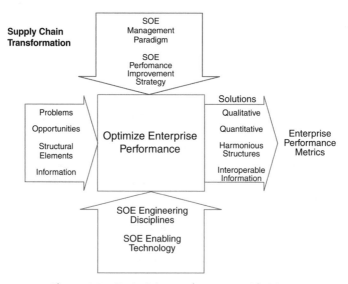

Figure 1.2 Optimizing performance with SOE.

course, by a host of other considerations that are accounted for in the details of SOE implementation.

What makes SOE different? Engineering information interoperability and embracing the value for openness and vendor independence are critical values.

What Is the Current Situation?

The current system software environment is the product of collaboration among mutually developed applications, hosted on a fixed configuration of structurally integrated platforms.

Undesirable qualities include:

- Tight coupling
- Brittle interfaces
- High maintenance
- Great difficulty in changing and upgrading

The desired state is a product of development from the enterprise view accomplished through collaboration among independently configured and structurally isolated platforms.

Acknowledged in this approach is the real-world situation that is fraught with legacy systems. Desirable qualities include more agile and less invasive solutions performing better to the benefit of the entire enterprise.

Figure 1.3 identifies engineering disciplines, advancing technologies, and structural elements in relative association. Effective enterprise integration requires an enterprise view, not silos or stovepipes. Optimizing enterprise performance requires refined engineering disciplines accompanied by advancing technology and with attention to certain structural elements (Figure 1.4).

Engineering Disciplines	Advancing Technologies	Structural Elements
Data Engineering	Enterprise Information Integration Metadata Management	Smart Data
Grid Engineering	Service-Oriented Architecture NetCentricity	Smart Grid
Process Engineering	Business Intelligence Modeling Enterprise Core Web Services	Smart Services

Figure 1.3 Starting point: effective enterprise integration requires an enterprise view.

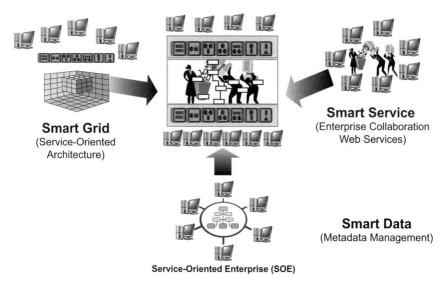

Smart Grid
(Service-Oriented
Architecture)

Smart Service
(Enterprise Collaboration
Web Services)

Smart Data
(Metadata Management)

Service-Oriented Enterprise (SOE)

Figure 1.4   Key structural elements.

SOE includes an icon. Subsequent description will introduce the elements comprising the icon.

1.4 NEW TECHNOLOGIES

Critical to the success of enterprise-wide strategy is ownership by the CEO in private enterprise. Leaders in the top positions of any organization command, lead, and integrate with the following elements: power, time constraints, focus, attributes, and outcomes. These elements can be seen in the following two case studies. In one case, Netscape built a successful e-business—Anthill.com—was successful, and sold for a profit. The second case, Fannie Mae, is a case study in critical success factors gone wrong.

Power translates into value-setting, business rules, resource allocation, and balance of consequences—that is, making certain organizational participants clearly realize the difference between positive behavior and deficient performance. Power was abused at Fannie Mae and even though the company was successful in the short term, the corruption of power led to its downfall. Time constraints force the identification of milestones, metrics, and outcomes in a meaningful period with associated metrics such as rate and frequency.

Netscape identified these constraints successfully in Anthill.com. However, over time Fannie Mae ignored these time constraints and failed. They got away from the critical success factors (CSFs) that made them successful originally. Focus describes the domain in which change and improvements are to be realized, leading to the formation of themes. The domain for Netscape and its theme for Anthill.com stayed focused on building a successful e-business. Fannie Mae, on the other hand, lost its focus and failed.

Attributes are the detailed characteristics that describe desired process and associated performance metrics such as cost and quality, in addition to previously mentioned budget constraints (resource allocation) and time. Netscape paid attention to cost and quality metrics; Fannie Mae lost site of these important metrics.

Outcomes are the end results that include everything from physical products, assets (including information assets), and performance results. Netscape sold Anthill.com for a profit and fulfilled its outcome objectives of turning a profit. Fannie Mae's end result, on the other hand, was bankruptcy, because, in the long term, it did not continue to pay attention to its information assets and performance results.

New technologies have been affording new and different strategies for years. Today, SOE is a strategy for developing business capability and helping the organization to achieve business agility and to create an information-sharing environment that can create new applications to support changes in goal requirements, increase the speed at which services and information can be shared to benefit others, and securely connect people and systems in real time.

Case Study: Building a Large-Scale E-Business from a Small Anthill

Dr. Rodger wrote a case entitled "Building a Large-Scale E-Business from a Small Anthill: A Case Study" to gain insight into how a company can incorporate important principles learned by other successful e-businesses. This study examined four essential principles used by Netscape Corporation for building a successful e-business as they relate to Anthill.com, which was a relative newcomer to the e-business world. Data were gathered via in-depth interviews with Anthill.com executives. It is hoped that other e-businesses will recognize the importance of closely following these principles to improve the opportunity to develop into a large-scale e-business.

We systematically analyzed the practices of Anthill.com, even though there was a paucity of guiding frameworks for successful e-commerce. We felt the framework was an appropriate first step for rigorous analysis of our case.

Principle 1: Create a compelling, living vision of products, technologies, and markets that is tightly linked to action.

The greatest strength of Netscape's vision was its ability to create a tight link between senior management's high-level view of the world and the products it delivered to the marketplace. Netscape's vision did more than map a path through the confusion of the Web's early years. It also mobilized the company's troops to develop and deliver an impressive range of client and server products in a very short period of time.

From humble beginnings, Curt Matsko and Scott Alexander envisioned developing a startup Internet business into a multimillion dollar enterprise. The two friends began the Anthill.com business from a two-bedroom apartment in 1999. Their vision was to bring as many small-to-medium businesses in secondary markets across the United States together into one e-business location. These secondary markets included cities having populations between 10,000 and 250,000 people. As a result of this vision, individuals are able to develop their business in a central Internet location that is highly visible to consumers.

Likewise, small e-business owners benefit from being able to place their business in a visible location without enormous cost or effort. With advances in Internet technology and hard work, Matsko and Alexander are beginning to realize their e-business vision. With Web access now very common and relatively inexpensive, Anthill.com is currently adding up to 1000 small business sites per week and is in the process of capturing over 10% of the entire small business market. This forward thinking is benefiting both consumers and retailers involved in e-business.

Principle 2: Hire and acquire managerial experience in addition to technical expertise.

Netscape's strategy of hiring experience was not restricted to the top ranks; it extended throughout the organization. Managers at every level tried to bring on board people who would hit the ground running. Netscape did not hire many green college graduates, fresh from studies in programming or marketing. Instead, it looked for people who had actually done these jobs.

Anthill was pleased with company growth but realized other resources were necessary for the future of the company. Matsko was sure that they would need an in-house lawyer to assist with legal issues and a possible move to bring the company public. "We didn't know what to expect, but we figured that rapid growth doesn't come without pain. I discussed with Scott the possibility of recruiting Dan Thurber for legal support. He is an outstanding and experienced attorney as well as a personal friend." Scott concurred with the choice. "I think he would make an important addition to our team. We not only need Dan, but should also approach his brother Brad. He knows the markets and is an exceptional manager" (S. Alexander, personal communication, November 5, 1999).

Since their company was made up of only the two partners, it seemed prudent to hire the management and legal experience they lacked. In addition, they knew the people they were hiring on a personal level: Matsko and Alexander thus followed this Netscape strategy. They brought in the Thurber brothers who have experiences in managing and providing legal advice for new businesses. Dan Thurber, a successful New York City attorney, was so impressed with the company's potential that he joined in 1999 while taking a 70% reduction in salary. His brother Brad quickly followed Dan's lead and joined. Brad emphasized the importance of his decision with the following comment, "the potential growth, stock options, and work environment at Anthill.com made it an easy decision for me to leave my current position as a successful stock consultant" (B. Thurber, personal communication, November 5, 1999). Substantial growth of Anthill.com has greatly increased the number of employees necessary to run the operation. Matsko developed a unique recruitment strategy that has enabled the company to add over 45 highly qualified employees. The recruitment strategy is unique at Anthill.com because it focuses on future growth and employee ownership, rather than traditional salaries and retirement plans. Employees at Anthill.com have bought into the vision, potential, and excitement of managing an e-business. "Employees that have joined the Anthill believe in our business model, and are eager to obtain stock ownership in the organization over competitive salaries which we cannot offer" (C. Matsko, personal communication, November 5, 1999).

Principle 3: Build the internal resources for a big company, while organizing like a small one.

Most start-up companies scale their systems to meet their current needs. In fact, they usually allow their systems to lag behind their growth. One of the biggest traps for an entrepreneur is to build an organizational structure in advance of sales, profits, and stable cash flow. Far too often, wildly optimistic sales projections do not materialize, the company gets overextended, and everything comes to a crashing halt.

Attracting attention to Anthill.com is a top priority. Management is responding by a method coined by Matsko as "travel and conquer." The method involves traveling the country and providing training seminars for small businesses to help them develop a variety of commerce sites. According to Matsko, "our clients need to know us on a personal basis and understand how we can affordably help them make money with the Web" (C. Matsko, personal communication, November 5, 1999).

This approach is the mainstay of Anthill.com. It continues to bring thousands of individuals to the Anthill.com colony. Likewise, this personal interaction makes Anthill.com have the coziness of a small business as it works to expand into a large-scale e-business. Since Anthill.com competes in an industry immersed in the technological revolution, it must invest much of its profits in technology resources and people with technical skills. Top executives Curt Matsko and Scott Alexander are investing in the future. Matsko underscores this by stating that "our company realizes the importance of staying abreast of the rapidly changing technologies associated with the Internet" (C. Matsko, personal communication, November 5, 1999).

What started out as a relatively simple Web server has emerged into a complex array of active server pages, Java, Pearl, and Netscape's SSL technologies. This change drastically increased the need to be able to attract technically competent employees. Recruiting the best technical minds is becoming more difficult as the competition for good technical people is at its highest in years in the information systems industry. Anthill.com executives spend much of their decision-making time devising strategies to attract talent. Matsko's plan is to recruit individuals with experience in both managerial as well as technical areas, as well as to develop unique business relationships with individuals who do not work directly for Anthill.com.

One unique strategy utilized by Anthill.com to improve business practices is the use of independent contractors and Anthill.com affiliates. Anthill.com currently has over 1000 active contractors (individuals who have purchased a commerce site from Anthill.com) who sell Anthill.com commerce sites to other businesses and individuals for profit. In addition, Anthill.com affiliates are compensated for each new business they bring to Anthill.com. The affiliates use an identification number when adding a business site to Anthill.com to receive compensation. The use of independent contractors and Anthill.com affiliates permits Anthill.com to develop a formable workforce similar to a large organization.

Principle 4: Build external relationships to compensate for limited internal resources.

Netscape would have been unable to keep up with the demands of Internet time without outside help. The company had a powerful vision, experienced leaders, and an organization geared toward fast growth, but ultimate success depended critically on a wide variety of external resources and relationships. These external assets compensated for Netscape's lack of scale in marketing, financing, and product development. Netscape was essentially able to exploit the Internet and other external resources to create a virtual workforce—people outside the organization who were working for free on the company's behalf.

According to Matsko, "individuals we train are equipped to promote the colony though a variety of means such as registering with search engines and maintaining promotions via traditional media" (C. Matsko, personal communication, October 15, 1999). As a result, there are thousands of individuals around the United States who work to attract attention to Anthill.com. Other methods of attracting attention to Anthill.com include mailers, online promotions, and face-to-face communications. For example, teams of Anthill.com employees meet daily with individual businesses. Brad Thurber emphasizes this facet of company growth: "Face-to-face communication helps us to personalize the Internet and our company. It is a viable and successful approach to business" (B. Thurber, personal communication, October 15, 1999). Additional external relationships have been developed with a number of Internet companies such as BedandBreakfast.com, LotteryUSA.com, Mapquest.com, CardSercicesInternational.com, sisna, FilmFrinzey.com, Astrologynet.net, Travel.com, and Barchart.com. These relationships help to enhance the quality services offered by Anthill.com without requiring additional resources.

According to Thurber, "we plan to develop additional relationships with additional Internet organizations, to help improve our site within our current budget constraints" (B.Thurber, personal communication, October 15, 1999).

Case Study: Fannie Mae

Fannie Mae is an example of how a once successful company got away from a smart data approach, with a strategy for change and improvement, and ended up as a prime contributor to the Wall Street fiasco of 2008, and how the adoption of a SOE model could make them successful. We postulate that the highest order activity shared by all enterprises independent of organization type and role is to "optimize performance." We also postulate that enterprise data is a most treasured asset, and that strategy focused on improving enterprise data is of equally high importance and aligned with enterprise performance optimization.

Generally, smart data is the product of engineering such that it presents itself to the right users at the right place and time in support of planning, problem solving, decision making, sense making, and predicting. At one time, Fannie Mae embraced these truths and was successful. In the near term, Fannie Mae ignored these core concepts of smart data as well as the methodology and algorithms that support the intended uses of smart data.

As a result, Fannie Mae has strayed from the smart data path and has become a liability as a government bailout. Fannie Mae forgot that smart data is interoperable and supports qualified users with diverse needs in an ubiquitous environment. It refused to follow a paradigm that produced advanced data engineering technologies, including modeling and metadata management, and smart application of open standards that account for credentialing and privileging as a dimension of security.

In a nutshell, this is why Fannie Mae finds itself in financial woes today. The following case was conducted in 1999 and shows Fannie Mae adopting many smart data concepts. Compare this environment with the one that Fannie Mae finds itself floundering in, a decade later, by ignoring the smart data paradigm.

Much like "Where's Waldo," can you point out the numerous smart data concepts that Fannie Mae ignored in the following case, Managing Radical Transformation at Fannie Mae: A Holistic Paradigm? [22] In this case, from 1999, we noted that change management is a critical issue in the current fast-paced business environment. Organizations are being bombarded with global business change, innovations in communications, and rapidly evolving information systems capabilities. Since there exists a paucity of rigorous and relevant literature on change management, we chose to embark on an in-depth case study to explore how one organization manages radical change. We developed a set of theoretical propositions to guide the effort and act as a theoretical lens.

Our specific research question is: How does an organization successfully manage projects that call for dramatic change to the way one normally conducts business? The case study allowed us to test the propositions in the business arena with one organization. It is hoped that this approach will offer insights and mechanisms to help other organizations effectively deal with change.

When we embarked on this study, we wanted to explore how organizations successfully manage dramatic change. We began by consulting the literature on transformation, BPR, and sociotechnical management. The literature helped us articulate five propositions relating to successful change management. We conducted an in-depth case study of one organization (Fannie Mae) involved in radical transformation to see how it coped. We found that each of the five propositions paralleled the case study in some way.

The case study provided a rich set of data that allowed us to delve deeper than the literature. We were able to show that these propositions are in fact applied in at least one case. We were also able to show that holistic management can increase information sharing and knowledge creation. Fannie Mae approaches training and development in a holistic manner. Top management is active and committed.

Resources are allocated. People are the central focus. Finally, every training course is aligned with the business it serves and education is customized for each person to maximize potential.

Each of the propositions is supported by the case. However, analysis of the data revealed more depth than what is currently in the literature. This study can be extended by modifying or rethinking the propositions based on what was uncovered by the case. The first proposition states that a systematic methodology will facilitate change efforts.

Fannie Mae uses the University Model to develop people, align classes to business objectives on the job, and enable equitable evaluation based on training. The model is also used to keep technical training flexible. A natural extension of this study is to explore the impact of these human and technical factors on change management.

People development concentrates on two dimensions—human and technical. Human development encompasses behaviors, personality traits, and attitudes. Technical development encompasses skills, knowledge, and competencies. For real change to occur, Fannie Mae has surmised that people must understand why change is important to the business. Hence, IT training is not limited to basic C++ or SYBASE. A curriculum is designed with a Fannie Mae slant; that is, it reflects what people will do back on the job. It also includes conflict resolution, creativity, communication, and teamwork training to develop the "soft" side of its technical people.

Personality tests are given to people to help customize training to the individual. Fannie Mae also believes that behavior can be shaped if it hires people with the right attitude. Technical development focuses on skills, knowledge, and competencies. Skill and knowledge development is pretty straightforward because the model helps match training with business. The second proposition states that top management support is critical to successful change. The case supports this proposition.

The CIO is the champion and visionary of training transformation at Fannie Mae. He hired Gus to help him implement his vision. The president is supportive of the University Model and budgets generously for IT training (approximately $7 million per year). Two other important factors are risk-taking and mistakes. Top management is happy if training hits the mark 20% of the time because of dramatic increases in people productivity. This means that 80% of the time mistakes are made. Allowing mistakes as a natural part of the learning process encourages risk-taking and innovation. Researchers can augment this study by gathering more data about each factor.

The third proposition states that a strategy-driven approach will facilitate transformation. Transformation at Fannie Mae focuses on its training paradigm because it believes that people are the key to success. The idea is to train information system (IS) professionals so that they can deliver whatever work, services, and products the business needs. Training works with business managers to develop a curriculum that is strategically aligned with the business needs of the enterprise. Exploration of training–business partnerships and strategic alignment of the curriculum can further transformation research.

The fourth proposition states that holistic management will facilitate change. The case brings out three factors that make up holistic management—equitable assessment, customized training, and adoption of a holistic approach. Holistic assessment is when everyone is evaluated based on the contribution to the enterprise. It is not subjective or trivial. It is performance based. Evaluation at Fannie Mae is based on the customized training a person receives, how⁻ the person uses it to deliver what the business needs, and best practices.

A group of managers and peers decide on what the best practices are for each job. Evaluation forms are uniform (except for management evaluation), which helps people perceive them as fair. Training is customized to the individual and what he/she must perform back on the job. Finally, Fannie Mae embraces a holistic management approach to IT training. The philosophy is top–down. Top management communicates to people that business value and understanding is important for everyone. People are rewarded for enterprise thinking, innovation, and creativity. The University Model guides the design of an aligned curriculum with business needs across functional areas and the enterprise. Human development is at the center of the philosophy. People do the work and can therefore make or break a holistic approach. The fifth proposition states that knowledge is created and retained by people.

Fannie Mae is "betting the farm" on this philosophy. Top management invests millions of dollars in people development. The goal is to create an environment that rewards value and encourages risk-taking. It is based on the notion that people are the conduit of information sharing. They create information and knowledge, pass it on to others, and retain what they believe is useful to them. "Knowledge is a tricky thing. It is not tangible. The inferential engine inside [a person's] head generates what we want. We don't know how it works, but we know that as we develop and challenge our people, information sharing and knowledge creation dramatically increase" (G. Crosetto, personal communication, October 29, 1999). Even though Fannie Mae is totally committed to the idea that people retention translates into knowledge creation and retention, it puts a lot of pressure on management. "It is much harder to manage autonomy than merely telling someone what to do. We have to allow people the freedom to be creative, but we have to make sure that what they create is valuable" (E. McWilliams, personal communication, October 29, 1999). "We are really managing chaos. Controls are minimal as compared to an autocracy. The time and effort we put into developing the people system is daunting, but the benefits are amazing" (G. Crosetto, personal communication, October 29, 1999).

Failure to adapt to a changing financial climate and complacency toward adopting a SOE model led Fannie Mae down the road to destruction and placed it among one of the major contributors to the 2008 Wall Street collapse fiasco. Surveying government and commercial enterprise customers for a number of years, and from having been directly engaged in related topics for more than 15 years, the authors have observed the following tendencies, in companies such as Fannie Mae, which have contributed to their downfall. First, these companies have deficiencies and omissions from a government and commercial data strategy. Second, the companies make advances in commercial enabling technology with gaps in application and implementation. Third, while there is continuous investment in information technology on a large scale, the increasing pressure on government and private enterprise for better use of scarce resources, as invested in information technology, can be improved by commercial off-the-shelf technologies that can accelerate adoption of smart data strategy. Finally, the companies forget that there are requirements for management and technical training, in order to keep the enterprise competitive.

Now, let us see how companies such as Honeywell and IBM apply these general considerations by developing a strategy that will address the following issues. It is generally accepted that "integrating" the elements of an enterprise is a good thing. Honeywell has integrated people with change and used smart data to accomplish this. It has produced corresponding benefits by developing process maps and continuous improvement. It is generally recognized that "information," whereby information = data facts + meaning, and information + context = knowledge, is an asset that can sometimes become noise. Therefore question everything and demand team ownership. It is generally accepted that information technology—infrastructure and software—has evolved into something that is useful and essential while costly, unwieldy, and unmanageable.

Honeywell realizes that IT is a necessary, but not a sufficient, enabler. It is generally accepted that "interoperability" among data, processes, interfaces, applications, taxonomies, policies, and social networks is highly desirable, including the latest semantic interoperability. Therefore Honeywell realizes that execution is the real difference between success and failure.

Case Study: Honeywell Business Process Reengineering (BPR)

We compare Fannie Mae to another example of an organization that embraced the smart data, SOE paradigm: Honeywell. In the case entitled "A BPR Case Study at Honeywell" [23], we embarked on a case study to explore one organization's experiences with radical change for the purpose of uncovering how it achieved success. The organization we examined was Honeywell Inc. in Phoenix, Arizona. From the interview data, we were able to devise a set of 10 lessons to help others transform successfully. Two important lessons stand out above the rest. First, execution of a carefully developed change plan separates the high performers from less successful BPR projects. Second, recognition that dealing with change is difficult and complicated is not enough. Top management should make change management a top priority and communicate the change vision across the organization.

From the case study, we developed a set of general lessons. The case experience allowed us to speak in-depth with people involved in enterprise transformation, which should make the lessons more practical.

> **Lesson 1**: People are the key enablers of change.
> Business processes are complex, but process mapping offers a comprehensive blueprint of the existing state. The blueprint enables systematic identification of opportunities for improvement. IT is complex, but vendors, consultants, and system designers can create models of the system. In contrast, people are unpredictable. They cannot be modeled or categorized universally.

However, people do the work and therefore must be trained, facilitated, and nurtured.

Lesson 2: Question everything.

Allowing people to question the way things are done is imperative to change. Fail-safing provides a systematic approach to effectively question the status quo. People are encouraged to question the existing state.

Lesson 3: People need a systematic methodology to map processes.

Process mapping is the mechanism used to map and understand complex business processes. The systematic nature of the process mapping methodology keeps people focused and acts as a rallying point. Moreover, process mapping provides a common language for everyone involved in the project.

Lesson 4: Create team ownership and a culture of dissatisfaction.

Once a team perceives that they "own" a project, they tend to want to make it work. It becomes "their" project. In addition, management should encourage people to be dissatisfied with the way things are currently done. However, punishing people for complaining about ineffective work processes is an effective way to promote the status quo.

Lesson 5: Management attitude and behavior can squash projects.

If the managerial attitude remains that of "command and control" and/or management's behavior does not change, transformation will most likely fail. Success depends on facilitative management and visible and continuous support from the top. When Honeywell got its new president in 1996, the attitude toward criticism changed dramatically. The new president was not as accepting of casual criticism. Criticism of the status quo had to be based on well-thought-out ideas and presented with the logic behind the thinking. This drastically reduced the complaints about existing processes without justification.

Lesson 6: Bottom–up or empowered implementation is important.

While support from the top is critical, actual implementation should be carried out from the bottom–up. The idea of empowerment is to push decisions down to where the work is actually done. Process mapping and fail-safing are two systematic and proven methodologies that help support empowered teams.

Lesson 7: BPR must be business-driven and continuous.

Process improvements should be aligned with business objectives. Process mapping, fail-safing, and teaming should be based on what the business needs to change to become more successful. In this case, effective communication of ideas from top management throughout the enterprise is imperative. In addition, organizations should be wary of the "I've arrived" syndrome. Change is continuous and is never over.

Lesson 8: IT is a necessary, but not a sufficient, enabler.

IT is not a panacea. IT enables BPR by automating redesigned processes. However, information is for people. People work with people to produce products for other people. In addition, people need quick and easy access to quality information to help them make good decisions. Therefore IT needs to be designed to support the business and the production of products to be effective.

Lesson 9: Set stretch goals.

> Goals should be set a little higher than what the team believes they can accomplish. Since teams have little experience with the new paradigm, goal setting will tend to be based on the past. Project managers should work with the team to help them develop stretch goals.

Lesson 10: Execution is the real difference between success and failure.

The Honeywell case introduces four powerful mechanisms to facilitate enterprise change. However, real change will not happen without a plan for change and aggressive execution of that plan. We believe this is where most organizations fail. We believe that execution fails in many cases because organizations are not willing to dedicate resources, time, and energy to the effort.

Structural Elements

Three structural elements are shown in Figure 1.4. These elements comprise a stable, technology-independent, structural description of the architecture. They represent a technology neutral description of the attributes or essential features that an integrated enterprise will exhibit. They do not describe an end state, that is, "do this and you will have an integrated enterprise."

As structural elements, they fall into the category of architectural features: each having associated design characteristics; each having varying, measurable degrees of performance, which, in implementation, can be traded off against other enterprise design and performance criteria (i.e., software system engineering).

Smart Data

Smart data is data that has been invested with explicit semantic content through formalization of metadata (by definition, characterization or modeling process). The term is intended to be broad in that "smartness" is a measurable quantity by degree, not a state of being. In other words, there are degrees of rigor, precision, accuracy, structure, and abstraction to which data can be formally described. It is intentionally broad in the sense that it applies to "data," being the grist that is crunched by applications, and to metadata, models (data, business process, and others), metamodels, and even SOA contracts.

Smart data is the product of a rigorous and published process that describes actionable information flowing across the enterprise. Actionable information includes data that is descriptive of events, phenomena, materials, processes, procedures, actions, applications, structures, relationships, or the data itself. The more structured the descriptive process, the more that data is reduced to descriptive relationships among fundamental abstract elements and the smarter the data.

Smart Grid

Smart grid is the physical infrastructure and protocol routine for intercommunication within the enterprise. Smart grid differs from a conventional interconnect structure in that it includes substantial pre-engineering of enterprise interfaces. The smart grid technology of today corresponds to SOA implemented through shared language such as web service protocols. This is in contrast to point-to-point or data hub architectures. These latter configurations should be considered dumb grids because they require substantial engineering of specific enterprise interfaces. Smart grid is technology neutral and defined independently from SOA or web services implementation. Characteristic of smart grid are:

- Shared interconnection network architecture with common entry and messaging methods
- Message management capability to ensure reliable delivery of data from originating source system to intended receptor system with notification of failure and recovery
- Information assurance controls that prevent intentional or unintentional corruption of the enterprise communication process
- Sufficient resources in the form of directories, routers, and so on to support the interconnectivity requirements of the enterprise

Smart Services

Smart services are synonymous with semantic services that are shared resources, configured as web services, which are considered assets available to the enterprise as a whole, regardless of physical ownership or economic model. Enterprise global repository (EGR) is essential to the SOE. EGR is a structured resource that provides access to metadata, process and data models, metamodels, and other constructs that describe the enterprise, including processes and data. The EGR is a build-time resource from which system developers access existing models for use in formalizing their own data and to which they post their completed products. The EGR is also a run-time resource, accessed as a web service, from which semantic translators access data models and the active translation of data from system to system takes place.

Three Engineering Disciplines

Three disciplines interact and underlie the structural elements. Enterprise integration (EI) technology gains will provide increasingly robust capabilities to implement the specific solutions within the SOE structure. As EI technology progresses, increasingly robust capabilities will be installed into the EI environments. Simultaneously, the structural element provides the conceptual background for technological advancement.

Data Engineering Data engineering is the practice of developing and documenting semantic content for enterprise data throughout the enterprise life cycle. It is an ongoing discipline that addresses new types or forms of data whenever in the life of the enterprise they are introduced. By investing in data engineering, the enterprise will see a reduction in data conflicts in enterprise operations. In addition, it will enable more cost efficient data reconciliation processes and products, which in turn can extend the "reach" of the enterprise into increasingly marginal (low-volume) applications.

Basic to the data engineering discipline is the need to semantically characterize all classes of data relevant to the enterprise (all data that is passed system to system) through a formal process. This process is practiced through various methods of formalization of metadata: for example, definition of data elements, information modeling, and metamodeling. It can be practiced at four levels.

1. **Unmanaged, Ad Hoc.** Data is defined locally on the fly.

2. **Systematic Data Definitions.** Data is defined on an element-by-element basis. This is basically the conduct of a data inventory, which is the necessary first step to higher levels of data engineering. In a bottom–up fashion, data elements are rigorously defined using a prescriptive process such as the International 11179 Standard. Accuracy and synchronization with physical sources and instances of data are largely manually achieved, as is configuration control and governance supported by a variety of data repository types, such as databases and spreadsheets.

3. **Formalized Information Modeling.** Data accuracy and synchronization are achieved at a higher level of abstraction through information modeling of the underlying metadata. Information modeling more fully describes data/metadata by describing the relationships between data elements as well as defining the data elements themselves. This increases the semantic content of the data enabling the interoperability of data by means of semantic mediation engines. Configuration management at this level requires a more complex data repository to maintain the models as well as data element definitions.

4. **Metadata-Driven Information Integration.** A model-driven architecture prevails for the enterprise. Supported by both a design-time metadata repository and run-time integration engine, this level facilitates enterprise integration by application and system developers as well as during operations. Both are achieved through extensive data engineering in a top–down fashion such that metadata and information models are integrated through metamodeling and similar means.

Grid Engineering   Grid engineering is the discipline that develops and evolves the smart grid architecture for an enterprise, consisting partially of adapting the generic SOA to specifics of the enterprise. This consists of selection of integration tools, processes, and standard protocols, such as the Web Services Description Language (WSDL) and Simple Object Access Protocol (SOAP). It also establishes the rules of engagement criteria for applications and systems to participate in the enterprise. It also prescribes the method for integrating legacy systems and applications into the enterprise.

With respect to the enterprise, selecting protocols and then broadcasting the selected protocols to system developers accomplishes grid engineering. However, grid engineering also entails the development and enforcement of an enterprise information assurance and security strategy. Future web service protocols that directly address this topic may simplify this task. Grid engineering also includes defining the service layers that connect enterprise systems to the grid. The attributes of a smart grid will not be attained through protocols alone. Every system that participates in the enterprise must provide functional capability as structured layers of services.

Grid engineering can be practiced at one of four levels.

1. **Unmanaged, Ad Hoc.** Legacy applications and corresponding business intelligence are integrated into the enterprise locally on the fly.

2. **Encapsulation and Objectification.** Legacy applications are adapted to a process of encapsulation and objectification. A service-oriented adaptor stack provides the "face to the enterprise." This stack includes translation and security engines. Modeling of information and process steps objectifies data and processes. Objectified data is maintained locally to the application. Configuration management is exerted locally mostly by application developers.

3. **Capture of Business Intelligence.** Business intelligence is captured in local applications and systems in a comprehensive fashion. A local content repository is used to store and manage business intelligence. Common local functions, such as troubleshooting and recovery, are selectively exported to the enterprise. Configuration management is exercised at the system level by system integrators.

4. **Externalization of Business Intelligence.** Business intelligence is captured out of local applications and systems in reference to enterprise level models and forms and exported to an enterprise repository. This enables asset mining by enterprise users. Through enterprise modeling, all business intelligence is derived from comprehensive unified models, resulting in a model-driven architecture. Configuration management is exercised at the enterprise level.

Process Engineering Process engineering is the practice of designing and documenting enterprise processes. Done properly, it enables process improvement while maintaining enterprise operations. At its most basic level, it involves development of the rules of interaction for the processes that comprise the enterprise and their enforcement. These rules of interaction, which apply to processes as well as data, are similar to the business process rules of today, but only apply to process characteristics that impact the execution of other processes within the enterprise.

Rules that constrain the internal operations of a business process application are excluded from the category of rules of interaction. Rules of interaction deal with formality in describing the outcomes of processes, not constraining how they work internally. Their purpose is to provide a human and machine interpretable characterization of what the process does so that other process designers can accurately anticipate the result as they design their process.

Process engineering can be practiced at one of four levels.

1. **Unmanaged, Ad Hoc.** Processes, legacy and otherwise, are integrated into enterprise business operations locally on an ad hoc basis.
2. **Business Rule Standardization.** Processes are integrated into the enterprise business operations through business rules that tightly constrain the manner in which the process is built within application software. Business rules are stored in various designer accessible repositories, registries, and spreadsheets. Accuracy and synchronization are issues that have to be worked out on an interface-by-interface basis. Manual governance (through meticulous checking of design rules) and manual configuration management prevail.
3. **Process Modeling.** Enterprise business processes are modeled with the process models stored on a generally accessible repository. Application developers generally post and subscribe to the repository. Process integration largely takes place at design-time and configuration management is exercised at the system level.
4. **Outcome-Driven Processes.** Enterprise business processes are modeled abstractly using enterprise level models as reference points. The enterprise business process repository enables process integration through orchestration or choreography engines at run-time as well as through design-time methods. Through enterprise modeling, all business processes are derived from comprehensive unified models, resulting in a model-driven architecture. Configuration management is exercised at the enterprise level.

Enterprise Integration

Enterprise integration has progressed from point-to-point connectivity, data hub connectivity, service-oriented architecture, enterprise service bus, and enterprise information interoperability manifesting in the service-oriented enterprise. Figure 1.5 also illustrates the combining of net-centric enterprise services and semantic interoperability aggregating into SOE, the complete paradigm.

Your world is full of artifacts representing each of these paradigms: some point-to-point, some hub-and-spoke, and some implementations toward SOE. Unfortunately, progress is a mishmash. Point-to-point is much more costly than hub-and-spoke, and hub-and-spoke architectures are more costly than SOE. The combination of conflicting legacy causes the enterprise to lope toward a state of entropy.

The U.S. government addresses enterprise-scale integration and performance optimization with different programs and with attempts at sharing knowledge, although the U.S. government is far from having a cohesive strategy that integrates the thinking and powers of three branches of government. New leadership must recognize the need for a common strategy and focus on America's public resource management that produces a common result—a secure and prosperous nation.

- Embodiment into functions
- Embodiment into outcomes
- Embodiment into things

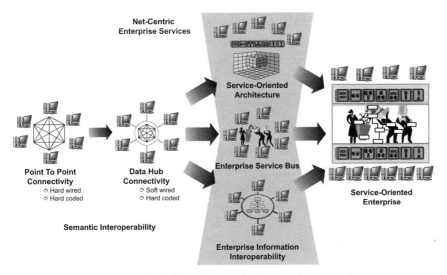

Figure 1.5 Evolution of enterprise integration strategies.

U.S. Army LTG General Jeffrey Sorenson, G-6 CIO, described the current system of systems strategy manifest in a program called Future Combat Systems (FCS) with the intention to produce a lighter weight vehicle that embodies all of the open interoperability characteristics with the ability to deliver line-of-site and beyond-line-of-site lethality to the enemy. According to Boeing, the prime contractor, "the Future Combat Systems (FCS) program is an Army modernization initiative designed to link soldiers to a wide range of weapons, sensors, and information systems by means of a mobile ad hoc network architecture that will enable unprecedented levels of joint interoperability, shared situational awareness and the ability to execute highly synchronized mission operations."

According to Wikipedia, "system of systems is a moniker for a collection of task-oriented or dedicated systems that pool their resources and capabilities together to obtain a new, more complex, "meta-system' which offers more functionality and performance than simply the sum of the constituent systems. Currently, systems of systems is a critical research discipline for which frames of reference, thought processes, quantitative analysis, tools, and design methods are incomplete. The methodology for defining, abstracting, modeling, and analyzing system of systems problems is typically referred to as system of systems engineering."

System of systems (SOS) is different from service-oriented enterprise (SOE) in that the context for SOE is, from beginning to end, the enterprise. SOS presumes standalone or ad hoc systems being brought together, leveraging metasystem characteristics to produce something stronger than the parts. However, we might argue that better results might come from end state designs that begin and end in the enterprise context.

"The U.S. Army could spend more than $300 billion to purchase and run its Future Combat Systems (FCS) family of armored vehicles, drones and communications networks during their expected multidecade life, according to a cost estimate prepared

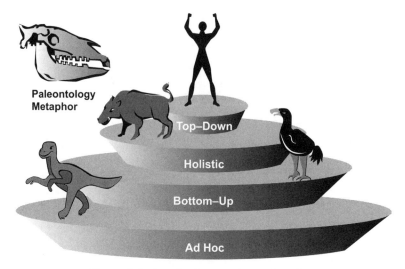

Figure 1.6   Levels of enterprise engineering.

by the U.S. defense secretary's Cost Analysis Improvement Group (CAIG)." Already, the program is woefully behind schedule and ahead in cost.

Figure 1.6 illustrates four progressions beginning with ad hoc, bottom–up progressing to holistic, and finally top–down. The references begin with dinosaurs and birds to warm-blooded mammals and finally humankind. The advance of animals from dinosaur reptiles, birds, four-legged mammals, to humankind paralleling ad hoc, bottom–up, holistic, and top–down remains an interesting metaphor. That's our sense of humor.

1.5   BREAKING FROM TRADITION WITH IMPROVED RESULTS

To break from tradition, executives must do some things differently in their pursuit of enterprise performance optimization. Ever present is possession of high value for data.

CEO Smart Data Strategy Work Breakdown Structure

What specifically are executives to do?

[A0] Optimize Enterprise Performance.

At what point in the order of business do CEOs or department secretaries address data strategy and attention to enterprise data? Observing the incoming Obama administration, we witnessed (1) attention to data details and (2) attention to staffing executive positions with those who understand processes and rules and can interpret data.

Propelled by events, incoming executives must grasp an operational environment that is fraught with problems and catastrophes. They must address strategy, tactics, and operational management issues in parallel, doing so with constrained bandwidth and capacity for change and improvement.

The national economy is collapsing. The nation is at war on two or more fronts. U.S. auto manufacturers are unprofitable and sales trend downward as products are uncompetitive. These issues are nontrivial, so how should executives factor attention to data strategy?

The CEO or president provides a brand of management and a set of guiding values and principles that are a part of what we call the *leadership and integration architecture*, which IT professionals might call the *control architecture*.

The leadership and integration architecture is a complete management system that contains process activities, inputs, controls, outputs, and enabling mechanisms needed by management to manage.

Did Obama have time to formally establish the leadership and integration architecture? He had a running start, but until his subordinate team is approved and on the job, it really can't be completed. Herein lays a fundamental flaw in our government process. One can attempt to manage intuitively, projecting personal charisma and the like, but that will only go so far in creating an illusion of management. Tim Geithner is an example of a consequence of being overwhelmed with too much, too fast. Look around at other departments and you will see DoD is in better shape because it has a carryover executive.

Part of the enterprise management system is structurally intact in mature enterprises, government, and commercial industry. It is inherited, though subject to change and improvement. Changing the structure requires deliberate planning, as often changing without a plan may result in a condition worse than before the change.

Therefore having a management approach is essential to undertaking the enterprise leadership position. Sooner than later, it is time to address the responsibilities strategically while managing day-to-day tactically.

[A1] Define Enterprise Outcomes.

Defining enterprise outcomes is a top priority as all resources and processes are committed to producing them.

Outcomes may be classified into the following types:

- Plans
- Results

  Problems solved

  Needs satisfied

  Opportunities exploited

  Products produced

  Services delivered

For our purposes, a plan contains a series of steps or milestones for guiding achievement accompanied by a schedule and resource allocations. How are plans different from processes and activities?

Enterprise processes and their component activities are recurring capabilities that may be engaged or applied to achieving plans. Plans are intended to address the production of results such as problems solved, needs satisfied, opportunities exploited, and products produced—all with associated measures and metrics.

Plans include attributes such as volume, rate, frequency, milestones, labor allocations, burn rates, and other resource allocations and utilizations resulting in costs, and accounting for assets, and value production.

Data are input to plans as well as input to processes. Data are outputs of plans and processes. Data are evidence of accomplishment.

[A2] Define Enterprise Controls.

We prefer calling enterprise controls the *enterprise leadership and integration architecture*. It accounts for management processes and their associated ` controls that permeate the organization as controls on subordinate processes. To a large extent, controls shape the brand of management.

As stated before, controls on government enterprise contain all of the laws and regulations as well as policies and guidances that may appear as memos and other communiqués such as visionary plans and programs.

One definition of *program* is "a system of projects or services intended to meet a public need." That is a complex idea. First, the existence of a program represents management's decision to commit considerable resources to produce results from an effort of considerable scope and scale. Therefore identifying a program as a part of a plan might constitute a statement of leadership intention, commitment, and control as it defines budget allocation and time constraints.

A strategy may be executed as a program whereby to accomplish it requires special resource commitment and management focus.

Smart data strategy would manifest programmatically by the following:

1. Executive policy statement emphasizing the importance of operational entities having a data strategy that contributes explicitly to the enterprise data strategy as defined in the policy.

2. Using smart data strategy as a catalyst for change and improvement as an enterprise initiative.

3. Adding and implementing smart data strategy enablement including adding new skill sets and new technologies.

[A3] Identify and Define Enterprise Inputs.

Enterprise inputs include accounting for all resources, capital, and materials that are consumed and transformed by the enterprise into intended outcomes. Under the smart data strategy approach, all resources would be accounted for by tracking their use and transformation through enterprise processes, concluding with classification into assets, results, cost, waste, and noise.

The activity of accounting for resource use, transformation, and results requires explicit data tracking and management. Refined data tracking can happen in the background, although at any time, information can be made explicit to support monitoring, planning, problem solving, decision making, sense making, and predicting.

Smart data strategy is specific in anticipating what data will be needed, when, where, and for what purposes, as well as who will have access based on credentialing and privileging.

[A4] Define Enterprise Processes.

Enterprise processes are modeled for all core functions. They are attributed to the lowest level necessary with respect for accounting for inputs, controls, outputs, and mechanisms.

[A5] Attribute Enterprise Processes with Enabling Mechanisms People and Technology.

Attributing enterprise processes is a deliberate and precise effort to assign people and technology enablement to accomplish the work needed to produce desired and required outcomes.

Herein lays a great opportunity. Improving processes and improving data quality (i.e., making data smarter) will improve automation and provide significant opportunity to reduce the requirement for large numbers of analysts. In addition, smart data will provide executives with the opportunity to refine resource deployment with greater precision.

[A6] Define Enterprise Metrics and Feedback.

Information measures are what are being measured and metrics are the specific units of measure and evaluation. All process metrics are accumulated in an enterprise performance repository, which is the basis for providing planned feedback to management and for further analysis.

Applying the Integrated Definition (IDEF) modeling technique, the enterprise management optimization process appears as in Figure 1.7. Omitted for simplicity are such things as capital and material inputs.

For enterprise management, there is typically a biannual annual (on-year and off-year) cycle. In the U.S. government, planning, budgeting, and funding cycles extend for multiple years with overlying processes and cycles that prescribe certain management products.

Since many commercial enterprises are dependent on government customers as prime contractors, subcontractors, and suppliers, their internal processes mirror their customers.

In government and commercial industry, certain management products are required as imposed by laws, regulations, or requirements from sources of capital. Requirements vary depending on organization type. For instance, public companies must comply with Sarbanes Oxley. Nonprofit organizations must comply with regulations governing their type while government organizations have federal requirements. Private enterprises may be thought to have more latitude, but they are constrained by sources of capital and by laws

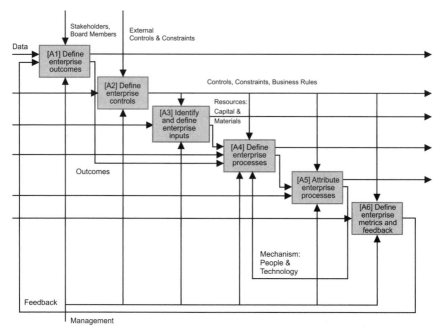

Figure 1.7 Optimize enterprise performance management process [A0].

for taxation, for instance. In a competitive environment, certifications emerge as self-imposed and imposed controls.

Where is the data in this diagram of management performance optimization? Data is input to and output from all processes. Data represents or measurably describes every control and every mechanism.

What data do executives need to determine to achieve the outcomes that the enterprise should produce? In an existing private enterprise, the nature of the business will suggest certain outcomes necessary to satisfy customer needs in return for which the business produces a profit. For a new business, the entrepreneur will anticipate customers and needs based on certain data from primary and secondary research. Even for an existing enterprise, primary and secondary research about customers and competitors is a continuous process.

For government enterprise, certain legislative givens identify outcomes, and, in addition, the executive branch of government working in concert with the legislative branch will generate new specifications for outcomes in response to constituent needs and wants.

In a large and complex enterprise, outcomes aggregate from subordinate processes. Outcomes specified by executives at the top of the enterprise stimulate organization responses at various levels and activity.

REFERENCES

1. Michael D. Shear and Anita Kumar, "Obama Picks Technology and Performance Officers," *Washington Post*, April 19, 2009.

2. National Institutes of Health, Glossary. http://science.education.nih.gov/supplements/ nih4/technology/other/glossary.htm.

3. Max Stier, "Challenges for the New Chief Performance Officer," *Washington Post*, January 7, 2009.

4. Al Ries, "Metric Madness: The Answer to Mathematical Failure Seems to Be More Math. If You Run a Company by Numbers Alone, You'll Run It into the Ground," *Advertising Age*, May 4, 2009.

5. Jim Finkle, "IBM Plans Cloud Computing Services for 2009," Reuters, April 23, 2009.

6. Marcus Cake,"Applying Web 3.0 to Financial Markets and Economic Development," http://www.marcuscake.com/.

7. Bob Bragdon,"The Value of Data," *CSO Online*, April 1, 2008.

8. Michael C. Daconta, "Designing the Smart-Data Engine," http://74.125.95.132/search? q=cache:P34xt0K_BVgJ:web-services.gov/Designing%2520the%2520Smart-Data% 2520Enterprise.doc + Designing + the + Smart-Data + Enterprise.&cd=1&hl=en&ct= clnk&gl=us.

9. Michael C. Daconta, Leo J. Obrst, and Kevin T. Smith, *The Semantic Web*, Wiley, 2003.

10. "AIG Annual Report 2006," http://www.ezonlinedocuments.com/aig/2007/annual/ HTML1/aig_ar2006_0004.htm.

11. Tim Berners-Lee,"Worldwide Web Consortium," http://w3.org/Consortium/.

12. Larry Clinton,President, Internet Security Alliance, testimony before the House Subcommittee of Communications, Technology, and the Internet, May 1, 2009.

13. http://www.opensource.org.

14. http://www.ichnet.org/glossary.htm.

15. Sid Adelman, Larissa Moss, and Majid Abai, *Data Strategy*, Addison-Wesley, Reading, MA, 2005.

16. Steve Martin, *Picasso at Lapin Agile*, 40 Share Productions, 1996.

17. Terje Totland, Norwegian Institute of Science and Technology, Knowledge Systems Group, 1995 p. 1172.

18. Donald Hall, Wilbert Bailey, and Jim George, A Service-Oriented Enterprise Strategy for Enterprise Integration, *Business Integration Journal*, Vol 2, p 38–42, July/August 2005.

19. E. Altman, "Financial Ratios, Discriminant Analysis and the Prediction of Corporate Bankruptcy," *The Journal of Finance*, 23, 589–609, 1968.

20. J. A. Rodger, Ch. 9: Utilization of Data Mining Techniques To Detect and Predict Accounting Fraud: A Comparison of Neural Networks and Discriminant Analysis. In Parag C. Pendharkar (Ed.) *Managing Data Mining Technologies in Organizations: Techniques and Applications* Idea Group Publishing, Hershey, PA pp. 174–187, 2003.

21. D. B. Yoffee, and M.A. Cusumano, "Building a Company on Internet Time. Lessons from Netscape." *California Management Review*, **41**, 38–28, 1999.

22. D. J. Paper, R. Chang, and J. Rodger, "Managing Radical Transformation at FannieMae: A Holistic Paradigm." *TQM Magazine*, **14**(4), 475–789, 2003.

23. D. J. Paper, J. A. Rodger, and P.C. Pendharkar, "A BPR Case Study at Honeywell." *Business Process Management Journal*, **7**(2), 85–99, 2001.

24. Paul L. Francis, Director Acquisition and Sourcing, GAO Testimony Before the Subcommittee on Armed Services, House of Representatives, Defense Acquisitions Key Considerations for Planning Future Army Combat Systems, March 26, 2009.

Chapter **2**

Elements: Smart Data and Smart Data Strategy

He who asks is a fool for five minutes, but he who does not ask remains a fool forever.
—Chinese Proverb

2.1 PERFORMANCE OUTCOMES AND ATTRIBUTES

Smart data strategy begins with recognizing that managing the enterprise is all about the data. The enterprise is most likely an existing organization, although it can be a new one. It can be a government entity or commercial enterprise. It can be an organizational element of a larger enterprise, and it can be at the pinnacle. It is about the business of producing outcomes that the enterprise must achieve to be successful for constituents, customers, and stakeholders.

Commercial enterprises must be superior at what they do to stay in business: witness the commercial banks and U.S. automakers that failed at this test. Government enterprises must also be superior at delivering priority services to stay in business or else they become descoped, eliminated, or replaced. In the aggregate, failing government enterprise can result in societal breakdown. The stakes are high.

Attributes define the performance characteristics of the outcomes in measurable terms such that users and managers can gauge their success. It is imperative to have visibility about performance such that managers and stakeholders have predictive insight about alignment with plans and intended trends and targets.

Outcomes and attributes, like standards and measures, are defined and described as data. Under our approach, the process in doing this is deliberate and precise. Obviously, when new leadership assumes responsibility, they must scope, scale, and prioritize the effort based on (1) capacity for change and improvement, (2) legacy demand for change and improvement, (3) current priorities and sense of urgency, and (4) their own value adding vision and strategy.

The management approach commensurate with smart data and smart data strategy is holistic. Beginning the discussion with outcomes will force consideration for other things as well. For instance, to determine outcomes, you must first consider the inputs and input conditions. You must factor in the impact of constraints such as business rules, laws, and regulations. You must surely consider critical processes and activities, which are attributed with enabling people and technology.

Example Outcomes and Attributes

U. S. Department of Health and Human Services (HHS)

This discussion about HHS reveals how we think about performance optimization with a smart data focus.

When the Obama administration began, there were some fits and starts in appointing a department head for HHS. That cost some time, weeks or a couple of months, for the new leadership to take hold. Nonetheless, the president's agenda and the congressional agenda include attention to healthcare reform, and momentum was underway toward that subject. Former governor Kathleen Sebelius has been confirmed as Health and Human Services Secretary, though a Republican senator held up the process.

To optimize performance in the U.S. healthcare enterprise, one must recognize that the government is only one participant in the collaborative process that includes public and private healthcare professionals and their organizations, businesses, and individuals. Yet, from the federal executive perspective, the president appoints a department head for HHS and that person is responsible for leading the collaboration to optimize healthcare performance for all U.S. citizens.

As we complete this book, commentators are discussing the administration's approach, posing questions such as "Is this a capitalist and free enterprise society, or are we moving toward socialism when it comes to healthcare?" That is a profound question and it is a sincere and serious one being waged in America today about a number of topics affecting all aspects of the federal government.

If you look at the HHS annual report for 2007–2012 and search for "outcomes," you will not find this term. What you will find is a discussion about how the department socialized the topics and the concerns of Americans through a workshop process that involved representatives from the healthcare community, including citizens. They produced a set of expectations that combine with executive leadership to guide what the department does in accordance with its legislative charter.

If you go to the HHS website on this day, you will see two main topics: Pistachio Recall and Health Reform. Obviously, the former is an immediate health threat and the latter is a long-term proposition, although both have one thing in common — each requires immediate action by the department. One outcome is to prevent further contamination and people getting sick from contaminated pistachios, just as the department had to do with peanuts the month before.

The data indicates that there is a problem inspecting nuts and ensuring their safety. Combine this with the previous year's onions-in-tacos and tomatoes-in-fast-food restaurants and the data would seem to indicate that there is an overall problem in inspecting foods. Yet, that is a very big leap without specific data facts.

Data-focused management operates in this way, however, and encourages exposing all of the data and how it aggregates to point to certain problems. When the problems are identified as to type and level of importance, they are systematically analyzed to determine solutions. This activity comes later, as now, in this example, we are in pursuit of identifying the department's planned and legislated outcomes.

The U.S. Department of Health and Human Services enterprise is a compilation of processes that have certain inputs: (1) rising healthcare premiums for U.S. citizens, (2) increasing drug costs, (3) people in increasing numbers without access to healthcare, (4) apparent lack of investment in preventative care, (5) healthcare quality disparities among the population, and (6) insufficient number of people entering the healthcare professions. These are all data points entering the HHS system. Most notable is that the United States spends more on healthcare than any other nation in the world and apparently gets much less in return.

Knowing that these things are the inputs into the HHS healthcare system, we will use the following label for the primary activity of the HHS: "Optimize U.S. Healthcare for Citizens."

Now that we see the input conditions, we can begin to develop some corresponding outputs or outcomes. Missing are the specific measures on the inputs:

- What are the average health insurance premiums for an individual today? What have they been in the past? At what rate are they increasing? How does the rate of increase compare to other pertinent economic factors?
- What are drugs costing today? What is their rate of increase? (Of course, one would want to know and understand the cost drivers.)
- What is the percentage of the U.S. population without access to affordable healthcare? What has it been in the past? What is the rate of expansion?
- What is the U. S. investment, public and private, in healthcare? What has it been in the past? What is the rate of investment?
- What is the investment in preventative healthcare, present and past, and associated rate?
- What is healthcare quality across the nation, averages and variations?
- What is the nation's present healthcare provider capacity versus capacity needed?

These are basic questions about the inputs that are needed to derive and define corresponding outcomes.

Corresponding outcomes might include the following:

- Targeted average premiums for individuals
- Targeted drug costs
- Targeted population with access to affordable healthcare
- Targeted U.S. investment, public and private, in healthcare
- Targeted U.S. healthcare quality
- Targeted healthcare provider capacity

Each of these requires specific metrics.

Is that all HHS does? The answer is no. These subjects happen to be the ones addressed by focus groups as a way of socializing the approach to planning.

Examining the department's strategic plan, we derived some additional outcomes. HHS identifies four major goals with associated objectives from which we derived outcomes.

1. Improved healthcare
   - Broadened insurance
   - Available insurance
   - Increased healthcare quality
   - Increased healthcare provider capacity
2. Improved public health
   - Disease prevention and limitation of disease spread
   - Prevention of environmental threats or limitation of injuries from environmental contamination incidents
   - Promotion of preventative care
   - Quick response to disasters and prevention of disasters
3. Promotion of economic independence and well-being
   - Promotion of total lifespan economic independence
   - Promotion of child and youth safety
   - Support for community development
   - Support for vulnerable populations
4. Advancement of scientific development
   - Support for health and behavioral science researchers
   - Increase in basic scientific knowledge
   - Supervision of applied research
   - Application of research results knowledge

Now that we have identified outcomes for which the department is responsible, we next define specific metrics. To complete this process, one would have to know the beginning states and the planned end states—not only the outcomes themselves as achievements, but also the associated costs and material profiles, so that the before and after pictures become clear. What do the laws and regulations say about the department's activities? Does the legislation and governance impose specific metrics regulating the budget and allocation to specific outcomes? Are there disparities between executive direction and legislative mandates? How much capital resource and material resource are allocated to HHS to accomplish its outcomes? Knowing all this is essential in determining the percentage of completion of each of the outcomes since they are also identified as goals.

Goals and objectives are often used synonymously, although we employ the term outcome and expect each outcome to have a specific quantitative measure or metric.

None of these outcomes say anything about lowering healthcare cost by automating patient records, and yet this is a major focus for the Obama administration. What is the current cost and expected savings? What is the investment required? Digitizing patient data has the potential for applying smart data and smart data strategy. Because the prospective insured are global, what consideration is given to international standards, for instance? This consideration belongs somewhere in the plan because its value aggregates as a part of a parent initiative, or such that it becomes a parent initiative.

Of course, these are all data points for which specific data is required, and you can see how accounting for data can cause executives and their organizations to be more specific, more transparent, and more accountable.

The government needs to go through this process and to identify outcomes that ultimately result in departments modeling their enterprises. They place quantitative measures on their primary outcomes and then begin the process of defining how work gets done to produce them.

Observe that this process does not begin by identifying a bureaucratic organization to determine how to keep them in business. That approach, which is close to the present practice, is rejected completely.

General Motors Corporation   What would be the first questions that Fritz Henderson should be asking about General Motors Corporation, besides how much time do we have? Based on what Fritz reads, we know he is drawn to adventure and intrigue, and that is different from being a former mortgage banking executive at GMAC.

General Motors (GM) is an automaker, a U.S. company that is competing in a global market with too many manufacturers, too many mediocre makes and models. The company has too much debt and is losing money, even though it is producing a few products that are world class.

GM says it has "100 years of innovation and that they are just getting started." You know that sounds true as they are starting over, but with a huge set of problems. It is not a zero-based start.

So the choices begin:

Outcome #1: Stay in business.

Outcome #2: Go out of business.

Outcome #3: If staying in business, doing what?

Inevitably, another outcome will be liquidation of assets that are not essential to a new strategy. If GM can stay in business, then the new business engineering process should supersede general operations; although keeping certain core operations moving is likely to become a part of the scenario. For all of this there is essential data.

2.2 POLICY AND BUSINESS RULES

All controls and constraints on enterprise process performance are expressed in the form of policies and business rules that may be externally and internally imposed. To account for this under our paradigm, we construct the enterprise leadership and integration architecture (ELIA) that in IT terms is called the control architecture.

ELIA is a model containing all of management processes and associated inputs, outputs, controls, and mechanisms that executives employ to manage the enterprise. ELIA flows down to each successive layer in the organization as a scaled replica focused on the missions of the respective organization to which it applies. This is the channel that provides continuity in management approach and accountability. It is the source from which resources are allocated to organizations and individuals to enable mechanisms to perform the work prescribed by processes.

Business rules establish conditions that guide and otherwise constrain performance as intended. Business rules may be embedded as methodologies and algorithms. To a large extent, rule implementation and enforcement can be automated, although some measure is left to intellectual intervention and interpretation.

A policy is a deliberate plan of action to guide decisions and achieve rational outcome(s). The term may apply to government, private sector organizations and groups, and individuals. A policy can also be a formal statement of principles that govern actions. Business rules are policies by which a business is run.

Such policies appear in a variety of forms, sometimes contractual. Many govern business automation; however, the method of delivery and appearance is in words and not executable code, for instance. That is an area worth exploring for improvement.

Policies are often embedded as business rules in software. When two organization entities subscribe to using the same software for certain functional or process execution, they may be agreeing to employ or to adopt the same business rules. However, since software is adjustable and adaptable, certainty about agreement cannot be guaranteed without deliberate attention and configuration management.

Smart data is engineered with methodology and associated algorithms such that policies and business rules are implemented with a higher state of automation, and such that planning, problem solving, decision making, and predicting are more autonomic as in self-regulation.

Information technologists use the term *data governance*. One definition is based on an IBM and Data Governance Institute combined definition: "'Data governance' is a quality control discipline for assessing, managing, using, improving, monitoring, maintaining, and protecting organizational information. It is a system of decision rights and accountabilities for information-related processes, executed according to agreed-upon models which describe who can take what actions with what information, and when, under what circumstances, using what methods."

We believe that before you get into the details of data governance, it is essential for executives to specify their data requirements and to know their data. Then the IT specialists can convene the collaboration needed to manage the details.

The smart data approach and a SOE model adoption should be incorporated directly into the organization's governance process, with the objective of maximizing SOE's value for internal transformation efforts within any organization. A smart data strategy toward viewing costs, at the beginning of a software project, would make a difference in cost saving outcomes and optimizing performance.

Case Study Example: An Empirical Study of the Cobb–Douglas Production Function Properties of Software Development Effort

An example of breaking from tradition with improved performance results, by adopting a smart data strategy toward costs, was shown in the paper entitled "An Empirical Study of the Cobb–Douglas Production Function Properties of Software Development Effort [1]. In this case study, we examine whether software development effort exhibits Cobb–Douglas functional form with respect to team size and software size [2].

This is also an example of how smart data is engineered with methodology and associated algorithms in such a way that policies and business rules are implemented with a higher state of automation, and planning, problem solving, decision making, and predicting are more autonomic such as in self-regulation.

We empirically test these relationships using a real-world software engineering data set containing over 500 software projects. The results of our experiments indicate that the hypothesized Cobb–Douglas function form for software development effort with respect to team size and software size is true. We also find as increasing returns to scale relationship between software size and team size with software development effort. There are over a dozen software effort estimation models and none of these models reliably predict software development effort.

There are two reasons why traditional software effort estimation models do not predict software effort accurately. The first reason has to do with the selection of a model that appropriately captures the underlying software development production process. The second reason may be the fact that the variables used in software effort prediction models do not capture all of the actual software development effort drivers.

Traditional software cost estimation models predict either software development effort or software development cost. Among the popular models for predicting software effort is the common cost model (COCOMO). Using a smart data approach, we have shown that software size and team size have a Cobb–Douglas relationship with software development effort.

A Cobb–Douglas production function is a convex production function, which means that for a fixed software size and team size a unique minimum software development effort exists. We have also shown that an increasing returns to scale economy exists, which means that when software size and team size are considered in software effort estimation models, nonlinear models will provide a better estimate than linear models.

Additionally, for large sized projects, linear models may be prone to large errors. The increasing returns to scale relationship indicates that team size and software size should be managed carefully by project managers because doubling the team size and software size will more than double the software development effort.

For projects facing schedule and budget overruns, managers must exercise caution before increasing the team size to achieve desired goals. We believe that the traditional COCOMO may be extended by adding additional multiplicative terms, so that the resulting COCOMO may be consistent with the Cobb–Douglas model and together they can optimize performance. It is likely that such a model may provide better cost estimates. While future research needs to address this issue, we feel that we have provided better cost estimates by breaking from tradition.

2.3 EXPECTATIONS: MANAGERIAL AND TECHNICAL

Smart data and smart data strategy have managerial and technical dimensions. Defining outcomes and attributes tends to be a managerial-oriented activity, whereas defining aspects of automated implementation and operations is more technical oriented.

Evidence that management has implemented smart data strategy will be the presence of the following:

- Enterprise smart data strategy
- Enterprise leadership and integration architecture (including data governance)
- Enterprise process models
- Enterprise data model
- Enterprise data accountability
- Resource commitment to implementation

Evidence that management smart data strategy has been implemented and is operational includes the following:

- IT professionals are performing the right work the right way.
- Information technology organization staffing is current with the right skills and proficiencies.
- State-of-the-art mechanisms for data exchange are being utilized.
- Highly proficient metadata management is occurring.
- Technology infrastructure is enabled, featuring open interoperability with less dependency on rigid standards.

- IT metrics are aligned with enterprise performance optimization metrics.
- Improved operational footprint and savings exist.

2.4 CAPACITY FOR CHANGE AND IMPROVEMENT

While America watches a new administration address the need for economic improvement, the public and government executives are cognizant that investment and government expenditure is being made on a large scale, such that it may be approaching a boundary of tenability.

How much, how fast, and how far can the government and private enterprise partnership go in making changes and improvements? Knowing well that it takes resource investment to make changes that are intended to produce reward on which to capitalize and otherwise exploit, what are the data and associated metrics as planned and as realized along a dynamic continuum?

There is interplay between government change and industry change that makes the process of change management even more complex to understand and to manage. The capacity for change and improvement for an enterprise is determined and constrained by executive bandwidth that comprises the chief executive officer and the executive team. The capacity for change and improvement is constrained by capital and by the intellectual and professional capacity of the staff. It is affected by the degree to which the organization has been modeled for optimal performance and otherwise being prepared for the subject period of performance or the starting position.

Effective organizations are designed to ensure a level of healthy autonomy that is subject to cyclical leadership inspiration and guidance. Such guidance comes in the form of plans that are executed as programs or initiatives that involve allocating resources to organizations for managing the achievement of specified outcomes. Outcomes may be routine products of defined processes as well as special outcomes from new or modified processes. What changes most is the environment in which outcomes are produced.

Elements in Determining Capacity for Change and Improvement

- Executive and management time
- Operations discretionary time/surge capacity
- Professional staff capability
- Cash flow
- Access to capital
- Customer/constituent situation
- Executive performance optimization infrastructure
- Planning, budgeting, funding cycle
- Number and criticality of deficiencies (deficiency profile)

- Competitor/threat situation
- Schedule and event drivers

2.5 ITERATION VERSUS BIG BANG

Enterprises cannot accomplish all that they can envision, need, and want at the same time; therefore smart data strategy and the associated pursuit of enterprise performance optimization must be iterative. It must happen with a plan and schedule that is within the scope of capacity for change and improvement.

The enabling technology platform on which SOE and smart data is based permits individual organizations in the enterprise to make improvements as they are ready. Entities should not be placed in the position of doing nothing while waiting for another improvement. The goal is to promote change and improvement that can happen under local control.

However, the process must also make dependencies visible and consequences of change visible, such that higher levels of management can intervene as necessary to manage collaboration and to leverage for maximum improvement. Herein lies the opportunity to apply smart data with associated methods and algorithms.

Case Study Example: Performance of Genetic Algorithm Based Artificial Neural Network for Different Crossover Operators

The following case illustrates these points, in which we examine the performance of a genetic algorithm (GA) based artificial neural network (ANN) for different crossover operators in order to provide a method for executives to exert local control while promoting change and improvement [3].

GA based ANNs are used in several classification and forecasting applications. Among several GA design operators, crossover plays an important role for convergence to the global heuristic solution.

Several crossover operators exist, and selection of a crossover operator is an important design issue confronted by most researchers. The current case investigates the impact of different crossover operators on the performance of GA based ANNs and shows how smart data and smart data strategy have both managerial and technical dimensions.

Defining outcomes, such as correct classification and forecasting, as well as identifying attributes tends to be a managerial-oriented activity, whereas defining the aspects of automated implementation and operations is more technical oriented.

The classification problem of assigning several observations into different disjoint groups plays an important role in business decision making and the development of an enterprise smart data strategy. A binary classification problem is a subset of classification problems. It is a problem where the data are restricted to one of two disjoint groups. Binary classification problems (also called a two-group discriminant analysis problem) have wide applicability in problems ranging from credit scoring, default prediction, and direct marketing to applications in the finance and medical domains.

To provide managers with an opportunity for enterprise leadership through integration architecture and to provide an enterprise process model, we must be aware of the several approaches that are proposed for solving various classification problems. The approaches can be categorized as linear and nonlinear discriminant analysis approaches.

The linear approaches use a line or a plane to separate the two groups. Among the popular approaches for linear classification models are statistical discriminant analysis models (Fisher's discriminant analysis, logit, probit) and nonparametric discriminant analysis models such as genetic algorithm/artificial neural network based linear discriminant models.

Nonlinear approaches used for discriminant analysis fall into two categories: the connectionist approaches employing some form of artificial neural network (ANN) learning algorithm, and the inductive learning models, where the discriminant function is expressed in symbolic form using rules, decision trees, and so on. The back-propagation ANN is the most commonly used connectionist scheme for nonlinear discriminant analysis. Various induction algorithms have been suggested for classification. Some of the most popular among them are CART, ID3, and CN2.

It is through supervised learning approaches that enterprise data accountability and resource commitment to implementation can be utilized by managers to make important decisions. Among the popular supervised learning approaches for non-linear binary classification problems are ANN, genetic programming, and ID3/C4.5.

All of these approaches, however, have been criticized for overfitting the training data set. However, it has recently been shown that adding random noise during the ANN training may alleviate the problem of overfitting the training data set at the expense of a lower learning performance on the training data set.

In the current case, we use the principles of evolution to train an ANN. Specifically, we use nonbinary genetic algorithms (GAs) to learn the connection weights in an ANN. GAs are general-purpose evolutionary algorithms that can be used for optimization. When compared to traditional optimization methods, a GA provides heuristic optimal solutions. Although heuristic optimal solutions are less attractive when traditional optimization approaches are likely to find better solutions, they may be attractive when finding an optimal solution has a chance of overfitting the training data set. Furthermore, the global and parallel nature of genetic search makes finding heuristic optimal solutions efficient when compared to the traditional local search based hill climbing and gradient descent optimization approaches, such as back-propagation.

For complex search spaces, a problem that is easy for GA may be extremely difficult for steepest ascent optimization approaches. GA based learning of connection weights for an ANN has received some attention in the computer science and operations research literature. Most studies, however, use forecasting and function learning domain as an application and many studies use binary GAs. Binary GA representation has been criticized for longer convergence times and lack of solution accuracy.

Recent studies have shown that GA based learning of connection weights is a promising approach when compared to gradient descent approaches, such as back-propagation and GRG2, and other heuristic approaches, such as simulated annealing and tabu heuristic search.

Unlike some of the previous studies, we use nonbinary GA representation. We specifically investigate the impact of different types of design parameters (crossover operators), group distribution characteristics, and group dispersion on learning and predictive performance of GA based ANN.

The contributions of our case are the following: we study the learning and predictive performance (with special interest in overfitting) of GA based ANN on training and unseen test cases under different data characteristics, rather than focus on the training/learning performance of GA based ANN for forecasting and function approximation problems.

We investigate the predictive performance of GA based ANN for classification problems, which has received little attention in the literature. We investigate the performance of different crossover operators on the learning and predictive performance of GA based ANN for classification. Most studies in the past used only one type of crossover and benchmarked the performance of GA based ANN with other approaches.

Unlike studies in the past that were limited with few data sets and functions, we conduct extensive experiments to increase external validity of our study. Finally, we use nonbinary representation since there is evidence in the literature that nonbinary representation is intuitively appealing and more efficient in terms of the use of computer memory and convergence times. The results of our study should therefore, be interpreted in the realm of nonbinary GAs.

In summary, we utilize smart data sets to provide evidence that a management smart data strategy is implemented and operational by demonstrating that (1) IT professionals are performing the right work the right way, (2) information technology organization staffing is current with the right skills and proficiencies, (3) state-of-the-art mechanisms for data exchange are being utilized, and (4) highly proficient metadata management is occurring.

The case also shows ways of enabling technology infrastructure featuring open interoperability with less dependency on rigid standards by alignment of IT metrics with enterprise performance optimization metrics and providing an improved operational footprint and savings. ANNs and GAs were developed to mimic some of the phenomena observed in biology. The biological metaphor for ANNs is the human brain and the biological metaphor for GAs is evolution of a species.

An ANN consists of different sets of neurons or nodes and the connections between the neurons. Each connection between two nodes in different sets is assigned a weight that shows the strength of the connection. A connection with a positive weight is called an excitatory connection and a connection with a negative weight is called an inhibitory connection. The network of neurons and their connections is called the architecture of the ANN. Let A = {N1;N2;N3}, B = {N4;N5;N6}, and C = {N7} be three sets of nodes for an ANN. Set A is called the set of input nodes, set B is called the set of hidden nodes, and set C is called the set of output nodes. Information is processed at each node in an ANN.

For example, at a hidden node, the incoming signal vector (input) from the three nodes in the input set is multiplied by the strength of each connection and is added up.

The result is passed through an activation function and the outcome is the activation for the node. In the back-propagation algorithm based learning, the strengths of connections are randomly chosen.

Based on the initial set of randomly chosen weights, the algorithm tries to minimize the square root of the mean-square error. Most supervised learning applications use back-propagation as the learning algorithm. Back-propagation as a method has several limitations.

One limitation is that learning time tends to be slow during neural network training using back-propagation. Learning time increases with the increase in the size of the problem. A second limitation occurs in the degree of difficulty of the training data itself. Researchers have attempted to accelerate the learning that takes place with back-propagation.

One study used variations in the learning rate and corresponding step size to decrease the learning time. Another study used various second-order techniques, which use a second derivative in the optimization process to utilize information related not only to the slope of the objective function but also to its curvature. A few studies used least-squares optimization techniques.

All of these techniques might have offered improvements over the basic back-propagation method. Since neural network training is a nonlinear problem, the GRG2 nonlinear optimizer was tested as an alternative to back-propagation. GRG2 is a FORTRAN program that is used to solve nonlinear constrained optimization problems by the generalized reduced gradient method. GRG2 evaluates the objective function and gradient at points required by the algorithm. The weights of the interconnectivity arcs and the bias values of the nodes are the decision variables.

These variables were initialized to zero. Training inputs were supplied to the network, which allowed the net inputs and activation levels of the succeeding nodes to be computed until activation levels were computed for the output layer. The sum of the squares of the errors for all of the training patterns was the value of the objective function. GRG2 allows the user to choose the method for generating search instructions. The Fletcher–Reeves formula (a conjugate gradient method) and the Broyden–Fletcher–Goldfarb–Shannon method (a variable metric technique), both using the GRG2 software, find solutions faster than back-propagation. The major limitation of back-propagation is its scalability.

As the size of the training problem increases, the training time increases nonlinearly. GRG2 had scalability problems as well, but to a lesser extent than back-propagation. The degree of difficulty in training data has also been studied. One study introduced an induction method called *feature construction* to help increase the accuracy in classification, as well as the learning time of the neural network.

Feature construction is a different way of representing the training data prior to input to the neural network. Instead of using raw data as training data, higher level characteristics, or features, are constructed from the raw data. These features are then input to the neural network.

For example, if the purpose of the neural network application is to determine the financial risk of a corporation, instead of using raw accounting data, the features of liquidity, probability, and cash flow could be used for more efficient learning by the neural network. This is a perfect example of why and how executives need to adopt the smart data paradigm for developing an enterprise smart data strategy, for building an enterprise leadership, for sustaining integration architecture, for following the enterprise process models, and for ensuring enterprise data accountability and resource commitment to implementation.

To construct the features, a feature construction algorithm called FC was used. FC uses the original data but builds new representations of that data. Training data that is difficult for a neural network to learn characteristically exhibits a high degree of dispersion. Feature construction can lead to a reduced degree of dispersion in the search space in which learning occurs. Feature construction was combined with back-propagation and then compared to back-propagation alone to determine deferences in learning performance. The feature construction/back-propagation hybrid performed faster than back-propagation alone.

The increase in speed is due to the fact that feature construction effectively reduces the size of the search space. The result is that instead of improving the back-propagation algorithm itself, scalability is made to be less of an issue. Also, by choosing features carefully, the number of peaks in the search space can be reduced. These peaks are often a source of difficulty for back-propagation. In addition to the increased speed of the feature construction/back-propagation hybrid, the accuracy of the hybrid was higher as well.

Limits of Tenability

There are several important elements in determining capacity for change and improvement in management of data. These include executive and staff time, operations discretionary time/surge capacity, cash flow, access to capital, customer/constituent situation, executive performance optimization infrastructure, planning, budgeting, funding cycle, number and criticality of deficiencies (deficiency profile), competitor/threat situation, and schedule and event drivers.

The proper use of artificial intelligence (AI) methods such as GA and ANN can help managers make optimal use of resources as well as managing these elements of change management. In this case we will compare various AI methods and show how they can be applied to managerial tasks such as forecasting and classification [4].

GAs are based on Charles Darwin's biological theory of survival of the fittest, which means that, in a species, the characteristics of the fittest members of the population will be passed to future generations. Over time, the species will grow to be increasingly well adapted for survival. This analogy can be applied to computer search problems. When trying to solve a mathematical problem or function, there may not be one correct answer. The problem then becomes one of finding an optimal solution.

A GA will find a solution set of increasingly optimal solutions to the problem. The first step in a GA is to randomly generate a population of numbers or binary strings, which are analogous to chromosomes in a biological connotation. Each chromosome is a vector of parameters, or genes, each of which represents a partial solution to the problem. The possible values of each gene are analogous to alleles. The chromosomes are then each substituted for a variable in a function to generate a solution. A fitness value is calculated for each solution, depending on how well the solution solves the function.

The chromosomes are each put into a weighted pool, where the weight of each number is its fitness value. Through random selection, members of the pool are selected for the mating process. The solutions with the highest fitness value have the highest probability of being selected. The selected members are then mated in pairs, through a reproduction crossover method. Each mate contributes to a portion of the offspring. A new generation is created, and the process starts over again, continuing in an iterative fashion until convergence to a solution occurs.

Convergence is realized when there is no change in the population from one iteration to the next. Often, at this point, the members of the population are exactly the same. GAs differ from the traditional back-propagation technique in several ways. One difference is that GAs optimize the trade-off between exploring new search points and exploiting the information already discovered. A second difference is that GAs have the property of implicit parallelism. In other words, the GA has the effect of extensively searching the schemas of the domain search space.

A schema is a way of representing a group of solutions that share similar characteristics. When studying the fittest members of the population, the most useful information is found by those traits that make them similar. A template that represents a member of the population is created, whereby each allele is a consensus value of the members being compared. If the values of each member's allele are all over-whelmingly similar, then the template takes that value. If the values of each member's allele differ, the template takes a wildcard value, represented by as asterisk (*). The order of a schema is a count of the alleles that are not wildcard symbols, giving a measure of specificity. Another difference is that GAs are randomized algorithms.

Random numbers are used to generate the results for any given operation. Finally, since GAs generate several solutions simultaneously, there is a lesser chance of convergence to a local maximum. The performance of a GA depends on a number of factors, including its population size, the cardinality of the alphabet chosen to represent the organisms, the complexity of the evaluation function, and the fitness function. There is a separate problem—identifying the domains that are suitable for GAs. The domain characteristics of problems that are suitable for GAs may be identifiable by predicting the behavior of GAs in those domains in terms of the fitness function. There are some problems that are easy for GAs to solve, but not for traditional search techniques. GA-easy implies that, within the given problem, the schema with the highest fitness value in every complete set of competing schemata contains the optimum.

Problems that are not GA-easy will cause the GA to be misled, resulting in a possible failure to find the global optimum before converging to a suboptimal solution. However, even though a problem may be GA-easy, it may not be optimizable by basic hill-climbing techniques, such as bit-setting optimization and steepest ascent optimization.

Because GAs begin with a population of strings, they are thought to have an unfair advantage over basic hill-climbing techniques. A comparison of the techniques shows that the GA was more effective in the search with fewer string evaluations than the single hill climbers. The reason for this result could be that each of the hill climbers searches a different space than the other hill climbers, and there is no crossover technique to distribute the good searches among the hill climbers.

GAs are parallel search techniques that start with a set of random potential solutions and use special search operators (evaluation, selection, crossover, and mutation) to bias the search toward the promising solutions. At any given time, unlike any optimization approach, a GA has several promising potential solutions (equal to population size) as opposed to one optimal solution. Each population member in a GA is a potential solution. A GA starts with a random set of the population. An evaluation operator is then applied to evaluate the fitness of each individual. In the case of learning connection weights for ANN for classification, the evaluation function is the number of correctly classified cases.

A selection operator is then applied to select the population members with higher fitness (so that they can be assigned higher probability for survival). Under the selection operator, individual population members may be born and be allowed to live or to die. Several selection operators are reported in the literature; the operators are proportionate reproduction, ranking selection, tournament selection, and steady-state selection.

Among the popular selection operators are ranking and tournament selection. Both ranking and tournament selection maintain strong population fitness growth potential under normal conditions. The tournament selection operator, however, requires lower computational overhead. In tournament selection, a random pair of individuals is selected and the member with the better fitness of the two is admitted to the pool of individuals for further genetic processing.

The process is repeated in a way that the population size remains constant and the best individual in the population always survives. For our case, we used the tournament selection operator. After the selection operator is applied, the new population special operators, called crossover and mutation, are applied with a certain probability.

For applying the crossover operator, the status of each population member is determined. Each population member is assigned a status as a survivor or a non-survivor. The number of population members equal to survivor status is approximately equal to population size × (1 − probability of crossover). The number of nonsurviving members is approximately equal to population size × probability of crossover. The nonsurviving members in a population are then replaced by applying crossover operators to randomly selected surviving members.

Several crossover operators exist. We describe and use three different crossover operators in our case. The crossover operators used in our research are the *one-point crossover* in which two surviving parents and a crossover point are randomly selected. For each parent, the genes on the right-hand side of the crossover point are exchanged to produce two children. The second operator is the *uniform crossover*. In the uniform crossover, two surviving parents are randomly selected and exchanging the genes in the two parents produces two children; probability of exchanging any given gene in a parent is 0.5.

Thus for every gene in a parent, a pseudorandom number is generated. If the value of the pseudorandom number is greater than 0.5, the genes are exchanged, otherwise they are not exchanged.

If we have two random surviving parents P1 and P2 (as shown in one-point crossover section) then a child C1 can be produced. The third operator is the *arithmetic crossover*, which consists of producing children in a way that every gene in a child is a convex combination of genes from its two parents. Given the two parents P1 and P2 (as illustrated before), a child C1 can be produced. Arithmetic crossover ensures that every gene in the child is bounded by the respective genes from both the parents.

Unlike uniform and one-point crossover, arithmetic crossover provides some local/hill-climbing search (if the parents are on the opposite side of the hill) capability for a GA. Arithmetic crossover is a popular crossover operator when GA is used for optimization. A mutation operator randomly picks a gene in a surviving population member (with the probability equal to probability of mutation) and replaces it with a real random number.

Since GAs are population based search procedures, at convergence of population fitness, there are several promising solutions. Unlike the traditional ANN where only one set of weights exists, GA has several sets of weights (equal to the population size). Since one of the objectives of our research is to minimize overfitting and to increase predictive accuracy on holdout data sets, we do not use the best fitness population member from the training data set to predict the group membership for the holdout sample. The holdout data set contains data that are not used for training ANN, but it has similar properties (kurtosis, group means, etc.) as that of the training data. We use the availability of several potential solutions to minimize the impact of overfitting on the training data set.

In order to select the population member (set of weights) to predict the group membership for the holdout sample, we identify all the population members that have a similar set of weights as that of the best fitness population member on the training data set. For the holdout sample, we select a population member that is the average of all the vectors. This aggregation reduces the chances of overfitting, where the best fitness population member from the training data is used for the holdout sample.

In our experiments, we use all three crossover operators (one at a time) and investigate the performance of a GA when different crossover operators are used. Thus, based on the crossover operator, we have three different types of GAs: genetic algorithm with arithmetic crossover called GA(A), genetic algorithm with uniform crossover operator called GA(U), and genetic algorithm with one-point crossover operator called GA(O). Our ANN architecture consists of four input nodes (three inputs + one threshold), six hidden nodes (five hidden + one threshold), and one output node. We benchmark the performance of our GA based training of ANN with the results of a back-propagation algorithm based ANN. For our architecture, we have a population member defining length of 26((3 inputs + 1 threshold) × 5 hidden + (5 hidden + 1 threshold) × 1 output).

Data distribution characteristics determine the learning and predictive performance of different techniques for classification. Specifically, researchers found that variance heterogeneity and group distribution kurtosis, the peakedness or flatness of the graph of a frequency distribution especially with respect to the concentration of values near mean as compared with the normal distribution, and their interactions affect the learning and predictive performance of the different techniques used for classification.

Please keep in mind as you read the following hypotheses that these AI techniques can optimize performance for executive and staff time, operations discretionary time/surge capacity, cash flow, access to capital, customer/constituent situation, executive performance optimization infrastructure, planning, budgeting, funding cycle, number and criticality of deficiencies (deficiency profile), competitor/threat situation, and schedule and event drivers. With these potential smart data applications in mind we propose the following hypotheses:

> *H1: The group variance heterogeneity will have an impact on both the learning and predictive performance of the different techniques.*

> *H2: The group distribution kurtosis will have on impact on both the learning and predictive performance of the different techniques.*

> *H3: The interaction of group variance heterogeneity and group distribution kurtosis will have an impact on both learning and predictive performance of the different techniques.*

A back-propagation ANN uses a gradient descent algorithm to minimize RMS, and a minimization problem is an unconstrained minimization of a convex function. Gradient approaches are known to find the optimal solutions given the appropriate initial starting position. GAs, on the other hand, are general-purpose optimization methods that use survival of the fittest strategy to find heuristic solutions. For unimodal convex minimization based optimization problems, GAs are likely to underperform the gradient based approaches. This suggests that GA based ANN performance during the training phase will be lower than the back-propagation based ANN performance. This leads to our fourth hypothesis:

> *H4: The back-propagation based ANN will have higher performance than the GA based ANN during the training phase.*

The arithmetic crossover, as described previously, incorporates some hill-climbing capabilities (when the two parents are on the opposite sides of the hill). Several researchers, because of the low disruption of schema (a specific pattern of genes in a population member), have used arithmetic crossover for the optimization problems. In our case, a GA using arithmetic crossover (because of its hill-climbing nature arising from convex combination of two parents on the opposite side of hill) will have better performance than uniform and one-point crossover GAs during the learning phase. This leads to our fifth hypothesis:

> *H5: The arithmetic crossover GA based ANN will have higher performance than the uniform crossover and one-point crossover GA based ANNs during the training phase.*

It has been observed that adding random noise to the connection weights during the back-propagation of ANN decreases the performance of ANN during the training phases but improves the predictive performance of ANN. One argument was that the gradient descent algorithms show high training performance and have a tendency to overfit the training data sets (overfitting is sometimes referred to as learning noise in the training data).

Adding random noise to the connection weights during the training phase prevents the ANN from overfitting the training data and improves its performance on the test data. GA based ANN has a tendency not to overfit the training data. The reason for overfitting is that GA works with a population of potential solutions as opposed to the one optimal solution approach used by most other techniques.

At convergence, the GA has a population of members that have similar fitness. The procedure to select potential candidates ensures that the population member used for the holdout sample is selected from a diverse population that helps avoid overfitting the training data. This approach helps GA based ANN to learn the general patterns and avoid overlearning the training data. The generalized learning approach, without overlearning the training data, is likely to perform better on the test data set when compared to back-propagation ANN. This leads to our sixth hypothesis:

H6: The GA based ANNs will outperform the back-propagation based ANN in the classification of unseen cases.

Crossover consists of exchanging information from the two parents to produce children. Different crossover operators can be used to produce children. Uniform crossover, because of its higher search space, has a potential to produce children that are less likely to contain the traits of any one parent. However, for two promising parents, one-point crossover because of its restricted ways to cross over, is likely to produce promising children. It has been found that uniform crossover underperforms one-point crossover and we propose the upper bound heuristic for the probability of the solution (schema) disruption for one-point crossover and uniform crossover. This leads to the following hypothesis:

H7: A uniform crossover GA will underperform a single-point crossover GA for both learning and prediction.

Smart data is a product of data engineering discipline and advanced technology. Making data smart enables a smart data strategy. Today, many organizations do not have a data strategy, much less a smart one, and the subject of data is not a part of executives' lexicon.

At a high level, smart data is applied as part of the systems engineering discipline. Yet, this subject is very much integral to enterprise management and therefore management and information science. Modern executives must have a hybrid command of these various disciplines. Whether you lead a commercial enterprise or a government enterprise, there is one activity that is at the top of the hierarchy in importance for which all CEOs and senior executives are responsible and accountable, and that is to optimize performance. Optimizing performance means applying scarce resources to business processes under constraint and transforming them into highest yield and best use outcomes by managing people and enabling technical mechanisms (technology).

All enterprises exist for a purpose, that is expressed in mission and value statements, goals, and objectives, otherwise summarized into business plans. Once desired outcomes are identified, leaders organize resources into functions. Functions identify the work that needs to be done to produce outcomes. How the work is accomplished is defined as processes, where process activities constitute proprietary differentiation.

Proprietary differentiation or a unique way of accomplishing things is achieved through a variety of means that begin with creative leadership:

- Selecting the right customers to service
- Selecting the right things to do
- Organizing activities
- Attributing activities with application of a superior mix of people and technology
- Applying scarce resources in an optimal manner
- Structuring the balance of consequences such that doing the right things the right way is rewarded and deviations are dissuaded
- Ensuring that customers receive valuable results
- Assuring stakeholders that the enterprise is performing optimally.

Data is at the heart of each of these management activities and that is why we focus on data as a principal contributor to optimizing performance.

There are so many moving parts in an enterprise that it is unfathomable that modern executives would attempt to manage without techniques to keep track of them, but many do. That condition is unacceptable in the highly automated world of the 21st century because resources are exceedingly scarce and risks are too high to operate intuitively.

One of the greatest barriers to realizing the full benefit of automation is from people inserting themselves with manual intervention, usually unintentionally. They need and want data to perform better, but it can't because they did not plan and prepare sufficiently for the moment of need. For enterprises to perform optimally, executives must insist on better data planning, preparation, and engineering.

For our case, we use data sets that have been used previously for comparing a number of techniques for classification. The data sets consist of 1200 data samples. Each data sample consists of three attributes and has 100 observations equally split between two groups. The data varies with respect to type of the distribution, determined through the kurtosis, and variance–covariance homogeneity (dispersion). The second data set used for this case is a real-life data set for prediction of bankruptcy filing for different firms.

We use two sources to identify a sample of firms that filed for bankruptcy. The first source was the *1991 Bankruptcy Yearbook and Almanac*, which lists bankruptcy listings in 1989 and 1990. The second source for bankruptcy filings was the Lexis/Nexis database. We used the following strategy to screen the firms for our bankruptcy data:

1. All filings for financial companies and real estate investment trusts were excluded.
2. Only filings by publicly traded firms were considered.
3. Data on financial ratios (identified later) are publicly available for 3 years prior to bankruptcy filing.

Our final sample had bankruptcy filings from 1987 to 1992. The financial data, for the firms identified in the sample, was obtained from the Compustat database. Since we use classification models to predict bankruptcy, we used data that would be publicly available at the time of filing. For each firm that filed for bankruptcy, we identified a nonbankrupt firm with the same four-digit SIC code and total assets similar in size to the bankrupt firm.

All the nonbankrupt firms in the final sample were publicly traded for 4 years prior and 3 years subsequent to the year of filing by the bankrupt firm. To predict bankruptcy, we used five financial ratios. These financial ratios (predictor variables) were *Earnings Before Interest and Taxes/Interest Expense, Earnings Before Interest and Taxes/Assets, Current Assets/Current Liabilities, Retained Earnings/Assets,* and *Market Value of Equity/Book Value of Debt.* The final data set contained 100 total firms with 50 firms that had filed for bankruptcy and 50 firms that did not file for bankruptcy. We took 50 bankrupt and 50 nonbankrupt firms and randomly split them into two data sets. The first data set was the training data set that contained 25 bankrupt firms and 25 nonbankrupt firms (a total of 50 firms).

The second data set was a holdout data set that contained the remaining 50 firms from the total original sample of 100 firms. We compared the three different types of GAs with back-propagation based ANN (referred to as NN in the tables) and genetic programming (GP) approaches. After initial experimentation, we used a population size of 100, a crossover rate of 0.3, and a mutation rate of 0.1 for our GA implementation. Table 2.1 illustrates the training results of our experiments. Tables 2.2 and 2.3 illustrate the results of three-way ANOVA for training performance of different techniques. From Table 2.2, it can be seen that hypothesis 1 (variance heterogeneity) ($F = 2739:30$) and hypothesis 2 (distribution kurtosis) ($F = 22:87$) are supported (at the 0.01 level of significance) for learning (training) performance of the techniques.

Furthermore, the interactions between variance heterogeneity and kurtosis and between technique and variance heterogeneity were significant (at the level of significance = 0.01), supporting as well hypothesis 3. From Table 2.3, it can be seen that hypothesis 4 is supported (at the 0.01 level of significance) with NN versus GA(O) ($F = 235:20$), NN versus GA(U) ($F = 256:50$), and NN versus GA(A) ($F = 206:42$).

TABLE 2.1 Training Performance

Variance Heterogeneity		Group Means			GA(A)	GA(O)	GA(U)	NN	GP
Group 1	Group 2	Group 1	Group 2	Kurtosis	Mean	Mean	Mean	Mean	Mean
1	1	0	0.5	−1	76.62	76.62	76.38	80.28	78.08
1	1	0	0.5	0	77.22	76.90	76.60	80.18	79.78
1	1	0	0.5	1	77.46	77.46	77.36	80.32	79.42
1	1	0	0.5	3	78.58	78.18	78.16	81.80	79.60
1	2	0	0.6	−1	85.58	85.04	84.78	92.38	83.54
1	2	0	0.6	0	83.92	83.62	83.44	89.00	82.46
1	2	0	0.6	1	83.78	83.58	82.80	88.14	81.37
1	2	0	0.6	3	83.46	82.96	82.86	86.72	81.12
1	4	0	0.8	−1	94.82	94.90	94.58	96.78	93.92
1	4	0	0.8	0	93.14	92.82	92.84	96.50	90.72
1	4	0	0.8	1	90.56	90.28	90.48	95.08	90.44
1	4	0	0.8	3	91.68	91.48	91.56	93.16	90.08

No support to a very weak support (level of significance = 0.1) was found for hypothesis 5 (GA(A) vs. GA(O) and GA(A) vs. GA(U)) ($F = 2:76$).

The results of testing the different techniques on the holdout samples illustrates that GA based ANNs outperform both back-propagation neural network and genetic programming. The difference in the performance between GA based ANN and back propagation ANN decreases as the group variance heterogeneity increases. The results of three-way ANOVA on the holdout sample experiments show that there is support (at the 0.01 level of significance) for hypothesis 1 (variance heterogeneity) ($F = 3907:66$) and hypothesis 2 (distribution kurtosis) ($F = 25:61$) for holdout test data sets as well. Similar to the training results, for hypothesis 3, the interaction

TABLE 2.2 Correct Classification ANOVA Summary Table for Training Data

Source	Sum of Squares	Degree of Freedom (df)	Mean Square	F Ratio	P > F
Main effect					
Distribution (D)	1314.04	3	438.01	22.87	0.0001[a]
Variance (V)	104,928.34	2	52464.17	2739.30	0.0001[a]
Technique (T)	7455.30	4	1863.82	97.32	0.0001[a]
Two-way-interaction effect					
$D \times T$	161.15	12	13.43	0.70	0.7518
$D \times V$	2312.50	6	385.42	20.12	0.0001[a]
$T \times V$	1846.56	8	230.82	12.05	0.0001[a]
Three way interaction effect					
$D \times T \times V$	524.99	24	21.87	1.14	0.2868

[a] Signicant at 0.01 level of significance.

TABLE 2.3 Overall Pairwise Comparisons on Training Data Sets

Contrast	df	Sum of Squares	F Value	P > F
NN vs. GP	1	5100.56	266.31	0.0001[a]
NN vs. GA(O)	1	4504.68	235.20	0.0001[a]
GP vs. GA(O)	1	18.50	0.97	0.3258
NN vs. GA(U)	1	4912.65	256.50	0.0001[a]
GP vs. GA(U)	1	1.76	0.09	0.7616
GA(A) vs. GA(O)	1	18.50	0.97	0.3258
NN vs. GA(A)	1	3945.81	206.02	0.0001[a]
GP vs. GA(A)	1	74.00	3.86	0.0494[b]
GA(A) vs. GA(U)	1	52.92	2.76	0.096
GA(O) vs. GA(U)	1	8.84	0.46	0.4969

[a] Significant at 0.01 level of significance.
[b] Significant at 0.05 level of significance.

effects of distribution kurtosis and variance heterogeneity were significant (at the 0.01 level of significance) as well. However, unlike the training results, the interaction effect of distribution kurtosis and technique was significant at the 0.05 level of significance.

Furthermore Table 2.4 shows that hypothesis 6 is supported at the 0.01 level of significance and the GA based ANN outperforms back-propagation based ANN (NN vs. GA(O) ($F = 411:08$), NN vs. GA(U) ($F = 379:05$), and NN vs. GA(A) ($F = 366:80$)). No support was found for hypothesis 7 (GA(U) vs. GA(O)) both in Tables 2.3 and 2.4. GA based ANNs performed better or similar to back-propagation ANN on the holdout sample. For training, except for GA(A), which performed similar to the back-propagation algorithm, GA(O) and GA(U) and GP did not perform well. GP had the worst performance on the holdout sample.

TABLE 2.4 Overall Pairwise Comparisons on Holdout Test Data Sets

Contrast	df	Sum of Squares	F Value	P > F
NN vs. GP	1	3474.80	167.92	0.0001[a]
NN vs. GA(O)	1	8506.69	411.08	0.0001[a]
GP vs. GA(O)	1	22855.14	1104.47	0.0001[a]
NN vs. GA(U)	1	7843.85	379.05	0.0001[b]
GP vs. GA(U)	1	21760.08	1051.55	0.0001[a]
GA(A) vs. GA(O)	1	26.11	1.26	0.2614
NN vs. GA(A)	1	7590.27	366.80	0.0001[a]
GP vs. GA(A)	1	21336.33	1031.07	0.0001[a]
GA(A) vs. GA(U)	1	2.08	0.10	0.7510
GA(O) vs. GA(U)	1	13.44	0.65	0.4203

[a] Significant at the 0.01 level of significance.
[b] Significant at the 0.05 level of significance.

Smart data is not just plain old vanilla-flavored data, and it may not appear to the executive or user any differently than data they have seen in the past. The difference is that smart data will be more reliable, timely, accurate, and complete. Therefore smart data is actionable. It will be the product of a more cost-effective approach to delivering higher quality information.

Smart data is engineered for high states of interoperability among disparate users, which includes semantic mediation that translates meaning from one enterprise lexicon to another despite differences. This is essential when conducting commerce or delivering critical human services, such as the health services described in this case, where inaccuracies are unacceptable and subject to legal liability. Our case shows that GA based ANN training follows the basic premises of smart data because this method shows resistance toward overfitting in a binary classification problem.

We think that this is a strong facet, and researchers and practitioners should use GA based ANN when a higher predictive performance is desired. To understand the difference between outdated methods of data analysis and smart data requires (1) knowing how information is exchanged by the enterprise today, including the support footprint and costs, and (2) knowing how information is engineered in a smart data paradigm by contrast. Achieving this knowledge is the intent of this chapter.

Our case provides a glimpse about what should be expected from smart data technology so that it can be used for decision making. Our case follows the smart data paradigm because it is one of the first to compare the performance of different crossover operators related to design of GA based ANN. This technique can then be applied to the important elements that are necessary to determine capacity for change and improvement in management of data, including executive and staff time, operations discretionary time/surge capacity, cash flow, access to capital, customer/constituent situation, executive performance optimization infrastructure, planning, budgeting, funding cycle, number and criticality of deficiencies (deficiency profile), competitor/threat situation, and schedule and event drivers.

The proper use of AI methods such as GAs and ANNs can help managers make optimal use of resources as well as manage the elements of change management. We postulate that smart data and a smart data strategy are essential elements for optimizing enterprise performance. Optimizing enterprise performance is the most basic executive responsibility. In our quest to find higher learning performance and to optimize our solution, no significant difference was found between the different crossover operators. However, we believe that for larger networks crossover may play a vital role during learning (since GA(A) vs. GA(U) were significantly different at the 0.1 level of significance during the training phase). In our case, we kept the ANN architecture constant. ANN architecture plays an important role during the training and predictive performance. In cases where higher learning performance is desired, hybrid approaches might have potential.

For example, the best fitness population member from the GA based ANN can be used as the initial set of weights for the back-propagation algorithm and higher learning performance may be obtained. The hybrid approaches may have some merit when compared to random initialization of weights for an ANN. Future research may focus on the impact of network architecture on the training and predictive performance of GA trained ANN. Hybrid approaches deserve merit for future investigation as well.

Using the Department of Defense (DOD) as an example, each military service represents a unique brand of warfighting and providing for the common defense. The Joint Chiefs consider scenarios based on combining these unique brands for optimal effect.

The same can be said about the Department of Homeland Security, where each agency is responsible for a unique brand and contribution to securing the homeland.

What are the data profiles for these departments and their agencies? What are the key operational and performance metrics to deliver specific outcomes?

These are questions that must be answered precisely under a smart data strategy at every level of the organization to ensure accountability. These questions extend throughout the enterprise to every contractor and contract and supply chain. Smart data strategy is the common thread.

As stated previously, the enabling technology platform on which SOE and smart data is based permits individuals and functional areas of the organization within the enterprise to make improvements as they are ready. Entities should not be placed in the position of doing nothing while waiting for another. The goal is to promote change and improvement that can happen with local control.

Optimizing Performance and Breaking from Tradition

An example of optimizing performance and breaking from tradition, with improved results, is evidenced by adopting a smart data strategy toward costs, as was shown by pendharkar and Rodger in "The Relationship Between Software Development Team Size and Software Development Cost" [4]. We show that most of the software development cost is related to the programmers' salaries and recent evidence shows that team-related factors can affect project performance and its cost. Selecting an appropriate team size for software development projects continues to be a challenge for most project managers.

Larger teams represent better distribution of skills but lead to higher communication and coordination costs. Smaller teams lead to lower communication and coordination costs but may result in programming biases that make software maintenance costly. Smaller teams, due to homogeneous programmer backgrounds, may lack the problem-solving expertise for large size complex projects. In addition to programmer skills, heterogeneity in team member personalities may be desirable in certain phases of the Systems development life cycle (SDLC). Additionally, it has been emphasized that diverse expertise in IT and business application experience is always desirable in any project.

Software development project tools provide advanced coordination and communication capabilities and are often used to control communication and coordination costs for large sized teams. Without the use of software development tools, it is easy to show that team communication requirements increase exponentially with the increase in team size. Since most projects use advanced software development tools, the assumption of exponential increase in software development cost with increase in team size will seldom hold true in the real world. Even a linear relationship between software development cost and team size may be considered a mute point, at best.

Understanding the relationship and impact of team size on software development cost is important because most managers, when confronted with strict deadlines, are tempted to increase team size. Among the reasons for increasing team size are the following:

1. There is pressure to meet the schedule, as competitive advantage of the technology decreases with time.
2. Assigning one person to a project does not mean linear completion time.
3. Dividing work into small subprojects allows many people to work on the project simultaneously, which reduces the project completion time.

Although increasing team size for strict deadlines appears to be a good strategy, it is not really clear if it works well. For example, increasing team size will more likely increase the number of defects found by the team and may not necessarily mean the project will be completed on time. In fact, most software quality professionals admit that team size should be larger in the beginning or middle of the project than at the end of the project. The larger team sizes in the beginning of the project allow for strict testing of the software product in early phases of SDLC.

In this case, we study the relationship between team size and software development cost. Since larger size projects are likely to contain large team size and vice versa, we use software project size in function points (FPs) as a contextual variable. We use real-world data from 540 software projects for our data analysis. Our data were obtained from the International Software Benchmarking Standards Group (ISBSG) and is publicly available.

We use production economics theory as a guideline for testing the relationship between team size and software cost. We use software effort as a surrogate measure for software cost because once the software effort is known software cost can be computed by multiplying software effort with average hourly programmer wage. The existing literature in production function analysis provides three models that can allow us to test the relationship between team size and software effort.

These three models are the linear regression model, the log-linear model to test nonlinear relationships, and the nonparametric data envelopment analysis (DEA) model. Linear and log-linear models assume that the variables used in the models—regular values for linear regression and log-transformed values for log-linear models—are normally distributed. Using the SPSS software package, we conducted a Kolmogorov–Smirnov (KS) normality test for three variables and their natural logarithm transformed values to check if the variables and their transformations were normally distributed.

The KS method tests the null hypothesis that there is no difference between the variable distribution and normal distribution. Since a perfectly normal distribution has skew and kurtosis values both equal to zero, we report the skewness and kurtosis values for the variable distributions using the KS test.

The results of our normality tests indicate that, except for the log-transformed software effort variable, all the variable distributions were nonparametric. Since most of our variables violated normal distribution assumptions for linear and log-linear models, we did not use these models for our analysis. We decided to choose the last remaining nonparametric DEA model for our analysis. The DEA model does not assume normal distribution.

It has been proposed that the DEA approach can be used for production function estimation. It has also been argued that the DEA approach is superior to parametric approaches, as it does not impose a particular form on the production function and assumes a monotonically increasing and convex relationship between inputs and outputs. Monotonicity and convexity are standard economic production function assumptions.

To test the relationship between team size and software development cost, we use the nonparametric DEA statistical testing approach. We calculated the inefficiencies, θ_i^C and θ_i^B, for each ith software project. We assume that the reader is familiar with these models. For a single input of team size variable a single contextual variable of software size in FP, and a single output variable of software effort, was used Central Contractor Registry (CCR).

We used a software package called Frontier Analyst Professional by Banxia Software Inc. for our analysis. As shown in Figure 2.1, the software provides the option for running both models.

We test the hypothesis that relationship between team size and software effort is constant returns to scale. We assume nonparametric exponential and half-normal distributions to test our hypothesis. For data where skewness is a large positive number and all the values of variables are positive, half-normal distribution is the

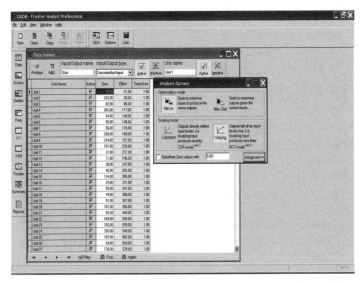

Figure 2.1 DEA data analysis in Frontier Analyst Professional.

preferred distribution. The exponential distribution is a common nonparametric distribution for positive data values. Since all three variables in our study are positively skewed and all the values take positive values, we test our hypothesis under these two different distributions. The F-ratio (df $= 1080$) value of 1.30 for the exponential distribution was significant at $p = 0.01$, and the F-ratio (df $= 540$) value of 1.54 for the half-normal distribution was significant at $p = 0.01$.

The null hypothesis of constant returns to scale was rejected. The results indicate the existence of variable returns to scale economy, which means that proportional change in team size will mean higher or lower proportional change in software effort. The constant returns to scale (CRS) means that proportional change in team size will mean the same proportional change in software effort. Using the variable returns to scale model instead of the CRS model will lead to improvement in software cost estimation. This improvement can be estimated. For our data set this cost reduction formula can be written as follows.

Total Cost Reduction $=$ Wage Rate

$$\times \text{ Total Effort} \left[1 - \sum_{i=1}^{540} \left(\frac{\theta_i^B}{\theta_i^C} \times \frac{\text{Actual Effort for } i\text{th Project}}{\text{Total Effort}} \right) \right]$$

At a wage rate of \$65/hour and the total effort of 2,817,557 worker hours, a maximum cost saving of \$184.15 million can be achieved.

Through the inputs of smart data from global software projects, we used DEA analysis technique processes to determine optimized performance metrics for team size and costs. We have investigated the relationship between team size and software cost and found a nonlinear variable returns to scale economy exists between team size and software development cost. This seems to indicate that increasing team size does not linearly increase software development cost. The variable returns to scale relationship indicates that team size should carefully be managed by project managers because its impact is not easy to predict. If there are team communication or coordination problems, the managers can assume increasing returns to scale economy, which would mean that increasing the team size by a certain proportion would require more than a proportional increase in resources. Under this assumption, increasing team size for projects facing schedule overrun may not be a good idea. However, if team members are known to efficiently communicate and coordinate their problems, then managers can assume decreasing returns to scale economy, where proportional increase in team size would mean less than proportional increase in the software cost.

The variable returns to scale relationship between team size and software cost also indicates that when the team size variable is used to predict software development cost, using nonlinear software prediction models provides a better estimate than using linear regression models. Furthermore, the actual saving resulting from nonlinear regression models may be on the order of several million dollars.

Our study does have some limitations that are worth mentioning to the reader. Since we did not screen our projects based on different programming languages and use of computer aided software engineering (CASE) tools, our conclusions are not specific to projects that use certain programming languages or CASE tools. The use of object-oriented programming and CASE tools might allow managers to control the cost escalation due to increase in team size. Another limitation of our study is due to unavailable information on team members' experience. The ISBSG data set does not provide this information. Team member personalities and their experience and knowledge can all impact software development cost. Selecting team members based on their experience and personality type may provide project managers with another avenue to control software development cost.

The DEA model used in our study has some limitations as well. Since DEA uses extreme points to compute efficiencies, errors or noise in data can have a significant impact on the results. Additionally, for large data sets, the linear programming based DEA model can be computationally intensive.

As stated earlier, smart data and smart data strategy have managerial and technical dimensions. Defining outcomes and attributes tends to be a managerial-oriented activity, whereas defining aspects of automated implementation and operations is more technical oriented. Both of these dimensions are demonstrated in the paper entitled, "A Field Study Utilizing the Medical Companion Healthcare Module (MCHM) and Smart Data to Measure Importance, Effectiveness, Suitability and Ease of Use in Mobile Medical Informatics (MMI) Systems" [5].

The purpose of this paper was to offer a field study example of a smart data approach to the Medical Companion Healthcare Module (MCHM) residing on the Palm Personal Digital Assistant (PDA) and utilizing the Palm operating system. Kelly's personal construct theory (PCT) was used as a framework to explore the end-user perceptions of information system design and development (ISDD) activities.

This research utilizes the PCT as a methodology to understand the Subject Matter Expert SME's reluctance to accept and use new Mobile Medical Informatics MMI systems in the military. Smart data was collected and analyzed for measuring the importance, effectiveness, suitability, and ease of use of MMI systems. We postulate that smart data and a smart data strategy are essential elements for optimizing enterprise performance.

Optimizing enterprise performance is the most basic executive responsibility. Smart data is a product of data engineering discipline and advanced technology. Making data smart enables a smart data strategy. Today, many organizations do not have a data strategy, much less a smart one, and the subject of data is not a part of executives' lexicon.

Previously on page 103 we said "At a high level, smart data is applied as part of the systems engineering discipline. Yet, this subject is very much integral to enterprise management and therefore management and information science. Modern executives must have a hybrid command of these various disciplines. Whether you lead a commercial enterprise or a government enterprise, there is one activity that is at the top of the hierarchy in importance for which all CEOs and senior executives are responsible and accountable, and that is to optimize performance. Optimizing performance means applying scarce resources to business processes under constraint and transforming them into highest yield and best use outcomes by managing people and enabling technical mechanisms (technology)." Now, here is a case that demonstrates that this is true.

Data Is at the Heart of Creative Leadership

Documentation of medical encounters is a military requirement. In the course of conducting a clinical encounter, staff personnel triage, diagnose, and treat up to a thousand patients in a day. Smart data collection and the resulting documentation of these encounters is a moral and legal requirement. Since paper dissolves in tropical environments and paper-based documentation is labor intensive, an alternative method of smart data documentation was investigated. There are further challenges in the rapid compilation of data from paper for reporting purposes and trend analysis. This is particularly true with multiple and dispersed medical staff contingencies. Compilation and computation of the data for statistical morbidity and trend analysis can take weeks. In addition, paper-based systems are notoriously problem laden when it comes to legibility, interpretation, and errors. The MCHM was developed as a smart data prototype, to address the medical encounter documentation requirement and decrease transcription errors.

The field study mission called for the deployment of a subset of the staff of a Navy Fleet Hospital to conduct programs and collect data on programs such as optometry, physical therapy, dental, family practice, and veterinary services in Thailand. However, the overall responsibility was with the Royal Thai Military and Department of Public Health.

Emphasis was placed on the importance of capturing end-user feedback associated with this research and development (R&D) product. Total training time was approximately 30–45 minutes. All end-users were given the opportunity to access the system, create a patient encounter, and finalize the encounter. At the conclusion of the training session, end-users were provided with a performance-based test script and asked to create a patient record and finalize the encounter. All 30 participants, in the training, successfully completed the training and achieved a passing score on the performance-based competency test.

In this section, we describe the marshalling of assets and staff at the respective areas in preparation for employment. Thirteen Tungsten Palm computers with integrated keyboards, five portable expandable keyboards, one laptop used as the in-theater server, two personal laptops, and support gear were packed in two hardened Pelican cases and carried aboard a commercial aircraft from Pennsylvania to Thailand without incident. The equipment was set up and tested upon arrival in a Thai hotel using a Thai power converter and U.S. surge protectors. No problems were experienced during the initial testing.

Usually, it took one IT engineer 15 minutes to be fully operational, with the server up and running and a systems test performed. Another member of the IT team would assist the triage personnel with establishing the patient data flow process, delivering the patient data collection sheets and the color stickers that would be placed on the sheets, and informing the triage personnel where the IT team would be located for further assistance. The data collection usually started between 0800 and 0830 and lasted until the last patient was seen, usually around 1600–1800 hours. The site layouts were different at each location, and the clinical treatment areas, pharmacy, and lens-making areas were positioned based on available space and power considerations.

Users involved in the employment phase of the data collection included two experts from physical therapy, one expert from general medicine, two veterinarians, one licensed practical nurse, one optometrist, one dentist, one pharmacy technician, two computer specialists, one observer, and one respondent of undetermined status.

All of the patient demographic information and chief complaints were collected in the triage area on the patient collection form. Thai Military or Thai Public Health, along with U.S. Navy personnel, staffed the initial entry point. From that point, at each site, patients were initially screened and sent to one of several functional areas, including general medicine, physical therapy, dental, optometry, and the pharmacy. Palm PDAs with the MCHM installed were located at each of these functional areas, with three to five PDAs located in the general medicine area, one PDA located in ophthalmology, one PDA in dental, and one PDA in physical therapy. The veterinary treatment area was provided with two PDAs, with one of the PDAs used by the veterinary team that would travel to outlying areas.

Following treatment in each area, data was entered into the PDAs, including narrative comments. Several times each day, this data was downloaded to a laptop computer, utilizing small PDA disks. This provided the HCA Commander, Commanding Officer Fleet Naval Hospital, and local Thai public health officials with near real-time information regarding the total patients processed at that point. Before departing the site, the IT team was able to deliver all of the customized reports of the day's activities available on the web-based server to the personnel mentioned previously. This was of valuable interest to the mentioned parties, because the HCA Commander could provide his superiors with a recap of activities at the site, and the medical personnel could see the results of their work.

At the conclusion of each session, these data were downloaded and transmitted, via dial up or broadband connection, to the U.S. server. In this manner, the data could be accessed via the Internet, for those who had established user accounts. The data were presented in several reporting formats and provided near real-time patient information data for U.S. and Thai military personnel and Thai public health officials.

The average file size was 549.89 kb, with the largest file being 907 kb and the smallest being 396 kb. The average transmission time for dial-up access was 19 minutes 41 seconds, with the longest time being 26 minutes at a connection speed of 36.6 bps for a 907 kb file, and the shortest time being 5 minutes at a connection speed of 56.6 bps for a 721 kb file. The average transmission time for broadband connection was 3 minutes for file sizes of 579 kb and 470 kb.

Following the exercise, each of the known users participated in structured interviews and completed surveys. The users' opinions and insights regarding the operation of the MCHM were recorded and will be used for future development of the system. At the conclusion of the final data collection, the Healthcare Administrator (HCA) and the Officer in Charge (OIC) of the Fleet Naval Hospital, along with pertinent Thai officials, were provided a detailed roll-up of the total patient encounters captured throughout the nine supported data collection sites. This data would be used to inform their superiors and proper personnel of the outcome and success of the data collection. These data are still available on the website. Presently, users are querying this data for reporting purposes.

The team's task was to test and evaluate the MCHM and to measure its capability to collect smart data. The team functioned as "active participants" in the project. According to Creswell [6], a qualitative approach is necessary when research must be undertaken in a natural setting where the researcher is an instrument of smart data collection, data is analyzed inductively, the focus is on the meaning of participants, and a process needs to be described. Qualitative researchers want to interpret phenomena in terms of the meanings people bring to them [7]. Selection of a qualitative approach should flow from research questions that attempt to answer why or how [8], so that initial forays into the topic describe what is going on [6].

The intention of the research was to test and evaluate an MCHM that streamlines and automates the discovery disclosure process of smart data. The team was also responsible for the education and training of end-users to assist them in using the system to develop their own processes based on their own needs. From a problem solution perspective, the team wanted to better understand how new technology is implemented in support of a complex process or discovery disclosure in an environment unaccustomed to change. With this knowledge, we could then derive better solutions for improving information flow related to discovery disclosure in a medical environment. The MCHM system not only required dramatic changes to the existing technology infrastructure, it also required a paradigm shift in the way end-users interact with technology. The MCHM system allows end-users direct control over the design of their processes without the need for computer programming.

As a result of the collection of this smart data, both the end-users and management had to rethink the way they operated. Change is difficult in any setting, but radical change is even more difficult to manage. In pursuing the answers to the research question we wanted to record the people, political, technical, and managerial dynamics of the smart data involved in the case. We entered into the project with an understanding that the MCHM testing and evaluation needed to be completed on schedule and that what was learned from the smart data collection and our participation would be used to facilitate process change.

A survey was designed to investigate the smart data collection utility of the MCHM. This exploratory instrument was offered to MC decision makers. The survey was specifically designed to initially assess the importance, effectiveness, and suitability of the MC as used in the data collection and the HM. Implementation was successful, if success was measured as a function of use, and the MCHM project overall was considered to be a success.

A review of the smart data supplied by the surveys suggests that decision makers consistently found the MCHM important for completion of the commanders' missions. For example, 10 SMEs had favorable comments on the importance of MC tasks that were performed. Nine respondents had favorable comments on the importance of the task.

Ten decision makers commented favorably on the difference made by the MCHM. All 10 of those sampled agreed that there was a desire for this technology, and seven SMEs had favorable general MCHM importance comments.

The initial impression given by the respondents to the survey indicates that the MCHM has smart data collection capabilities. This conclusion can be drawn from the preponderance of favorable comments by the decision makers regarding importance, effectiveness, and suitability. The MCHM met several areas of importance in addressing the requirements of the military medical mission. It was demonstrably effective in documentation of mass-patient encounters. It was felt to be more suitable than the current paper-based status quo. The possibility of further hardening—adapting the unit for use in other extreme environments—is reasonable. Finally, the initial and sustaining costs are under consideration.

According to our research, importance is the highest weighted factor of smart data collection. The importance of the MCHM has been sustained by the surveys. The decision makers surveyed recommended that the MCHM continue to be developed. Their comments lend credence to the importance of the MCHM in the tasks performed, the importance of the tasks, the difference made by the MCHM, the importance of technology, and favorable MCHM importance comments in general. These factors alone would warrant further consideration of the MCHM prototype for experimentation in other exercises. It is strongly recommended that the MCHM be identified and marketed to the sponsors of all exercises worldwide for potential use and experimentation.

Based on the surveys, the MCHM performed effectively. However, with minor changes to the screens and pick lists, the tool was perceived to have extraordinary potential effectiveness. The decision makers gave favorable comments about their perceptions that the MCHM performed as advertised. The decision makers also commented favorably on the effectiveness of the actual MCHM performance and the ease of training. Several decision makers offered unfavorable comments concerning the ability to personally see results and the overall effectiveness of the MCHM. To this end, it is recommended that a panel of subject matter experts be established in a 3-day session to review and study the screens and pick lists and recommend necessary changes.

Generally, the top three suitability factors are usability, compatibility, and cost. Several unfavorable perceptions related to these factors were reported by the decision makers. Problems regarding security, battery life, use in other environments, and technical support were identified. Usability can be associated with effectiveness as detailed earlier. Compatibility must be addressed in detail. In order to move the MCHM closer to operational deployment, it must be integrated with the Navy Medical Enterprise Architecture. MCHM must be able to access and accept data from legacy applications and share data with appropriate systems.

The preceding case study illustrates the fact that all enterprises exist for a purpose, that is expressed in mission and value statements, goals, and objectives, otherwise summarized into business plans. Once desired outcomes are identified, leaders organize resources into functions. In this case, according to our research, importance is the highest weighted factor of smart data collection, followed by effectiveness and suitability. Functions identify the work that needs to be done to produce outcomes. How the work is accomplished is defined as processes, where process activities constitute proprietary differentiation.

In this case example, proprietary differentiation or a unique way of accomplishing things is achieved through a variety of ways that begin with creative leadership in the military command. The MCHM was used to select the right customers to service, that is, the Thai people. Selecting the right things to do by performing humanitarian services and organizing these activities into a superior mix of people and technology, while applying scarce resources in an optimal manner, lead to the collection of smart data that pointed out the importance, effectiveness, and suitability of the system for delivering optimal healthcare performance.

By structuring the balance of consequences such that doing the right things the right way is rewarded and deviations are dissuaded, the MCHM ensured that the Thai customers received valuable results and assured stakeholders that the military enterprise is performing optimally in the delivery of medical services. Data is at the heart of each of these management activities and that is why we focus on smart data as a principal contributor to optimizing performance.

Difference Between Smart and Dumb Data Example

There are limitations with capital budgeting problems that use the traditional approach and assume the future will be an extrapolation of the present. We will provide an artificial intelligence (AI) and simulation method for capital budgeting that is a smart data approach to show executives the right way to use data for increasing interoperability and enterprise performance optimization.

An example of dumb data is the traditional accounting methods that led to the 2008 Wall Street debacle. This accounting information provides little help for reducing costs and improving productivity and quality and may even be harmful. These systems do not produce accurate product costs for pricing, sourcing, product mix, and responses to competition. The system encourages managers to contract to the short-term cycle of the monthly profit and loss statement. This accounting data is dumb data that is too aggregated, too late, and too distorted to be relevant for managers' planning and control decisions. Therefore smart data is *actionable*. It will be the product of a more cost effective approach to delivering higher quality information. Smart data is engineered for high states of interoperability among disparate users, which includes semantic mediation that translates meaning from one enterprise lexicon to another despite differences.

Activity based costing (ABC), which is similar to cost management accounting practices, utilizes smart data because it allocates costs to the products and services that use them. Knowing the costs of activities supports efforts to improve processes. Once activities can be traced to individual products or services, then additional strategic smart data is made available. The effects of delays and inefficiencies become readily apparent and the company can focus on reducing hidden costs.

To understand the difference between dumb data and smart data, the executive must (1) know how information is exchanged by the enterprise today, including the support footprint and costs, and (2) know how information is engineered in a smart data paradigm by contrast. We want to provide a glimpse about what should be expected from smart data technology so that it can be used in decision making.

Costing systems that stress direct-labor performance measures reinforce the tendency for companies to invest primarily in short-term cost-reducing or labor-saving projects. This dumb data approach biases against approaches that develop people or customer satisfaction without reducing current costs.

We postulate that smart data and a smart data strategy are essential elements for optimizing enterprise performance. The adoption of the discounted cash flow approach for evaluating capital investment projects is a major innovation in optimizing cost management accounting practices. In this example of how executives should use smart data, we describe an information technology capital budgeting (ITCB) problem, show that the ITCB problem can be modeled as a 0–1 knapsack optimization problem, and propose two different simulated annealing (SA) heuristic solution smart data procedures to solve the ITCB problem.

Using several simulations, we empirically compare the performance of two SA heuristic procedures with the performance of two well-known traditional ranking methods for capital budgeting. Our results indicate that the information technology (IT) investments selected using the SA heuristics smart data approach have higher after-tax profits than the IT investments selected using the two traditional ranking methods.

The capital budgeting problem is a problem of selecting a set of capital expenditures that satisfy certain financial and resource constraints and has been studied extensively in the finance and decision sciences literature. The capital budgeting problem is often formulated as a binary integer programming formulation, where the decision variable takes a value of 0 indicating that a potential investment is not chosen for funding or 1 indicating that a potential investment is chosen for funding.

When the number of potential investments is large, solving the integer capital budgeting problem is time consuming. Under such circumstances, researchers have used soft constrained programming, problem decomposition, and investment clusters to lower the time to solve the capital budgeting problem.

The type of capital budgeting problem considered in our research is sometimes called a hard capital rationing problem in the literature. In a hard capital rationing problem, a decision maker has to select the best project mix among several competing projects such that a budget constraint is satisfied. In the case of hard capital rationing, budget constraint is defined such that no borrowing, lending, or forward carrying of cash from one period to another is allowed. The hard rationing capital budgeting problem is a very common problem occurring in manufacturing, healthcare, and information technology industries. Capital budgeting in IT is slightly different from the traditional capital budgeting problem as certain capital investments are mandatory; for example, a local telephone company is required to provide service to anyone in the area who desires a telephone service.

Additionally, the capital investments in IT are sizable; that is, more capacity or bandwidth can be added in a project, and only certain IT capital investments can be depreciated. Most telecommunications and Internet service providers find that it is necessary to invest in new technologies and expand their services to compete with their competitors.

Furthermore, these IT companies are also subject to the pressure of not borrowing heavily for these investments as any unused capacity is wasted, which leads to a high opportunity cost of not investing in competing interests within the company. Considering small technology life cycles and mandatory capacity expansion, the total cash outlay in IT projects is usually greater than the budget available and not all the IT projects are always funded. Different IT investments have various tax implications.

For example, in the United States, certain hardware and software investments can be depreciated, but personnel and outsourcing investments represent nondepreciable expense. In a survey of various U.S. companies, approximately 27% of IT budget is allocated to hardware, 42% is allocated to personnel, 7% is allocated to software, 10% is allocated to outsourcing, and the remainder to other IT activities. In the case of multinational firms and the presence of market imperfections, a firm may be able to take tax advantages due to market imperfections. Thus, considering the U.S. tax code and its implications, the ITCB problem is exacerbated owing to nonlinear interactions between different IT projects. The ITCB problem with tax interactions is slightly different from traditional non-IT capital budgeting problems with tax interactions.

The primary difference is due to the fact that the U.S. Internal Revenue Service (IRS) provides a maximum upper limit on the IT depreciation expense deduction. This upper limit requires that all depreciable IT projects be treated collectively and not as standalone depreciable investments. For example, assuming an IRS upper limit of $25,000, if there are five depreciable IT investments and the combined depreciation expense of four IT investments is $25,000 then the fifth IT investment cannot be depreciated.

This upper limit restriction leads to a discrete objective function in a binary integer capital budgeting formulation, which cannot easily be solved using traditional integer programming software. In this section, we study an ITCB problem with tax-induced interactions of IT investments and propose two simulated annealing heuristic procedures to solve the ITCB problem. Because the ITCB problem considered in our research does not have a clearly defined objective function, we cannot use conventional optimization procedures for benchmarking. Thus for the purpose of benchmarking, we use two simple ranking methods that can be used to solve the ITCB problem.

Using the ranking method (RM), all potential projects can be ranked in either ascending order of their investment values or descending order of their investment values before the projects are selected in a particular order until the budget is exhausted. The RM can be broken down into two submethods—the A-Rank method and the D-Rank method. If projects are ranked in the ascending order of their investment values, before their selection using the budget constraint, then the ranking method is called the ascending ranking method (A-Rank). If projects are ranked in the descending order of their investment values, before their selection using the budget constraint, then the ranking method is called the descending ranking method

TABLE 2.5   Descriptive Statistics for *Information Week's* Survey of the Top 500 Firms

Variable	Mean (millions)	Standard Deviation (millions)	Skewness	Kurtosis
Annual revenues	$8,016.6	12,846.26	4.91	32.14
IT budget	$168.97	323.44	6.10	54.90

(D-Rank). Both A-Rank and D-Rank methods do not always lead to optimal solutions because they ignore tax considerations in IT projects, but these methods do provide flexibility to solve the ITCB problem without a clear definition of an objective function.

Given restrictions posed by ranking methods, and difficulty in using integer programming models, we find that heuristic approaches are best suited for solving the ITCB problem. The heuristic approaches provide us with the flexibility to model the objective function appropriately, define accounting rules, and find the solution to the ITCB problem in polynomial time.

Therefore we developed two simulated annealing (SA) procedures for solving the ITCB problem. Our first SA procedure utilizes a feasibility restoration component. In our second procedure, we eliminate the "feasibility restoration" step from the first SA procedure. The feasibility restoration always maintains a feasible solution. In the event when a solution is not feasible, failure occurs in our second SA procedure and annealing schedule is omitted.

We call our first SA procedure the feasibility restoring SA (FRSA) procedure, and the second procedure the simple SA (SSA) procedure. We use an object-oriented programming language C ++ to build a simulation model for generating a real-world set of IT projects, revenues, and IT budgets. The probability density functions for revenues, IT budget, and individual project types were developed using the data from an *Information Week* survey of Compustat's top 500 firms. The descriptive statistics for revenues and IT budgets are summarized in Table 2.5. Table 2.6 illustrates the descriptive statistics for the natural log-transformed variables. The correlation between log transformed revenue and IT budget variables was 0.722.

The simulation model used base e antilogarithm exponent transformation of Revenue and IT Budget variables to compute the actual values of revenues and IT budget for analysis. Individual hardware, personnel, software, and other category projects were generated with normal distributions with means of $4.56 million, $7.09 million, $1.182 million, and $4.05 million, respectively. The standard deviations of 10% were considered for each of the categories.

TABLE 2.6   Descriptive Statistics for Natural Log-Transformed Variables

Variable	ln(Mean)	ln(SD)	Skewness	Kurtosis
Annual revenues	22.20	1.04	0.42	−0.02
IT budget	18.04	1.38	−0.32	1.53

The value of corporate tax was kept constant at 35%. The value of the IT Budget variable was randomly generated. The expected value of the Revenue variable was generated using an expectation formula. The actual values of the IT budget and revenues were then generated using the antilogarithm function. A prespecified number of projects were generated for funding.

Four different approaches—the A-Rank, the D-Rank, the FRSA, and the SSA— were used to make four different IT project funding selections. We provide a simple working example to illustrate the concept of project generation and selection using the A-Rank and D-Rank methods. We assume 10 simulation replications and generation of 10 random projects for each simulation replication. For 10 simulation replications, we first generate 10 random IT budgets. Using our simulation model, these 10 random IT budgets are {17.05, 16.47, 18.45, 18.55, 18.58, 15.52, 16.29, 17.52, 17.23, 17.66}. The expected 10 revenues of each of the IT budgets are computed using the expectation equation mentioned previously. These 10 expected revenue values are {21.66, 21.34, 22.42, 22.47, 22.49, 20.82, 21.24, 21.91, 21.75, 21.99}. As the 10 random values represent the log-transformed values, we use the base *e* antilogarithm exponent transformation to obtain actual values of IT budget and revenue. These values are represented in millions of dollars in the following sets:

IT Budget 5 {25.39, 14.22, 102.97, 113.80, 117.27, 5.49, 11.87, 40.62, 30.40, 46.75}

Revenues 5 {2555.02, 1863.53, 5472.97, 5779.01, 5874.12, 1111.33, 1689.67, 3299.57, 2817.92, 3560.75}

For the first replication, we use the first elements from the IT Budget and Revenues set. These elements represent an IT budget of $25.39 million and revenues of $2.55 billion. For the first replication, we generate hardware, personnel, software, and other projects with an approximate probability of 27%, 42%, 7%, and 24%, respectively. These probabilities were chosen due to previous studies. The actual computer implementation can generate 10 random numbers between the interval (0, 1), and based on the value of each random number, we can allocate the project type using the following four rules:

Rule 1: IF random number A (0, 0.27], THEN project 5 hardware.

Rule 2: IF random number A [0.27, 0.69), THEN project 5 personnel.

Rule 3: IF random number A [0.69, 0.76), THEN project 5 software.

Rule 4: IF random number A [0.76, 1), THEN project 5 other.

In our case, the 10 values of random vector were {0.819, 0.848, 0.513, 0.653, 0.034, 0.199, 0.596, 0.722, 0.62, 0.26}. Using the four rules, we have three hardware projects, four personnel projects, one software project, and two other projects to be funded. The individual values for each project belonging to each category were generated using a normal distribution with the means and standard deviations for each category reported above. We report the individual project values, in millions of dollars, of each of the 10 projects in the following sets for each category:

Hardware 5 {4.451, 4.701, 3.713}

Personnel 5 {7.45, 8.089, 7.761, 8.286}

Software 5 {1.237}

Other 5 {4.09, 4.27}

The total funding requested for all the projects is about $54.05 million and the IT budget is $25.39 million. Thus the projects selected by the D-Rank method are {8.286, 8.089, 7.761}, and those selected by the A-Rank method are {1.237, 3.713, 4.09, 4.27, 4.451, 4.701}, respectively. Assuming that other non-IT-related costs are constant and ignoring these costs and adding a depreciation tax shield for software projects selected using the A-Rank method, we have after-tax profits for projects selected by D-Rank and A-Rank methods of $1,645.07 and $1,646.19 million, respectively. We can repeat this procedure for the other nine values in the IT budget and revenue sets to obtain results for all 10 simulations.

Our actual simulation experiment, for a given complexity level, consisted of a company with one randomly generated IT budget value, revenue value, projects to be funded, and solution of the ITCB problem using four different approaches—the A-Rank approach, the D-Rank approach, the FRSA approach, and the SSA approach. We consider two different levels of complexity. In the first level of complexity, the number of projects to be considered for funding was considered constant and equal to 40. Ten projects within each of the four categories (hardware, personnel, software, and other) were generated with particular means and standard deviations mentioned previously.

The worst-case solution search space complexity for the first level of complexity was well over one trillion unique combinations. In the second level of complexity, the value of projects to be considered for funding was considered constant and equal to 60. Fifteen projects within each of the four categories were randomly generated. The worst-case solution search space complexity for the second level of complexity was over one million trillion unique combinations. For each of the two levels of complexity 50 different random simulation experiments were conducted.

For our statistical tests on means, we use a simple pairwise difference in means *t*-test. There are other multiple comparison procedures that are available, including Tukey's test and Scheffé's test. The multiple comparison procedures are typically more suitable as post hoc tests after an analysis of variance (ANOVA) for some planned hypothesis is conducted. The multiple comparison procedures also assume that the pairwise comparisons are not independent.

Given independent simulation runs, we find the pairwise differences in means *t*-test suitable for our study. For the first level complexity, the average required funding for all 40 projects was $168.8 million and the average IT budget was about

TABLE 2.7   Descriptive Statistics of After-Tax Profit for the Two Approaches

Method	Mean Profit (millions)	SD Profit (millions)	Mean Projects Selected	SD (projects)
A-Rank	$3,708.07	$2,685.45	25.26	12.08
D-Rank	$3,708.42	$2,684.95	16.96	15.64
SSA	$3,740.77	$2,715.49	8.22	2.79
FRSA	$3,740.76	$2,715.51	8.22	2.76

$157.86 million. Tables 2.7 and 2.8 illustrate the results for the first level of complexity. The results indicate that the FRSA and SSA approaches outperformed the A-Rank and the D-Rank approaches in terms of selecting the projects that maximize the after-tax profit. The pairwise differences in means comparison for after-tax profits between SSA and A-Rank, between SSA and D-Rank, between FRSA and A-Rank, and between FRSA and D-Rank were significant at the 0.01 level of significance.

The pairwise differences in means comparisons between A-Rank and D-Rank, and FRSA and SSA were not significant. One of the reasons for lack of significant differences in means, between A-Rank and D-rank, may be because of a set of common projects that were selected by both the A-Rank and the D-Rank selection approaches. On average, both the A-Rank and the D-Rank methods selected at least one common project.

A similar behavior of SSA and FRSA indicates that the "feasibility restoration" procedure in FRSA does not appear to provide any significant advantages over the traditional SSA approach. Given the very similar behavior of FRSA and SSA, we only plot the differences between SSA and A-Rank, and between SSA and D-Rank after-tax profits for each of 50 simulation runs. Figures 2.2 and 2.3 illustrate these differences in after-tax profits. The negative values in the figures indicate that either the A-Rank or the D-Rank outperformed the SSA procedure. The figures illustrate that the A-Rank and the D-Rank methods outperform the SSA method a few times, but overall the SSA method outperforms the A-Rank and the D-Rank methods several times by large amounts.

TABLE 2.8   Pairwise Difference in means t-Test Results for After-Tax Profits

Mean SSA	Mean FRSA	Mean A-Rank	Mean D-Rank	t-Value	P > F
3740.77	3740.76			0.09	0.922
3740.77		3708.07		6.65	0.000[a]
3740.77			3708.42	6.42	0.000[a]
	3740.76	3708.07		6.65	0.000[a]
	3740.76		3708.42	6.45	0.000[a]
		3708.07	3708.42	1.51	0.137

[a] Significant at 99% confidence level.

The Difference (SSA – A-Rank) Profits

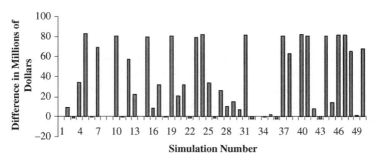

Figure 2.2   Differences in after-tax profit for the simple simulated annealing (SSA) and the A-Rank methods.

The Difference (SSA – D-Rank) Profits

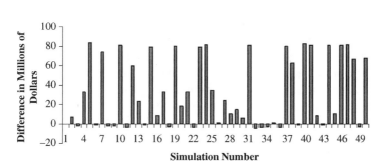

Figure 2.3   Differences in after-tax profit for the simple simulated annealing (SSA) and the D-Rank methods.

Table 2.9 shows the average number of projects selected by each of the four procedures. The average number of projects selected by the ranking methods was higher than the average number of projects selected by the two SA methods. As the ranking methods select the projects from low budget to high budget (A-Rank) or high

TABLE 2.9   Pairwise Difference in Means t-Test Results for the Number of Selected Projects

Mean SSA	Mean FRSA	Mean A-Rank	Mean D-Rank	t-Value	$P > F$
8.22	8.22			0	1.000
8.22		25.26		12.11	0.000[a]
8.22			16.96	4.40	0.000[a]
	8.22	25.26		12.09	0.000[a]
	8.22		16.96	4.40	0.000[a]
		25.26	16.96	10.58	0.000[a]

[a] Significant at 99% confidence level.

budget to low budget (D-Rank) until the IT budget constraint is satisfied, these ranking methods tend to be biased toward either smaller projects (A-Rank) or larger projects (D-Rank). The SA approaches do not have this bias and show preference toward the projects that may increase the after-tax profits.

The pairwise differences in means comparisons, illustrated in Table 2.9, for number of project selections were significant for the SSA and the A-Rank methods, the SSA and the D-Rank methods, the FRSA and the A-Rank methods, and the FRSA and the D-Rank methods, respectively. Significant differences in means were also observed, for the average number of selected projects, between the A-Rank and the D-rank methods. The A-Rank method, because of its bias toward selecting smaller investment projects first, always selects more projects than the SSA method and the D-Rank method. No significant differences in means were observed between the SSA and the FRSA procedures.

This case has described the ITCB problem as a 0–1 knapsack optimization problem. Given that a 0–1 knapsack optimization problem is NP-hard, we proposed two SA heuristic procedures to solve the ITCB problem. We also developed a simulation model that used probability distribution functions for large Fortune 500 companies and tested the performance of the two SA procedures against two ranking-based selection procedures. Our experiments indicate that the two smart data SA heuristic procedures — FRSA and SSA—outperform the traditional A-Rank and the D-Rank selection methods for IT capital budgeting.

Among the major contributions of our research were the use of corporate taxation and depreciation for selecting a portfolio of profitable IT investments. Given that the depreciation expense from a few IT investments can be used as a tax shield and may improve profitability of an organization, IT managers may benefit from the proposed approach for selecting IT investments.

Our approach is flexible and can incorporate any changes in tax laws. There are several potential improvements and extensions for our case. First, we used after-tax profit maximization for our objective function in the ITCB problem. In an event where management may not desire to consider the tax-induced interactions in their decision making, the objective function can be changed to maximize the combined net present value (NPV) of selected projects. Second, the two SA procedures are only some of the heuristic procedures that can be used to solve the ITCB problem.

Among other popular polynomial time heuristic procedures are the genetic algorithm (GA) and the tabu search (TS) procedures. Both GA and TS procedures require higher volatile memory than the SSA procedures used in this research and may add to the computational overhead. Finally, we only considered a single investment budgeting period in our case. This single investment budgeting period is a simplification of the real world IT capital budgeting, where IT investments are often scaled or delayed over multiple periods.

Future research is needed for developing techniques and algorithms for solving a multiperiod ITCB problem. This case demonstrates how utilization of smart data AI methods such as SSA and FRSA can outperform traditional capital budgeting methods such as A-Rank and D-Rank methods.

It is our contention that optimizing enterprise performance is the most basic executive responsibility, and that smart data is a product of the data engineering discipline and advanced technology. Making data smart, such as employing a better algorithm, enables a smart data strategy for capital budgeting. Today, many organizations do not have a data strategy, much less a smart one, and the subject of data is not a part of executives' lexicon.

At a high level, smart data is applied as part of the systems engineering discipline. Yet, this subject is very much integral to enterprise management and therefore management and information science. Modern executives must have a hybrid command of these various disciplines, as the previous case demonstrates. Whether you lead a commercial enterprise or a government enterprise, there is one activity that is at the top of the hierarchy in importance for which all CEOs and senior executives are responsible and accountable, and that is to optimize performance. Optimizing performance means applying scarce resources to business processes under constraint and transforming them into highest yield and best use outcomes by managing people and enabling technical mechanisms (technology). In our case, we used artificial intelligence and a superior method to solve the ITCB problem.

Our method allows executives to differentiate among IT projects based on a proven superior AI method.

Proprietary differentiation or a unique way of accomplishing things is achieved through a variety of means that begin with creative leadership:

- Selecting the right customers to service
- Selecting the right things to do
- Organizing activities
- Attributing activities with application of a superior mix of people and technology
- Applying scarce resources in an optimal manner
- Structuring the balance of consequences such that doing the right things the right way is rewarded and deviations are dissuaded
- Ensuring that customers receive valuable results
- Assuring stakeholders that the enterprise is performing optimally.

Smart data is at the heart of each of these management activities and that is why we focus on data as a principal contributor to optimizing performance.

There are so many moving parts in an enterprise that it is unfathomable that modern executives would attempt to manage without techniques, such as AI, to keep track of them, but many do. That condition is unacceptable in the highly automated world of the twenty-first century because resources are exceedingly scarce and risks are too high to operate intuitively.

One of the greatest barriers to realizing the full benefit of automation is from people inserting themselves with manual intervention, usually unintentionally. They need and want data to perform better, but it can't because they did not plan and prepare sufficiently for the moment of need. For enterprises to perform optimally, executives must insist on better data planning, preparation, and engineering. They must be cognizant of the smarter ways to conduct business such as adapting AI methods such as SA to outperform traditional capital budgeting problems.

REFERENCES

1. Parag C. Pendharkar, James A. Rodger, and Girish H. Subramanian,"An Empirical Study of the Cobb–Douglas Production Function Properties of Software Development Effort," *Information and Software Technology*, **50**(11), 1181–1188, 2008.
2. Parag C. Pendharkar, Girish H. Subramanian, and James A. Rodger, "A Probabilistic Model for Predicting Software Development Effort," *IEEE Transactions on Software Engineering*, **31**(7), 615–624, 2005.
3. Parag C. Pendharkar and James A. Rodger, " An Empirical Study of Impact of Crossover Operators on the Performance of Non-binary Genetic Algorithm Based Neural Approaches for Classification," *Computers & Operations Research*, **31**, 481–498, 2004.
4. Parag C. Pendharkar and James A. Rodger, "The Relationship Between Software Development Team Size and Software Development Cost," *Communications of the ACM*, **52**(1), 141–144, 2009.
5. James A. Rodger, Pankaj Pankaj, and Micki Hyde, " A Field Study Utilizing the Medical Companion Healthcare Module (MCHM) and Smart Data to Measure Importance, Effectiveness, Suitability and Ease of Use in Mobile Medical Informatics (MMI) Systems," *Academic e-Journal eJ.D.E.* ISSN, 1776–2960, 2009.
6. J. W. Creswell, *Qualitative Inquiry and Research Design: Choosing Among Five Traditions.* Sage, Thousand Oaks, CA, 1998.
7. Y. S. Lincoln, "Emerging Criteria for Quality in Qualitative and Interpretive Research." *Qualitative Inquiry*, **1**, 275–289, 1995.
8. R. K. Yin, *Case Study Research: Design and Methods.* 2nd Edition. Sage, Thousand Oaks, CA, 1994.

Chapter **3**

Barriers: Overcoming Hurdles and Reaching a New Performance Trajectory

Chains of habit are too light to be felt until they are too heavy to be broken.
— Warren Buffet

3.1 BARRIERS

Identifying hurdles toward reaching a new performance trajectory is most direct. To defeat inhibiting behavior, we must make clear what is getting in the way. A characteristic of smart data strategy is to leap beyond legacy problems to achieve a new performance trajectory. We don't want to level off at a new plane; instead, we want to push the performance trend upward with momentum.

To accomplish this, consider two elements: (1) executive and organizational inhibitors and (2) technical inhibitors.

The following list contains the principal executive and organizational inhibitors that we encounter as barriers to implementing effective smart data strategy:

- Don't understand it
- Executives aren't on board
- Reluctance to change
- Limited discretion and bandwidth to change
- Threat from and to stovepipe behavior
- Lack of awareness of ability to improve anticipation and prediction

Smart Data: Enterprise Performance Optimization Strategy, by James A. George and James A. Rodger
Copyright © 2010 John Wiley & Sons, Inc.

- Fear of transparency
- Facts bear the wrong answers

Second, there are technical inhibitors:

- Legacy infrastructure based on outmoded ideas
- Deficient skill, knowledge, and experience needed to leap beyond the legacy
- Lacking a strategy such as smart data to escape from legacy performance

3.2 OVERCOMING BARRIERS

Understand the Strategy

Smart data strategy begins when executives declare that they need and want total enterprise performance represented as fully attributed process models featuring data inputs and outputs, controls, and mechanisms. They must want the models activated in a real-time executive support system.

Knowledgeable executives need, want, and expect their enterprise data to be presented as an operational model that aggregates to a top line view and decomposes into details that explain every enterprise function within the domain of responsibility, and extending to supply chain and community of participant dependencies, including customers and sources of capital.

Knowledgeable executives will articulate and share an inventory of routine questions for which actionable data must be available at all times; and executives will share dynamic scenarios enabling what-if calculations and measurable effect of alternative decisions and choices.

Smart data strategy requires data engineering such that data is open and inter-operable. Smart data strategy is dependent on state-of-the-art smart grid and smart services as depicted in the service-oriented enterprise (SOE) paradigm and as part of cloud computing services and World Wide Web progression.

Smart data is a product of current technologies, and superiority is gained from progress in applying emerging technologies as suggested in the draft concept of operations and operational architecture described herein.

Smart data strategy is implemented with the introduction of the executive enterprise performance optimization system (EPOS) and associated management interface.

Become Aware of the Possibilities and Get on Board

To remove or mitigate these barriers, the executive needs convincing. Our approach is to demonstrate that data is on the critical path to enterprise performance optimization, and the CEO can achieve this best with a smart data strategy. The principal idea is to focus executives' attention on data that is essential to enterprise performance and executives' ability to execute their primary responsibilities.

Executives should expect more from IT services than infrastructure alone. They need analytical capability to model and simulate conditions in response to a dynamic environment. They need the ability to anticipate and predict performance with high certainty in the near term and with increased visibility over the longer term.

Executives taking ownership for data that is needed to optimize enterprise performance may be considered invasive because it disturbs the status quo in information technology management where the data is buried. Ask for answers to such requests as:

- Let's see the enterprise process model.
- Let's see the data inputs.
- Let's see how processes transform data inputs into higher yield outputs.
- Let's see the system in real-time performance reporting mode.
- Show the value of data outputs compared with associated costs.
- Show how much is spent on enterprise application interface maintenance.
- Show how much is spent on data exchange and business transactions.
- Show what will happen when certain variables are changed such as changing the employment mix, adding or changing equipment, tools, or systems.
- Show the material effect of changes in business rules, policies, laws, and regulations.
- Show the enterprise data model.

Smart data is intended to make enterprise performance visible and observable, and to enable executives to try what-if scenarios and conditions to evaluate possible decisions and alternatives.

Reluctance to Change

Technically, risk to change is minimal because smart data implementation is intended to leap beyond the IT legacy into a new world of open data exchange and transactions. Changing IT from old culture to smart data culture requires managing, executive leadership, and collaboration with enabling technologists who possess contemporary knowledge, skill, and proficiency.

The data engineering discipline has progressed in that regard, and only those with current credentials are qualified to advance the idea. Therefore current staff must advance in professional development and may require augmentation to stimulate fresh thinking.

The cost to change expressed as capital and executive and organizational mind-share must be estimated based on having a clear picture of what it will take to move from the present situation to an improved state. In addition, the enterprise must estimate its own bandwidth given all of the priorities and activities demanding attention. However, if data is not made a top priority, there can be no meaningful change.

Limited Discretion and Bandwidth to Change

Another way to say this is that if you find yourself short of bandwidth to attend to enterprise data essential to performance optimization, reorder your priorities.

What must change?

First, executive leadership must present the right questions and information requests and insist on smart data solutions. Executives must take time to address the enterprise smart data requirements and associated strategy.

Second, information technology management habits must change. Make an inventory of enterprise assets and describe how they are accounted for and expressed in current enterprise data for the purpose of reviewing with executives how this can be improved. Address the sufficiency of current data in supporting executives' planning, problem solving, decision making, sense making, and predicting.

Discuss with executives how to make predictive capability a strategic differentiator. This assumes that IT will truly catalyze the focus with creative and concrete smart data-centric ideas.

Have a frank and precise discussion with executives to identify key questions to be answered and (1) decisions to be made routinely and (2) decisions to be made for unique situations with accompanying standards for timeliness, accuracy, frequency, rate, completeness, cost, and quality.

What is the current enterprise performance optimization portfolio? Asked differently, what systems and tools do executives need to manage the enterprise optimally? How well are data requirements being addressed today—with the answers being (1) from the executive user's view, (2) from the competitive view, and (3) from the state-of-the-art technology view?

IT must stop serving up the same deficient solutions. Stovepipe behavior must be addressed head on as it is simply not acceptable in the culture of enterprise performance optimization with smart data strategy. IT cannot wage the battle alone as executives must lead the charge to change.

Threat from Stovepipe Behavior U.S. Government Generic Case Example

This example is based on a pattern of problems that appear in multiple government agencies.

As part of a department, the agency has five divisions. In the year 2000, the department established a data center based on software and architecture popular at that time, point-to-point or likely hub-and-spoke. All divisions are supposed to use the data center, although over time, each division has developed its own data collection and reporting system. Divisions send their data to the data center, although sometimes users go to the data center for information and sometimes not. Configuration management has degraded over time, so divisions have adopted the adage that "the only data they can believe is that which they create themselves."

As for sharing with other divisions, personnel in a division prefer not to because each division claims that it has special regulations and charters that makes it responsible for its data, and personnel in the division don't want to lose control.

Part of the problem here is that regulators invent and impose new rules on an ad hoc basis and do not consider implementation in a holistic enterprise. Therefore they legislate part of the problem and must become a part of the solution. Everyone must address what we call the balance of consequences problem.

Therefore, in this example, there are five fiefdoms with stovepiped data and a central data hub that may or may not have some or all of the data.

Now it is 2010 and the 10-year old data center has grown technically obsolete. The department executive plans to address the "data center" problem, but in the meantime, divisions have increasing demands and are impatient.

The department executive, who is a new appointee, says, "We need a data governance policy." A data governance policy might be a good thing to employ to provide direction to the situation. But what is the purpose of the policy? What is the desired outcome?

Fortunately, the Federal CIO Council has, at the direction of the Office of Management and Budget (OMB), identified this broad area in one of its top objectives: "Implement the Data Reference Model (DRM) as a common framework for managing and storing information across the Federal Government" [1]. With the outcome being interoperable IT solutions. Yet, this top level objective that is slated for accomplishment by 2010 will come too late to address the department's tactical needs immediately. (See Table 3.1)

The purpose of the policy and the desired outcome is a paradigm shift that allows for interoperable, optimized, and methodologically sound smart data that can be used between and across organizations, to improve performance, efficiencies, and effectiveness, while making the organization more competitive.

Investigate how to make the environment open and interoperable in an enterprise context. Is the current infrastructure suitable to support smart data and smart data strategy? What must change? What is the cost to change expressed in time to change and capital investment? What unique resources and capabilities are required? What are the expected improvements? How soon can improvements be realized?

Addressing smart data and smart data strategy development and improvement in an enterprise context will involve communications with other enterprise partners, and ultimately with customers. Smart data will become a catalyst for change and improvement as new improvement opportunities are discovered in the process. New technology partners may be needed.

Bear in mind that it is better to accomplish quality change than it is to attempt too much at once. Executives should follow through with a plan and schedule iterative improvements driven by the highest priority needs of the business.

Smart Storage

This strategy requires smart storage as part of the smart grid of interoperable services. A storage area network (SAN) is an architecture that fits our strategy. A SAN is used to attach remote computer storage devices to servers in such a way that the devices appear as locally attached to the operating system, much like our cloud computing

TABLE 3.1   Data Governance Needs and Requirements

Needs	Requirements
Shared data: Some data is common among all five divisions, and therefore common data must be accessible to all.	• Common data made accessible to all qualified users • Qualified users define
Shared data: More than one division can create data that will be shared among all credentialed and privileged users.	• Configuration management • Credentialing and privileging
Continue to use data center data that has relevant value: current and historical.	• Data quality • Legacy data strategy
Respect mission-specific autonomy while discouraging stovepipe behavior.	• Development of an open data exchange strategy leveraging model-driven data exchange • Development of an agency data model attributed with division user responsibilities and use case characteristics • Emphasize that all data is property of the department and agency with specific contributions from divisions • Establishment of a smart data exchange infrastructure that includes mapping from legacy systems to a neutral model for department/agency data exchange/sharing/and configuration management • Publication of a data governance policy that explains this for all user participants

example that is described in Chapter 4. The downside to this grid of virtual storage is that it remains complex and expensive.

Most storage networks use the small computer system interface (SCSI) protocol for communication between servers and disk drive devices. This enables creation of an information archipelago of disk arrays. A SAN literally builds a bridge between the islands of information in the gulag. The benefit of SANs is that they allow multiple servers to share the storage space on the disk arrays.

So how do SANs contribute to our smart data paradigm? For one thing, SANs facilitate an effective disaster recovery process because a SAN is a smart grid that can span a distant location containing a secondary storage array and facilitate storage replication over a much longer distance.

The goal for executives is to create an optimized, interoperable smart grid of storage for smart data that utilizes the correct methods and algorithms. This storage smart grid will be a high-level design for a centralized, fault-tolerant, and high-speed data storage solution that supports managing smart data across multiple departments and provides for fast backup of mission-critical data.

How can executives create a solution that meets the smart data storage needs of their organization? Here are some of the requirements of this smart data storage grid solution.

In order to effectively manage smart data, executives need to demonstrate their ability to use terminology associated with fiber channel and SAN technologies such as high-availability requirements, scalability requirements, performance requirements, and management requirements.

Furthermore, the executives need to recognize and discuss the features and benefits of various storage networking technologies, as well as present a compelling business case for a storage networking solution in terms of adding interoperability, optimization, and correct methods and algorithms that meet the smart data paradigm.

What are some relevant topics that executives need to address in this area? They must understand the storage market, storage network architectures, fiber channel technology, emerging technologies, storage networking applications, and storage management.

In order to allow their company to grow, executives must realize that they need a smart data grid of centralized data storage and management across the organization. This smart grid requires interoperability, reliability, scalability, and optimal performance of the smart data in their databases. Executives need to adopt a smart data approach to increase their ability to optimize the performance of their databases, apply algorithms to improve disk access speed and data retrieval, and improve interoperability by decreasing the time spent backing up and restoring data. This smart grid approach should lead to a dramatic increase in system reliability and manageability without incurring the cost of implementing redundant server complexes.

Today's executives require a smart grid of storage solutions that is interoperable, optimized, and can maintain adequate SAN performance under high traffic conditions. These executives must be able to add storage capacity quickly without taking applications offline. An interoperable, optimized smart grid storage network will allow IT to effectively apply algorithms and methods to manage and scale application performance, at both the SAN and application levels.

This smart data SAN network grid will strive for a configuration that is available at six-sigma levels and provide a solution that will allow it to quickly recover from component failures; it should also be fault-tolerant and ensure that available capacity is fully utilized. Connectivity, backup/recovery, and investment protection are also important considerations in a smart data grid strategy in order to improve interoperability, optimization, and utilization of correct methods.

Ability to Anticipate and Predict

To optimally satisfy the needs of enterprise constituents and customers, executives must be able to anticipate emerging needs and requirements, to get ahead of the demand curve. Our SAN example illustrates this need for anticipating emerging needs such as smart data storage.

In so doing, executives must improve their ability to predict the timing of events needed to satisfy demand in a timely, accurate, and complete manner at the optimal rate, frequency, cost, and quality. These are nontrivial requirements and this is why executives are compensated handsomely for their performance.

As we have witnessed in recent history, executives are not as well equipped to perform this responsibility as expected. We advocate high focus on smart data and an accompanying strategy and management technology portfolio to improve the situation.

Having studied management science and operations research, we often wonder throughout our management careers why there is so little evidence that executives apply the methods and algorithms that many have learned and some of us continue to teach. Upon investigation, a pattern appeared: while problem solving logic and formulations might apply, there is often insufficient data on which to operate. Data may be deficient in a variety of ways, ranging from "there isn't any" to a host of issues about quality.

Yet, mostly, deficiencies have to do with failing to prepare for and anticipate the need for data. Thus we suggest that executives raise questions and demand smart data to answer those questions in order to optimize enterprise performance.

Answering Questions that Management Does Not Believe Could Be Answered

When operating Talon Publishing & Research, a firm dedicated to answering questions that management did not believe could be answered, George conducted primary research for a new product initiative at GTE Telecom. The product was a telecommunications management system that was the brainchild of retired Air Force officers who envisioned applying what they had learned in the military combined with commercial off-the-shelf (COTS) technology to commercial needs about which they knew little. Before investing, GTE executives wanted to know the market for the system and details about how to price the product and associated services.

By employing a 20 page research instrument mailed to 5000 individuals selected from the prospective customer universe, and with a 20% response rate, GTE determined that it could sell 50 units with high certainty, and the research provided granular information related to pricing product and service features.

Before the survey, many executives were skeptical. "You will never get them to answer 20 pages of questions?" To make a long story short, success was based on presenting the right questions to a highly qualified audience for which the questions were relevant and important to them, and we promised to share the answers.

Research results-driven advertising and marketing instruments were rapidly prepared immediately upon completion of the analysis. Messages were closely aligned with what was learned about customers. Timely follow-up was directed to the hot prospect respondents and $80 million in new business was closed within 12 months with significant after-sales service of >$300 million.

The short and efficient selling cycle made believers of GTE management and they created the *GTE Intrapreneur Program* as a result, whereby the technique was routinely applied to developing ventures.

Today, companies employ this technique in an ever more sophisticated manner, leveraging social media, blogging, electronic white paper publishing, and surveying. The more the audience is bombarded by this surfeit of information, the more creative enterprises must be to entice and engage customers. Smart data is central to improving intelligent marketing as it leverages answers buried in information that is already available. Some call this *data mining*. Data mining is surely an aspect of smart data.

Executives were asking the following of questions:

- Who are our customers?
- How much will they buy?
- When will they buy?
- What must we offer them?
- How much should we charge?
- What after-sales services should we offer?
- Who are our competitors?
- What must we do to provide a superior offering?

The answers to these questions provide the basis for prediction with a degree of certainty. Every enterprise has questions like these for which quantitative answers are readily determinable. Notice also that these questions accompany other more routine business questions such as:

- How is our cash flow?
- What are the capital and material requirements?
- What are the enabling people and technology requirements?
- What are the processes involved?
- What are the performance metrics?
- What is the payback period and break-even?
- What is the return on investment?

Observe that many of these questions are not answered by routine business accounting, although some are. There is concern that application of general accounting principles and resulting reports are not predictive. They provide historical reporting from which analysis and comparison may serve as a retrospective report card, though not necessarily indicative about the future. A smart data and smart data strategy approach improves the ability of enterprises to prepare for and more accurately anticipate the future. The art of management becomes more of a science in this regard.

The U.S. Department of Defense has questions that are specific to its area of concern:

- What threats does the country face?
- What are the characteristics of the sources?
- What is the priority of these threats?
- How are the threats defined in terms of operational scenarios?
- What are the associated logistics footprints required in effective response?

- What is our current capacity in meeting and defeating these threats?
- How many threats can be managed at once based on various combinations?
- What is the operational use of military services in meeting these scenarios?
- What requirements are common among services that can manifest in shared enterprise services?
- What requirements are unique to the particular brands of military service?
- How does military strategy integrate with economic and diplomatic strategy to effect desired outcomes?

The art of war has already become more of a science, although the politics of war may make the management of war appear more artful. We know that the U.S. Department of Defense is pretty well advanced in answering many aspects of such questions, although we also know that resulting from complexities in operating in a democratic form of government, and with dependence on commercial enterprise, the questions and their answers are profoundly complex and dynamic. The same may be said for each department in the federal government and state and local governments.

The U.S. Department of Defense anticipates and plans for certain threats as expressed in scenarios and then stages resources accordingly, ready for deployment. As the needs arise, the department operates its massively complex organization to produce specific results up and down the chain of command and throughout the supply chain. Focusing on the right data at every level of the enterprise indicates success or deviations.

Why is it that year after year the U.S. Government Accountability Office reports that contractors are constantly off in their estimates, often by an order of magnitude? There is something inherently wrong about the defense acquisition system. If defense acquisition has highly visible problems that go without resolution, what makes Americans think that other federal government departments are performing any better, such as the Food and Drug Administration failing to address salmonella-tainted peanut butter? The answers to solving these problems are in following the data.

Commercial enterprise is no better. Why is it that after all of the years of publication from the likes of Peter Drucker and Michael Porter, for example, the rate of business failures is constantly high? We believe that the answer has something to do with repeating mistakes. Something is inherently wrong in the process and the way that we use and manage data.

Economist Brian Headd of the Small Business Administration said, "Businesses that have employees and that have good financing tend to survive longer" [2]. He reported that the larger the business, the greater access it has to capital. Such generalities provide no insight at all into what makes businesses tick. Surely all businesses are dependent on access to capital.

Access to capital has to do with risk. Lenders assess your risk based on ability of the business to say what it will do and then deliver results as planned. In commercial enterprise, success is based on anticipating customer demand, estimating addressable market share, and providing superior products and services at the right price and best value. Managing enterprise processes to produce desired outcomes is complex, although focusing on data that indicates the trajectory toward success in multiple dimensions makes it easier.

Importance of Transparency

Transparency is critical in garnering investment or access to capital. The more an enterprise is able to explain how its processes work to produce results, the easier it is to assess and mitigate risks.

On the lighter side, George worked for an advertising agency president, Tom Hall, of Ensslin & Hall Advertising, who personally interviewed clients before taking on their accounts. Basically, it went something like this. Hall would say, "We have demonstrated our ability to provide superior creative solutions for you as we have done for other clients with similar needs. We have been successful because they know that we get 51% of the vote when it comes to creative decisions. That means, even if you don't like it, we still make the final decision because we know what we are doing in our business, better than you know our business. Do you understand?" Since most of the clients knew the agency's reputation for producing results, they would nod their heads in agreement. Hall went on: "Now, since most businesses have a high failure rate, we expect to have our money in an interest bearing trust account where we have joint approval. We will get paid when you agree we have earned our fee, but the money cannot be withdrawn without both of our signatures." We share this story to emphasize the importance of transparency.

There are different types of transparency:

- Banking—disclosing information about lending and investment and underground banking for the purpose of revealing practices that are fraudulent, risk-filled, or discriminatory
- Corporate—enabling public access to corporate, political, and personal information
- Management—decision making is made public

Making everyone in the enterprise aware of the current situation and alignment with performance plans is one of the best ways to positively influence work performance. This means sharing the data and making it visible and actionable— the premise for smart data.

Facts Bear the Wrong Answers

Too frequently we encounter a situation in which we have been engaged to produce answers to questions and solutions to problems whereby the executives have already committed to a solution and result they expect. When the data facts do not support the expected conclusion, we sometimes hear:

- Wrong answer
- Can't be
- Hide the report

From this we learned that it is imperative to be at the right hand of the executive to become engaged in the thought process such that the pursuit is objective and based on sound logic, methods, and algorithms.

Executives are on a fast track to get the answers, and those providing data must be in a position to anticipate needs before the request. As such, the smart data engine in the enterprise must be engineered to anticipate and enabled to predict.

The following cases show the best and worst applications which were believed to be based on sound logic, methods, or algorithms. These cases represent the good, the bad, and the ugly when it comes to methods and algorithms.

A good smart data strategy begins when executives declare that they need and want total enterprise performance represented as fully attributed process models featuring data inputs and outputs, controls and mechanisms. As we stated previously, executives need, want, and expect their enterprise data presented as an operational model that aggregates to a top line view and decomposes into details that explain every enterprise function within the domain of responsibility, and extending to supply chain and community of participant dependencies.

The Good Example: Maximum Entropy Density Estimation Using a Genetic Algorithm

In this case we provide an alternative to unsupervised learning algorithms, neural networks, and support vector machine classification and clustering approaches that are kernel based and require sophisticated algorithms for density estimation [3].

- "Machine learning is a scientific discipline that is concerned with the design and development of algorithms that allow computers to learn based on data, such as from sensor data or databases. A major focus of machine learning research is to automatically learn to recognize complex patterns and make intelligent decisions based on data. Hence, machine learning is closely related to fields such as statistics, probability theory, data mining, pattern recognition, artificial intelligence, adaptive control, and theoretical computer science" [4].
- Neural networks are "computational data-analysis methods that mimic the function of a brain and have some learning or training capacity" [5].
- Support vector machine classification is "a new method of functionally classifying genes using gene expression data from DNA microarray hybridization experiments. The method is based on the theory of support vector machines (SVMs). We describe SVMs that use different similarity metrics including

a simple dot product of gene expression vectors, polynomial versions of the dot product, and a radial basis function" [6].

- "Clustering can be considered the most important 'unsupervised learning problem;' so, as every other problem of this kind, it deals with finding a structure in a collection of unlabeled data. A loose definition of clustering could be 'the process of organizing objects into groups whose members are similar in some way'. A cluster is therefore a collection of objects which are 'similar' between them and are 'dissimilar' to the objects belonging to other clusters" [7].

The density estimation problem is a nontrivial optimization problem and most of the existing density estimation algorithms provide locally optimal solutions. In this case, we use an entropy maximizing approach that uses a global search genetic algorithm to estimate densities for a given data set.

Unlike the traditional local search approaches, our approach uses global search and is more likely to provide solutions that are close to global optimum. Using a simulated data set, we compare the results of our approach with the maximum likelihood approach.

It is our intent to show how smart data can provide executives with answers to routine questions. Actionable data must be available at all times, and executives will share dynamic scenarios enabling what-if calculations and measurable effect of alternative decisions and choices.

Our case demonstrates how a smart data strategy to density estimation requires data engineering such that data is open, interoperable, and capable of optimizing through the correct choice of algorithms and methods.

Density estimation is a popular problem in data mining literature, and there are several data mining problems that lend themselves to density estimation. For example, several cluster analysis problems can be considered as multiple-density estimation problems, where each probability density function represents a cluster.

A few classification approaches such as radial basis function neural networks and support vector machines require transformation of nonlinearly separable input vectors from high-dimensional to low-dimensional vectors in feature space so that the inputs are linearly separable. Density or kernel estimation plays a vital role in this transformation of high-dimensional input vectors into the low-dimensional feature space vectors.

A density estimation problem can be defined as follows. Given a data set of size M of n-dimensional vectors $\{\mathbf{x}_1, \ldots, \mathbf{x}_M\}$ that are assumed to be obtained from an unknown probability distribution, estimate the probability density function (pdf) $p(\mathbf{x}_j)$, where $j = \{1, 2, \ldots, M\}$, that maps a vector $\mathbf{x}_j$ to the interval [0,1]. A density estimation problem is solved using parametric, nonparametric, and semiparametric approaches.

Parametric approaches assume a particular data distribution and fit a pdf on the given data set. Nonparametric approaches use data binning and histogram fitting, whereas the semiparametric approaches use maximum likelihood and expectation maximization approaches. Parametric approaches are relatively simple and don't fare well, particularly when the assumptions they make are violated.

Nonparametric approaches face problems of computational complexity. Semiparametric approaches combine the advantages of both parametric and non-parametric approaches and are more popular.

Semiparametric approaches are computationally efficient and provide locally optimal solutions. A popular semiparametric density estimation approach is the expectation maximization (EM) algorithm. The EM algorithm uses Gaussian mixture model and the maximum likelihood approach to estimate the pdf for a given data set. The EM algorithm is well known to converge efficiently to the local optimum.

Researchers from statistics and economics communities have recently shown interest in using maximum entropy (ME) measures for density estimation problems. For example, it has been shown that an encoding and decoding approach based stochastic gradient ascent method for entropy maximization can be used to estimate probability densities. It has also been proposed that a sequential updating procedure be used to compute maximum entropy densities subject to known moment constraints.

We may use a ME density approach to estimate error distribution in regression models. All researchers have reported good results with the use of ME measures. Others argue that ME densities have simple function forms that nest most of the commonly used pdfs, including the normal distribution, which is considered a special case as opposed to a limiting case. All the procedures proposed and used in these studies converge to the local optimum and suffer from similar computational issues as that of the EM algorithm.

While the ME approach for density estimation has shown promise, current approaches cannot be applied directly to clustering problems in data mining. These approaches assume either knowledge of data distribution or prior knowledge of movement constraints. Since a typical application of density estimation in the data mining literature is likely to be an unsupervised learning problem, such prior knowledge may not be available.

In this case, we use a global search heuristic genetic algorithm for density estimation. Genetic algorithms (GAs) are population based parallel search techniques that are likely to find "heuristic" optimal solutions to optimization problems. Unlike gradient search algorithms used in EM and ME density estimation, which are likely to get stuck in a local optimum, GAs are likely to provide solutions that are close to the global optimum. We use GAs to estimate pdfs on a simulated data set using both the maximum likelihood (ML) formulation and the ME formulation.

Semiparametric Density Estimation   Typical semiparametric density estimation consists of a finite mixture of the density models of k probability distributions, where each of the k distributions represents a cluster. The value of k is usually less than the number of data points. We use a Gaussian mixture model, where true pdf is considered to be a linear combination of k *basis functions.*

We use GAS for optimizing the nontrivial ML and ME optimization problems. GAs are general purpose, stochastic, parallel search approaches that are used for optimization problems [8]. We use floating point representation for our research.

All the optimization variables are represented as genes in a population member. We use single-point crossover and a single gene mutation.

Simulated Data and Experiments We generate a two-attribute simulated data set using simulations of three different normal distributions. We generate our data using three normal distributions with means of −1, 0, and 1. The standard deviations for all the distributions are considered equal to one. These three distributions represent three clusters. Figures 3.1 and 3.2 illustrate the data distributions and their contour plots. We generated 20 data points for each distribution with a total of 60 data points.

We test the accuracy of our ML and ME optimization on our simulated data set. Since the problem is that of cluster analysis, our data input vector contained the tuple $<x_1, x_2>$ and we specified the value of $k = 3$ in our experiments. We use the GA procedure to optimize the ML and ME functions. The GA parameters were set after initial experimentation as follows. Mutation rate was set to 0.1, crossover rate was set to 0.3, and the terminating iteration condition was set to 1000 learning generations. The cluster assignment for each data point in the sample was conducted after the optimization of the function and finding the maximum value of the likelihood.

Figure 3.1 shows the cluster means and standard deviations obtained for ME and ML objective functions. The cluster assignments from the algorithms were compared with the actual distribution from which a data point was generated to compute correct classification.

The preliminary results of our experiments indicate that ME approach, when compared to ML, appears to fare well. The standard deviations of the ML approach are lower than or equal to the ME approach and it appears that the ML approach is overly conservative.

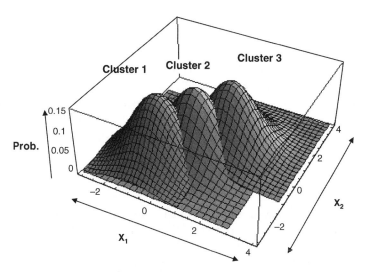

Figure 3.1 Data distributions for simulated data set with three clusters.

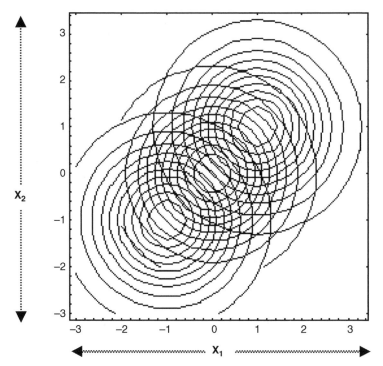

Figure 3.2   Contour plots illustrating the expected overlapping between clusters.

Conclusions   We used a heuristic and global search GA for a nontrivial density estimation problem. The objective function for the density estimation problem was derived from the ME and ML approaches. Using a simulated data set generated with two attributes and three different normal distributions, we tested our GA approach. Our results indicate that ME is a better objective function for the density estimation problem.

   Our case demonstrates how smart data is a product of current technologies, and superiority is gained from progress in applying emerging technologies, such as ME. Smart data strategy is implemented with the introduction of the executive enterprise performance optimization system (EPOS) and associated management interface utilizing algorithms and methods such as ME that outperform the more conservative ML approach.

The Bad Example: The Case of the Wrong Method

It is common, in academia, to write reviews and critiques of published works in order to improve quality in the editorial process [9]. It is also common for the author to have the opportunity to respond to these critiques [10]. In this case, we examine a published article on the adoption of administrative (TQM) innovation in IS departments, and we demonstrate that the dependent variables used in the study have problems related to content validity, that the techniques used for the data analysis are inappropriate, that one set of hypotheses are wrongly tested, that the unit of analysis is sometimes dangling, and that the binary transformation of one dependent variable results in loss of information and changes the underlying distribution of the dependent variable.

We believe that the study has internal validity problems and therefore we also believe that the results of the study may be flawed. We contend that the author didn't understand the method and that the journal executives weren't on board with the proper reviewing techniques to correct this problem. We also contend, in our case, that there was a lack of awareness on the part of the reviewers as to previous published manuscript flaws and the inability of the editor to improve anticipation and prediction of incorrect algorithms.

We conclude that the facts presented in the article bear the wrong answer due to the improper choice of methodology and that the reluctance of the editorial board to learn from previous studies, that were flawed due to a fear of transparency, led to limited discretion and bandwidth to change that threatened an existing stovepipe behavior within the journal hierarchy.

The case revolves around adoption of administrative innovation (TQM) in an information systems department (ISD). Using the diffusion of innovations and total quality management (TQM) literature, the author identified a set of independent variables (environmental, organizational, and task related), which can be used to explain variance in the two dependent variables (swiftness and intensity of TQM adoption). In this case, we show that making poor operationalizational and transformation of the dependent variables may be flawed.

Additionally, the unit of analysis is dangling and the techniques used for data analysis are inappropriate. The poor operationalization, measurement, and transformation of the dependent variables may limit the internal validity of the study.

Swiftness of Adoption of TQM   Swiftness of adoption was defined as follows: "Swiftness of adoption pertains to the relative timing of TQM adoption and measures when an IS unit adopts TQM in systems development." Later on, the author writes the following: "two measures of TQM adoption are used in this research: swiftness, which measures the relative earliness of adoption; and intensity, which measures the relative completeness of adoption." The author uses an ordinal scale to measure the swiftness of TQM.

"Swiftness of TQM adoption was measured by asking the respondents to indicate when TQM in systems development was formally initiated in their organization. A 6-point scale with the following values was used: *not yet started, <1 year ago, 1–2 years ago, 2–3 years ago, 3–5 years ago, and >5 years ago*. In addition, the scale included an option "initiated and abandoned."

We believe that the dependent variable—swiftness of adoption—has at least two problems. The first problem is related to the content validity of the dependent variable. Content validity ensures that the measure includes an adequate and representative set of items that would tap the concept. If swiftness is defined as relative earliness of adoption, then the question "When was TQM in systems development formally initiated in your organization?" may not capture relative earliness.

The term "relative" needs a benchmark for comparison. Relative measurements are common in multicriteria decision making models, where pairwise comparisons are used. One example of such a model is used in the analytic hierarchy process (AHP). The way the question is currently phrased, it only captures the *perceived* time since the formal adoption of TQM in the organization. The measure, thus, is an absolute measure.

Furthermore, if the unit of analysis is the information systems department (ISD), the question "When was TQM in systems development formally initiated in your organization?" appears to shift the unit of analysis from ISD to an organization. This leads to a dangling unit of analysis.

The second problem is that of the scale design. For example, categories on the scales are not mutually exclusive. If an organization has formally adopted TQM practices for 2 years, a respondent may choose one of the *1–2 year* or *2–3 year* options as 2 years appears in both intervals. Thus the measurement scale appears to be overlapping. We believe that relative earliness of adoption can only be measured on either a ratio scale or a pairwise comparison interval scale. The measurement of the dependent variable on either a ratio or a pairwise comparison interval scale is rare in IS surveys, however.

The author used Cox regression analysis for testing his hypotheses related to adoption swiftness. A justification for Cox regression and survival analysis was given by citing a reference in the literature. We believe that Cox regression analysis is not appropriate for the current research for the following two reasons:

1. Survey research is not longitudinal research.
2. Regression coefficients from Cox regression cannot be used to test the proposed hypotheses.

A description of Cox regression is as follows. Let

$N_i(t)$ = number of failures for subject i at or before time t

$Y_i(t)$ = 1 if subject is under observation, and 0 otherwise

$Z_i(t)$ = vector of time-dependent covariates for each individual

$\Lambda_i(t)$ = cumulative hazard function, which gives the expected number of events for the individual by time t

The Cox (proportional hazards) regression model can be written as follows:

$$d\Lambda_i(t) = d\Lambda_0(t)e^{\beta' Z_i(t)}Y_i(t)$$

Given several events over time t, the above model is fitted by estimating the regression coefficients βs and baseline cumulative hazard $\Lambda_0(t)$, so that the following partial likelihood is maximized for the product over all event times t of

$$\frac{e^{\beta' Z_i(t)}}{\sum_{j=1}^{n} Y_j(t) e^{\beta' Z_j(t)}}$$

where the individual i is the one failing at time t. The β coefficients are called the odds ratios, which express the increase of hazard in the presence of a covariate $Z_i(t)$.

In the presence of longitudinal data, Cox regression is an appropriate methodology to use. Cox regression is a popular approach for survival analysis of patients suffering from a disease and survival of mutual funds or when using Cox regression on medical data. The *time-dependent covariates* model examines the influence of patients' characteristics, which change over time, on the survival rates.

Since the respondents responded to the survey only once, no data was available across time and Cox regression analysis is therefore inappropriate. The problem is further aggravated when the significance of Cox regression coefficients is used to test the hypotheses. Cox regression coefficients provide information on the impact of a covariate in increasing the hazard (expected number of events), which is the dependent variable in the model. Using the significance results for testing hypotheses other than hazard increase is therefore inappropriate. The author uses these coefficients to test the level of significance of the impact of independent variables on the dependent variable (swiftness). Thus we believe that the set of hypotheses related to swiftness are wrongly tested. We believe that the reference cited for justifying survival analysis may suffer from similar problems as well.

Intensity of Adoption of TQM in Systems Development   The author offers three definitions for *intensity of TQM adoption*. The first one is "the number of these practices that have been adopted." The second definition is "intensity, which measures the relative completeness of adoption." And the third definition is "proportion of these practices that have been adopted." Based on the formula shown, the measurement of intensity appears to be an absolute measure and not a relative measure. In either case, the content validity problems from the first dependent variable apply to the second dependent variable as well.

The intensity of adoption is operationalized as percentage of responses that take a value of over 4 for 10 questions (measured on 7-point Likert scale). Unnoticed by the author, the formula used by the author assigns respondents to one of the 11 mutually exclusive and collective exhaustive categories. These categories are intensity of adoption taking one value from the set $A = \{0, 0.1, 0.2, 0.3, 0.4, 0.5, 0.6, 0.7, 0.8, 0.9, 1.0\}$. The creation of 11 categories transforms the interval scale to a categorical scale with some loss of information. Furthermore, the scale transformation changes the distribution of the dependent variable to the binomial distribution. In this section, we analytically illustrate the change in distribution and provide analytic models to estimate the mean and variance of the binomial distribution.

Let $B = \{p_0, p_1, \ldots, p_{10}\}$ denote the set of relative frequencies of 116 usable responses taking the values in the set A. According to the author, there were a total of 123 responses, with 81 respondents who were considered to be adopters, and 30 respondents who were considered to be nonadopters. Of the remaining responses, 7 responses were missing and 5 respondents indicated that they had abandoned TQM initiatives. Since the author used a sample size of 81 adopters in his analysis, it is fair to assume that all 81 adopters had at least one response that was greater than 4, on the 10 different questions that were asked to compute intensity. By the same token, all the other 35 respondents had all the responses below 4, in all the 10 questions measuring the multi-item dependent variable. In the absence of the actual survey data, the principle of maximum entropy can be used to demonstrate that the values of relative frequencies (probabilities) of the members of set B can be estimated by solving the following nonlinear program:

$$\text{Maximize} - \sum_{i=6}^{10} p_i \ln P_i$$

subject to

$$\sum_{i=5}^{10} P_i - \frac{81}{116} \quad \text{(for adopters)}$$

$$\sum_{i=6}^{4} P_i - \frac{35}{116} \quad \text{(for others)}$$

$$\sum_{i=0}^{10} p_i = 1$$

The solution to the above nonlinear program will provide estimates for the elements of set B, which provide the probability mass function of the dependent variable intensity. Thus the interpretation of a few elements of the set B is as follows:

$p_0 =$ probability that a respondent evaluates all 10 responses below or equal to 4.

$p_1 =$ probability that a respondent evaluates exactly 9 responses below or equal to 4 and 1 response higher than 4.

The values of the other elements can be similarly described.

Note 3.1   The entropy of the above nonlinear program is finite.

We believe that the transformation of the variable leads to a reduction in entropy (loss of information) in the dependent variable, which may lead to Type II errors (significant hypotheses that are not significant in reality). If the transformation is not used, then the average of all 10 variables can be used to measure the *perceived* intensity of adoption. This average variable, say, X, will be a continuous variable that will take values on the interval [1, 7]. The following lemma proves that using the transformed variable, taking discrete values from set A instead of variable X, leads to a reduction in entropy (loss of information).

Lemma 3.1 The entropy of a continuous variable X is higher than the transformed variable taking values in set A.

Proof: Let $f_X(x)$ be the probability density function of the variable X, which takes continuous values on the interval $[1, 7]$. Suppose that the interval is divided into a large number, K, of subintervals of equal length δ. Let $R(X)$ be the midpoint of the subinterval that contains X. If x_k is the midpoint of the kth subinterval, then $P[R = x_k] = P[x_k - \delta/2 < X < x_k + \delta/2] \approx f_X(x_k)\delta$, and thus entropy can be written as follows:

$$H_R \approx -\sum_{k=1}^{K} f_X(x_k)\delta \ln(f_X(x_k)\delta)$$

The above expression, after some simplification, can be written as

$$H_R \approx -\ln\delta - \sum_{k=1}^{K} F_X(x_k)\delta \ln(f_X(x_k))$$

For very small values of δ, the above expression for H_R approximates the true entropy of a continuous variable X. However, as $\delta \to 0$, $H_R \to \infty$. In other words, the entropy of the continuous variable X is infinity. Note 1 shows that the entropy of the transformed variable taking values in set A is finite. Thus the continuous variable X has higher entropy than the transformed variable taking categorical values in set A.

Since the author used a 7-point interval Likert scale and used a transformation to convert the results into a binary variable, the underlying distribution of the variables changed. Since there are 116 respondents, each responding to 10 questions related to the dependent variable, the responses can be considered to form a binomial distribution with values p (probability of scoring higher than 4) and $p' = 1 - p$ (probability of scoring lower than or equal to 4). For a binomial random variable X with n, p, and a sample space containing the number of questions with response values for 10 questions, $S_X = \{0, 1, 2, 3, \ldots, 10\}$, the probability mass function can be written as

$$P[X = k] = \binom{n}{k} p^k (1-p)^{n-k}, \quad k = 0, 1, 2, \ldots, 10.$$

Thus for the known value of p, the expected probability that a random respondent can have exactly one response over 4, and all the other 9 responses less than or equal to 4, can be calculated using the above formula ($n = 10$, $k = 1$). In fact, the expected value should be the same as the value of element p_1 in set B. For known values of elements in set B, the values p and p' for the binomial distribution can be estimated by solving the following nonlinear program:

$$\text{Minimize} \quad E = \sum_{i=0}^{10} \left[p_i - \binom{10}{i} p^i (p')^{10-i} \right]^2$$

subject to

$$p + p' = 1$$

The nonlinear program described above may be difficult to solve using traditional gradient based approaches, and global search approaches, such as genetic algorithms, may be used to estimate the optimal values of p and p'. Once the values of p and p' are known, the mean and variance of the binomial distribution can be estimated as mean $= 10p$ and variance $= 10pp'$.

The author used multiple regressions to study the significance of the impact of independent variables on the dependent variable. Multiple regressions require that the dependent variable be a continuous variable with the distribution of the dependent variable close to a normal distribution. Unfortunately, the dependent variable *intensity*, as used by the author, is not continuous and does not follow a normal distribution. While the author has used a transformation that was used previously in the literature, we believe the previous studies may suffer from similar problems.

The author used multiple regression for testing hypotheses related to the dependent variable *intensity of adoption*. Since the transformed dependent variable is discrete, the use of multiple regression is inappropriate. It is generally accepted that a continuous, interval-scaled dependent variable is required in multiple regression, as it is in bivariate regression. Interval scaling is also a requirement for the independent variables; however, *dummy variables*, such as the binary variable in our example, may be utilized. Given the change in the distribution of the dependent variable, loss of information as a result of scale transformation, and the inappropriate use of the technique for data analysis, the set of hypotheses related to intensity of adoption may have errors.

Conclusions   We believe that the author's study has internal validity problems. Among the reasons for our conclusion are problems with content validity, a dangling unit of analysis, the use of inappropriate techniques for data analysis, and problems related to scale design and change in data distribution due to the transformation of one of the dependent variables. The lack of theoretical grounding of different hypotheses exacerbates the internal validity problem.

To remove or mitigate these barriers, the journal executives needed convincing that there were flaws in their reviewing process. Our approach was to demonstrate that data is on a critical path to enterprise performance optimization, and the editorial board could achieve this best with a smart data strategy. The principal idea was to focus executives' attention on data that is essential to enterprise performance and executives' ability to execute their primary responsibilities. As a result, the editorial policies of the journal were changed and an optimization of manuscript quality was improved.

Although the journal executives had IT services and infrastructure available for their reviewers, it was apparent that they needed better analytical capabilities to model and simulate conditions in response to a dynamic environment. They needed the ability to predict performance with high certainty in the near term and with increased visibility over the longer term, not to continue to propagate their errors by a diffusion of errors that allowed the application of a longitudinal method such as Cox regression to be inappropriately applied to a snapshot survey environment.

It was only after the editorial executives took ownership for the data methods and algorithms, and realized that they needed to optimize enterprise performance, that they overcame the propagated errors that previously may have been considered to be invasive to the review process. The previous editorial review mind-set needed to be overcome in order to disturb the status quo in manuscript information management and reveal where the true meaning of the data methods and algorithms were buried.

The Ugly Instance

1. Lack of interoperability
2. Lack of optimization
3. Lack of correct methods and algorithms

Multiple deficiencies in smart data exist.

3.3 TOP–DOWN STRATEGY

The place to begin is with executives at every level of the enterprise recognizing and embracing their data—insisting that they have clear visibility into the data that they need to optimize performance, and into the data representing enterprise assets that they must protect. Executives need and want a holistic approach to data strategy and data management.

With the notion of smart data, the data should be engineered such that it addresses executive needs for problem solving, decision making, planning, sense making, and predicting. This means that when an executive or manager needs to refer to the data for answers, it is unnecessary to scramble to locate it or to judge its reliability or quality, as executive requirements are anticipated and the response preengineered as in data engineering.

Better still, the executive does not have to search for the answers because solutions and answers are presented to the executive, anticipating requirements. It is accompanied by methods and algorithms representing best analytics. Some describe this as an autonomic or self-managing system that is characteristic of smart data.

Data that is shared throughout the enterprise by multiple users and disciplines must be engineered such that it is available openly for seamless processing to those who are credentialed and privileged.

While there are a host of issues in the domain of smart grid and smart enterprise services associated with implementation, our focus is on anticipating data require-ments and engineering data for use throughout the enterprise. Since much data is created, stored, and maintained in a distributed environment, smart data strategy imposes common standards and strategy for members of the community/communities. It requires a system for smart data exchange.

Having a top–down strategy and direction trumps autonomous behavior that might deviate from and otherwise undermine what is best for the enterprise at each aggregate level of participation. However, this does not mean dumbing down data to the lowest level of interpretation. It does mean accommodating natural behavior and differences associated with highly autonomous and entrepreneurial operations. This should be true in government as well as in business.

Since organizations at the bottom of the enterprise hierarchy serve many higher order organizations, the smart data strategy must be easiest and most affordable to implement. For those reasons, we advocate open standards and high value for information interoperability.

Proprietary systems and software are a nemesis unless their data outputs are available in open standards for which access and use are minimally encumbered. The SOA approach makes great strides toward supporting this idea, although the top–down strategy is needed to reinforce the priority to implement the smart data strategy.

No Greater Priority than Optimizing Enterprise Performance

Attention to making data smart and using a smart data strategy is the most important priority because it is directly aligned with the single most important executive responsibility.

Elements of Top–Down Strategy

- Published smart data strategy with a description about engineering smart data
- Published compliance and expectations for enterprise member participants
- Implementation and operational policies
- System of sharing success stories and reporting deficiencies
- System for tracking and reporting changes and improvements
- Day-to-day reinforcement by top executives, principally through critical path usage

Executives experienced with enterprise resource planning (ERP) might observe or ask rhetorically, "Isn't that what we bought when we installed ERP?" Our response is that what you want is enterprise resource management (ERM) such that you can optimize enterprise performance.

What you have with most ERP packages is a monolithic software platform that grows and grows at incredible expense, for which interoperability is deficient unless everyone with which your enterprise works is on the exact version of your brand of software operating with synchronized changes and configuration management. We know from many examples that even within one corporate entity and with one software brand, this is difficult if not impossible to achieve.

We are suggesting that there are better technical solutions that will afford greater agility and data exchangeability and which are less costly and produce harmonious performance. The executive's specification for a smart data strategy must include this consideration.

Case Study: Technical Efficiency Ranking Using the Data Envelopment Analysis

The following case is a good example of how to improve optimization of human resources in a healthcare environment. In this case, we show that when nonlinear machine learning models are used for learning monotonic forecasting functions, it may be necessary to screen training data so that the resulting examples satisfy the monotonicity property [11].

We show how technical efficiency ranking using the data envelopment analysis (DEA) model might be useful to screen training data so that a subset of examples that satisfy the monotonicity property can be identified. Using real-life healthcare data, the managerial monotonicity assumption, and an artificial neural network (ANN) as a forecasting model, we illustrate that DEA-based screening of training data improves forecasting accuracy of the ANN.

Earlier, we made the point that smart data is central to improving intelligent optimization as it leverages answers buried in information that is already available. This case uses data mining, which is surely an aspect of smart data. The case study answers many of the questions that top executives asked, such as, who are our customers, how much will they buy, when will they buy, what must we offer them, how much should we charge, what after-sales services should we offer, who are our competitors, and what must we do to provide a superior offering?

The answers to these questions provide the basis for prediction with a degree of certainty. Every enterprise, including healthcare, has questions like these for which quantitative answers are readily determinable.

We have also pointed out that these questions accompany other more routine business questions such as how is our cash flow, what are the capital and material requirements, what are the enabling people and technology requirements, what are the processes involved, what are the performance metrics, what is the payback period and break-even, and what is the return on investment?

Observe that many of these questions are not answered in routine business accounting, although some are. There is concern that application of some methods and algorithms and the resulting reports are not predictive. They provide historical reporting from which analysis and comparison may serve as a retrospective report card, although not necessarily indicative about the future. A smart data and smart data strategy approach improves ability of enterprises to prepare for and more accurately anticipate the future. The art of management becomes more of a science in this regard whenever properties such as monotonicity are taken into consideration.

In certain forecasting problems it is very natural to assume that the forecasting function will satisfy the monotonicity property. For example, in financial forecasting models individual demand is shown to increase with one's income. In medicine, a patient toleration level for a drug increases with effectiveness of treatment. A few researchers in information systems have shown that a firm's financial performance increases with an increase in its information technology (IT) investment. Several models in economics, operations research, and transportation science are required to satisfy the monotonicity property.

The monotonicity property of the forecasting function makes it easy to aggregate forecasts. For example, assuming monotonicity of individual preferences relates individual preferences to predict market behavior and preserving monotonicity reduces overfitting problem in a nonlinear algorithm.

For a linear forecasting function, it is easy to check if the function satisfies the monotonicity property. Perhaps some measures such as positive values of regression coefficients can be used to ensure that the learned forecasting function satisfies the monotonicity property. When it is desired to use a nonlinear forecasting function, monotonicity can be preserved by (1) redesigning the learning algorithm so that the resulting forecasting function learned by the nonlinear algorithm preserves monotonicity, or (2) using the training data set that does not violate the monotonicity property.

There are several nonlinear machine learning forecasting algorithms available. The examples of nonlinear machine learning algorithms include artificial neural network (ANN), classification and regression tree (CART), and genetic programming (GP). All of these algorithms do not assume any particular functional form of the forecasting function and learn the nonlinear forecasting function based solely on the properties of the training data set.

Since the functional form of the machine learning algorithm depends on the training data, it may be necessary that the training data satisfy the monotonicity property. When training data violate the monotonicity property, some data screening may be required to create a training data subsample that does not violate monotonicity. In a small-size data set, such data screening may not be feasible due to very small resulting subsample size.

However, in mid- to large-size data sets, it may be possible to create a subsample that preserves monotonicity. In mid-size data sets, if a subsample that preserves monotonicity is too small, then perhaps a large subsample, not containing all the original training data, may be created with the subsample data that "weakly" preserves monotonicity. The term "weak" monotonicity means that most of the cases in the training subsample satisfy the monotonicity property, and a few cases may somewhat violate monotonicity.

The design of the data screening algorithm that may be used to create a subsample is an important issue. It is important that the screening algorithm should be nonlinear, preserve monotonicity, and not assume any particular nonlinear functional form. For multivariate input and one output, the data envelopment analysis (DEA) model seems to satisfy all the requirements.

The DEA model is a nonlinear (piecewise linear) and nonparametric model that is used to measure efficiency of production units. In this case, we use DEA as a data screening approach to create a subsample training data set that is "weakly" monotonic. We use the ANN as a nonlinear forecasting model. The contribution of our research is twofold: (1) we show how DEA can be used as a methodology to screen training cases, where forecasting models are subject to the managerial monotonicity assumption; and (2) we show how ANNs can be applied to forecast the number of employees in the healthcare industry. Empirical studies are conducted to compare the predictive performance of ANNs with learning from a set of DEA-based selected training cases, DEA-based rejected training cases, and a combined set of DEA-based selected and rejected training cases. We use the publicly available data from healthcare facilities in Pennsylvania to learn about and predict the number of employees based on a set of predictor variables.

DEA was a technique introduced for comparing efficiencies of decision making units (DMUs). The basic ratio DEA model seeks to determine a subset of k DMUs that determine the envelopment surface when all k DMUs consist of m inputs and s outputs. The envelopment surface was determined by solving k linear programming models (one for each DMU), where all k DMUs appear in the constraints of the linear programming model.

ANNs have been applied to numerous nonparametric and nonlinear classification and forecasting problems. In an ANN model, a neuron is an elemental processing unit that forms part of a larger network. There are two basic types of ANNs: a single-layer (of connections) network and a double-layer (of connections) network. A single-layer network, using the perceptron convergence procedure, represents a linear forecasting model.

A modification of the perceptron convergence procedure can be used to minimize the mean-square error between the actual and desired outputs in a two-layer network, which yields a nonlinear, multivariate forecasting model. The back-propagation learning algorithm, most commonly used to train multilayer networks, implements a gradient search to minimize the squared error between realized and desired outputs.

Figure 3.3 shows a three-layer network that can be used for multivariate fore-casting. The number of input layer nodes corresponds to the number of independent variables describing the data. The number of nodes in the hidden layer determines the complexity of the forecasting model and needs to be empirically determined to best suit the data being considered. While larger networks tend to overfit the data, too few hidden layer nodes can hinder learning of an adequate separation region. Although having more than one hidden layer provides no advantage in terms of forecasting accuracy, it can in certain cases provide for faster learning.

The network is initialized with small random values for the weights, and the back-propagation learning procedure is used to update the weights as the data are iteratively presented to the input-layer neurons. The weights are updated until some predeter-mined criterion is satisfied. The number of hidden layer neurons is chosen as twice the number of data inputs, a commonly used heuristic in the literature. Heuristics of "the more the better" could be used as a guide to select the number of hidden nodes in an ANN. Furthermore, a network with a higher number of hidden nodes can always be considered as a special case of the network, with additional nodes (in the case of an ANN with a higher number of hidden nodes) having connection weights taking values equal to zero.

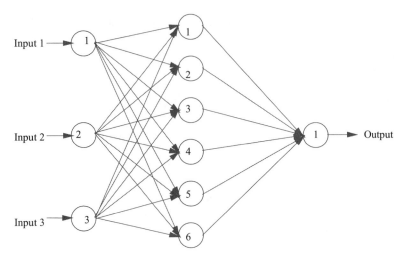

A three-layer network used in forecasting.

Figure 3.3 Three-layer network used in forecasting.

The disadvantages of having too many hidden nodes (two times the inputs and above) are excessive training time and memorization of training patterns. Since hidden nodes are one of the parameters that may play a role in ANN performance, we try two configurations—one with the number of hidden nodes equal to the number of inputs + 1, and the other with the number of hidden nodes equal to twice the number of inputs + 1. After some initial experimentation (experiment #1), we identify a better configuration and use it for our DEA-based data preprocessing experimentation (experiment #2). We do admit that better designs, in terms of selecting the number of hidden nodes, are possible but such issues are considered beyond the scope of the current research.

ANNs are data-driven techniques; thus high-quality training data for the ANN implementation cannot be overstressed. The quality and structure of the input data largely determine the success or failure of an ANN implementation. High-quality data do not necessarily mean data that are free of noise (e.g., errors, inaccuracies, outlying values).

In fact, it has been shown that ANNs are highly noise tolerant in comparison to other forecasting methods. A study comparing the forecasting performance of ANNs to linear regression under circumstances of varying data accuracy concluded that ANN-based forecasts were more robust as the data accuracy degraded.

Several studies point out the fact that, for ANNs, providing high-quality training input data is not as simple as "cleaning" inaccuracies from data or eliminating extreme outliers. On the contrary, it has been suggested that a valuable technique for improving the quality of training data is to *add* noise to the network inputs. The addition of noise is shown to strengthen the generalizability of ANNs, and the absence of noise in the training data forces the network to "memorize" specific patterns rather than abstract the essential features from the examples.

There are three essential requirements for high-quality training data (1) the data should adequately represent the fundamental features that the network must detect in order to obtain correct outputs; (2) the training data set should provide sufficient variation to allow generalization and discourage "memorization;" and (3) the training data should not contain contradictory examples (i.e., examples with identical inputs but different outputs).

In regard to the first and second requirements, we caution that pattern variations (noise), while helpful in the training process, must be controlled to avoid "swamping" the essential features in the input data. The need to find a balance between these two requirements has been described as the *noise-saturation dilemma*. The dilemma can be expressed in terms of communications theory as follows: some fixed dynamic range will always exist within which the components of a system can operate; that is, any signal below a certain baseline level will become lost in system noise, and any signal above a certain level will cause the components to become saturated at their maximum value. To successfully resolve the noise-saturation dilemma, an ANN must be able to process the dominating patterns without allowing the weaker patterns to become lost in system noise. At the same time, an ANN must be able to process weak patterns without allowing the dominating patterns to saturate the processing units.

There are several proposed approaches for training tasks in ANNs. The various approaches can be grouped into the following categories: (1) statistical and other data analysis techniques, (2) data transformation and preprocessing, (3) analysis of feedback obtained by interpreting the connection weights, and (4) hybrid techniques.

It has been proposed that statistical and numerical approaches can be useful in selecting and preprocessing training data for ANNs. One approach is to apply significant measures (correlation coefficients and plots) to assess the strength of the relationship between data elements; linear regression is used to examine the degree of contribution a candidate data element makes to the model as a whole. In cases where two data elements are highly correlated, one of the elements is eliminated.

Another approach is to calculate correlation coefficients between the error terms of a model prediction and each individual *unused* candidate variable. A high correlation between an unused variable and the error terms suggests that including that variable in the model might add explanatory power (resulting in improved ANN performance). Data aggregation approaches, such as the use of histograms, are proposed to facilitate inspection of the data. These approaches sometimes reveal unusual characteristics or complex relationships and outliers in the data. Stein pointed out that outliers and odd patterns in the data do not necessarily indicate that data should be eliminated or modified. Such patterns can be indicators of important relationships in the data, and further analysis of these areas can lead to improved input selections as the ANN model is refined.

Several other studies focused on training data transformation and preprocessing. The studies included both formal (applying a nonlinear transform to nonnormal data) and heuristic approaches. (If you are unsure about including a certain type of data, include it!). A transform was sometimes applied to data when the data did not approximate the normal curve. In general, the ANN performed better when the input data was normally distributed. Other formal techniques for preprocessing included the calculation of intermediate functions and trends. An example of an intermediate function is the use of ratios.

It is usually more meaningful and efficient to present a ratio, such as miles per gallon, to the network than to input the individual components, such as miles traveled and gallons consumed. Similarly, an algorithm is used to convert raw data into a form that captures a trend over time.

The following heuristics have been proposed for the use of binary and continuous data for training ANNs:

1. For naturally occurring groups, the binary categories are often the best method for determining the correlation.
2. Continuous-valued inputs should never be used to represent unique concepts.
3. For continuous value attributes, breaking them up into groups can be a mistake.

Among other heuristics is the "leave-k-out" technique, which involves the decision of how to handle input samples when relatively few examples are available. The procedure is to train several networks (each with a different subset), including most of the examples, and then test each network with a different subset.

In enumerating the three general requirements for high-quality training data above, it was stated that the training data should not contain contradictory examples (i.e., examples with identical inputs but different outputs). As a preprocessing issue, an algorithm was developed to deal with conflicting data. Conflicting data exists when two or more input patterns are either identical or very similar, but the two inputs have completely different outputs. Because the patterns being presented are contradictory, the ANN is unable to learn both patterns. In an example, 65% of the original data file consisted of conflicting data. The cause was not attributed to faulty data, but to an incomplete modeling process.

When a sufficient number of fields are not included in the model to distinguish between the two output categories, a conflicting-data situation exists and the network trains and tests poorly; however, the problem is not solved by eliminating portions of the data. The solution involves reexamining the data model and including the necessary fields, a tedious task when performed manually. To assist with this examination process, an algorithm was developed. With the use of the algorithm to reconstruct the training data, we can obtain a 25% increase in network performance.

A set of studies focused on improving the data model for ANN training by analyzing the feedback obtained by "interpreting" the connection weights of the network's processing elements. The nature of the ANN paradigm is such that the "meaning" of connection weights is opaque; therefore it is difficult to assess the relative importance of the input factors used by an ANN to arrive at its conclusions.

If such feedback were available, it might be possible to gain better insight into the types of inputs that could enhance the network effectiveness. Several techniques have been developed that attempt to provide an interpretation of the connection weights.

In one method, the partition is used to make judgments about the relative importance of the various inputs. The method consists of applying an equation that effectively partitions the middle-to-output layer weights of each middle (hidden-layer) node into components associated with each input node. It has been suggested that this method can make ANNs as effective for modeling applications as for pattern matching.

A more manually intensive method has been proposed for interpreting the weights that entails iteratively presenting specific isolated "features" to the trained network and assessing the network's "reaction." For example, after training the network to recognize seven letters of the alphabet, various features of the letters (such as the angular top of the letter A) were presented to the network, and the responses of each node were examined. Through this method, the discriminating features were identified. In both the above-mentioned methods, it was unclear to what degree the technique's applicability would extend to networks with different architectures and levels of complexity.

One of the most interesting methods of a weight-interpreting technique was the Knowledgetron (KT). The KT method uses a sophisticated algorithm to walk through the nodes of the network, interpret the connections, and generate a set of "if–then" rules to match the behavior of the network. These rules can then become the basis for a rule-based expert system. It is particularly noteworthy that the KT technique was able to process the classic "Iris" data set successfully. A set of only five relatively simple rules were generated from ANN trained on the Iris data. In fact, it was found that the generated rules actually outperformed the ANN from which they were generated. Because the Iris data set is not particularly complex, it would be premature to assume that the KT technique would scale to more complex, real-world data sets; however, its performance provides evidence that further research in this area would be worthwhile. From the perspective of the data issue, the production of rules provides a convenient mechanism for evaluating the importance and role of each input data element. This could form the basis for a "feedback-and-refine" iterative process to guide the selection and preprocessing of training data, which might result in improved ANN performance.

From the foregoing discussion, it can be observed that the range of techniques and possibilities for enhancing ANN effectiveness, even limiting the field to consider only the data component, is broad and complex. Many of these techniques require a level of expertise that may prohibit their use in practical situations; however, a growing body of research and applications in the area of hybrid intelligent systems could address this concern.

Many hybrid intelligent systems have been developed in recent years. Expert systems and ANNs have characteristics that can complement each other and form the basis for powerful hybrid systems. Several examples of expert systems used as frontends for ANNs have been proposed. These hybrid systems essentially automate the tedious work of preprocessing, including organizing data available from cases in a manner required by a specific ANN model. An expert system can also be used to select the best ANN paradigm for the data or to determine the optimal sample size of training and test sets.

Frameworks have been developed for expert system and hybrid technologies to assist in the improvement of ANN training. Among the experimental questions posed were those targeting the need for guidelines in the selection of the training data. Examining the example of character recognition, we see that an expert system is a natural fit for training.

Genetic algorithms (GAs) are another area of artificial intelligence (AI) application that can add value to the training and data selection tasks. Developments in hybrid systems of this type show that most of the work performed in this area focuses on the advantages of employing GAs to search large spaces and find an initial set of parameters for training an ANN. A number of GA techniques can be used to select optimal values for data and parameters used in the ANN training process.

Bayesian approaches have recently been used to select the best ANN architecture in a set of several candidate architectures. In particular, the Bayesian nonlinear regression model comparison procedure has been utilized to select an ANN architecture under which the training data set achieves the highest level of internal consistency.

Most of the techniques proposed in the literature attempt to increase the generalizability of ANN by removing inconsistent data or by adding random noise in the data. A few researchers illustrate different procedures used to identify the inputs that may significantly (in a statistical sense) be related to the output variable. While significant, none of the approaches focus on preserving the monotonicity property of the forecasting function learned by an ANN. In our case, we propose DEA-based prescreening of ANN data to preserve the monotonicity property. Our prescreening approach, while preserving monotonicity, doesn't hurt nonlinearity of the training data. We view DEA as a complement to other approaches to training data selection. When it is desired to increase generalizability of an ANN, DEA may not be an appropriate technique to use. However, when it is desired to preserve the monotonicity property, DEA may be a suitable approach for prescreening training data.

DEA models provide an efficient frontier, which is consistent with the monotonicity assumption. In fact, all the efficient DMUs (also called the reference set) lie on the efficient frontier, which satisfies the monotonicity property. Figure 3.4 illustrates several frontiers for a single-input/single-output combination. Frontier #1 is the efficient frontier. It can be shown that if DMUs lying on frontier #1 are taken out of the analysis, then frontier #2 becomes the next efficient frontier and so on. Assuming that all nine DMUs are used, it can be argued that DMUs lying on the efficient frontier #1 have an efficiency score equal to one and DMUs lying on frontier #2 have an efficiency score of less than one, but the score is higher than the DMUs lying on frontier #3.

ANNs use a least-squares error minimizing approach to learn nonlinear forecasting models. If only DMUs lying on frontier #1 are used for training, then it can be argued

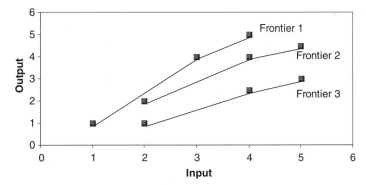

Figure 3.4 Set of piecewise linear frontiers.

that the resulting model is guaranteed to have a monotonicity property. In fact, the forecasting model will closely follow frontier #1. If all nine DMUs are used, then there is a likelihood that the monotonicity assumption may be violated and the resulting model may produce forecasts that are not consistent with the assumption of monotonicity.

Figure 3.5 illustrates two forecasting models. The first model (least-squares forecast #1) is an approximate model that may be learned by an ANN if DMUs lying on frontier #1 and frontier #2 are used for training. The second model (least-squares forecast #2) is an approximate model that may be learned by an ANN if all nine DMUs are used for training.

Let DMU_1, DMU_2, and DMU_3 represent the DMUs lying on frontier numbers 1, 2, and 3, respectively, and P(Monotonicity | DMU_i) represent the conditional probability that the monotonicity assumption is satisfied, if DMUs lying on frontier $i \in \{1, 2, 3\}$ are used for training ANN. In summary, we can conclude that when ANN forecasting models are subject to the managerial monotonicity assumption and the training data set is large, using DEA for selecting training data would reduce the ANN training time (as a result of fewer data) and reduce the probability of violation of the monotonicity

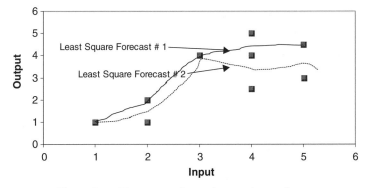

Figure 3.5 Two approximate forecasting surfaces.

assumption. The heuristic to remove the cases from the large data set is to remove the cases that have lower relative efficiency. This is similar to removing cases lying on frontier #3 in our univariate case example, and deleting lower efficiency outliers in the multivariate input case.

Human resource management represents significant expense to an organization. Among the responsibilities of a human resource department is the task of solving the problems of workforce utilization, organizational development, performance measurement, and adaptation to evolving business demands. The inappropriate management of human resources for a healthcare facility involves the risk of delays and the inability to deliver quality care. The delays and poor quality assurance translate into a weakened position of the company in terms of both cost and quality of care [12].

Among the factors that determine the human resource requirements in the healthcare industry are the number of patients, available physical resources and capacity, the type of hospital, and the types of services offered by the hospital. Because recruitment schedule and budget decisions are based on managerial estimates, we focus on the impact of the different factors on the human resource estimation. In reality, the human resource estimation depends on several complex variables (including the ones identified above), the relationships of which are often unclear. Given the lack of information on the interrelationships of the various variables, it becomes difficult to establish any specific parametric form of human resource requirement dynamics. Based on the review of ANN literature, we believe that an ANN model can be used to discover the nonparametric and nonlinear relationships among the various predictor variables.

Data on various hospitals were obtained from the Hospital Association of Pennsylvania, Pennsylvania Health Department, and the Pennsylvania Medical Society. The data set consisted of information on 275 hospitals throughout Pennsylvania.

Information about the following was collected from each hospital:

1. Hospital name
2. Ownership
3. Total beds
4. Employees (full-time equivalents)
5. Emergency services visits
6. Top diagnostic-related group
7. Admissions
8. Outpatient visits
9. Average daily census
10. Other information (e.g., surgical combination, nursing home admissions and long-term care, special services)

Not all 275 hospitals contained complete information. From our analysis, we found that approximately 188 hospitals contained the complete data. A preliminary analysis

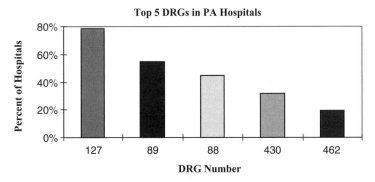

Figure 3.6   Percentage of hospitals offering top five DRG services.

was undertaken on the collected data. First, we plotted the percentage of hospitals that provide the top five diagnostic related group (DRG) services.

Figure 3.6 illustrates a bar chart of this analysis. Note that many hospitals provide more than one DRG-related service; thus the percentages do not add up to 100%. We further plotted the ratio of admissions per employee and the beds available per employee. Figures 3.7 and 3.8 show the results of plots for 188 hospitals.

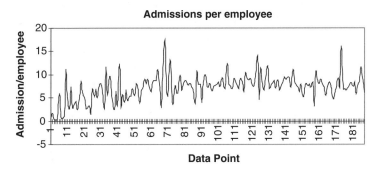

Figure 3.7   Number of admissions per employee.

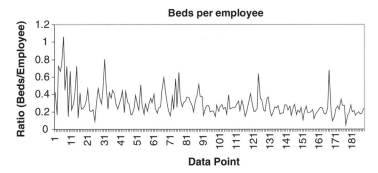

Figure 3.8   Number of beds per employee.

The mean and standard deviation for the number of admissions per employee were 7.12 and 2.74, respectively. The mean and standard deviation for number of beds per employee were 0.29 and 0.14, respectively. The higher standard deviation of number of admissions per employee compared to the number of beds per employee shows that some hospitals have lower admissions per employee compared to other hospitals. The differences in hospitals in terms of number of beds per employee are relatively low.

Among the set of factors that lead to the demand of human resources are the admissions, physical capacity (beds in our case), special ambulatory services, type of hospital, and the top DRG service. In other words, the employee requirement may be represented in the following functional form:

$$\text{Number of employees} = f(\text{admissions, number of beds, number of emergency}$$
$$\text{service visits, hospital ownership, top DRG})$$

where $f(.)$ maps the independent variables to the dependent variable. The total number of employees is a monotonic function of admissions, number of beds, and number of emergency service visits. The hospital ownership and top DRG are binary variables that were operationalized as 0 and 1. For the not-for-profit hospitals the value of 1 was used and for profit hospitals a value of 0 was used. In the case of top DRG, when top DRG was 127 (about 80% of the time) a value of 1 was used for input; a value of 0 was used for input when the top DRG was not 127.

Since there are only three independent variables that take continuous values, and the dependent variable is considered to have a monotonic relationship with these three independent variables, four variables are used to compute technical efficiency using DEA. These four variables provide a one output (number of employees) and three inputs (admissions, number of beds, and number of emergency visits) combination for computing efficiency using DEA [13].

In our ANN analysis, 175 hospitals were selected from 188 hospitals. Thirteen hospitals were discarded since the top DRG in these hospitals was different from the top DRGs listed in Figure 3.6. All the selected 175 hospitals contained one of the top five DRGs identified in Figure 3.6 as the major DRG. We conducted two different experiments. Our first experiment consisted of randomly dividing the data into two sets. The first set, consisting of 100 hospitals, was used to train the neural network, while the second, comprising 75 hospitals, was used to test the performance of the trained neural network.

In our second experiment, we took the first set of 100 hospitals and computed the relative efficiency of each hospital using DEA. The of 100 hospitals was then divided into two sets of 50 hospitals. The first set of 50 hospitals was called "efficient" and the other set of hospitals was called "inefficient" set. The "efficient" set contained the top 50 highest-ranking hospitals based on their DEA efficiency score.

It has been suggested that training data for nonparametric models should be at least 10 times the number of input variables. Since we have five inputs, a minimum of

50 training examples were necessary to have reasonable learning of connection weights. Since we had only six efficient cases (efficiency of 100%) and a minimum of 50 hospitals for training were desirable, we selected the top 50 hospitals (sorted in descending order of their relative efficiency) in our "efficient" set. The word efficient in the context of DEA means DMUs with efficiency of 100%.

Since the "efficient" set in our cases doesn't contain all DMUs with 100% efficiency, we use quotation marks to identify that the word "efficient" has a little different meaning than the word efficient when used in the context of DEA. The same line of reasoning applies for the word inefficient (efficiency < 100%) when used in the context of DEA and "inefficient" when used to represent the 50 hospitals that had the lowest efficiency rating. The performance of "efficient" and "inefficient" hospitals was then tested on the set of 75 hospitals in the test set. Among the design issues were the following:

1. *Normalization of Dependent Variable.* Because we were using the logistic activation function $f(x) = 1/(1 + e^{-x})$, it can be shown that $\lim_{x \to +\infty} f(x) = 1$ and $\lim_{x \to -\infty} f(x) = 0$. We normalized our dependent variable, the number of employees (y), so that $y \in (0,1)$, where 0 stands for zero employees and 0.9 was the largest number of employees in the sample.

2. *Training Sample Size.* Our initial decision on a training set sample size of 100 and a test set of 75 was determined on the heuristic of the training set size being ≥ 10 times the number of independent variables. Because we had five independent variables, we selected the training set sample size of over 50.

3. *Learning, Generalizability, and Overlearning Issues.* The network convergence criteria (stopping criteria) and learning rate determine how well and how quickly a network learns. A lower learning rate increases the time it takes for the network to converge, but it does find a better solution. The learning rate was set to 0.08. The convergence criteria were set as follows:

```
IF (|Actual_Output - Predicted_Output| ≤ 0.1 for all
examples)
OR Training iterations ≥ Maximum iterations THEN
Convergence = Yes
  ELSE
  Convergence = No
```

We selected the above-mentioned convergence criteria to account for the high variability of the dependent variable. A more strict convergence criterion was possible; however, an issue arose regarding overfitting the network on training data. Evidence in the literature shows that overfitting minimizes the sum-of-square error in the training set at the expense of the performance on the test set. We believed that the above-mentioned convergence criteria can make the network learning more generalizable. It is important to note that convergence (stopping) criteria should not be confused with the procedure used to learn the weights. A standard back-propagation algorithm was used to learn the weights. The back-propagation algorithm uses minimization of the sum of squares as the optimization criteria. The maximum number of iterations was set to 5,000.

4. *Network Structural Issues.* The network structure that we chose for our study was similar to the one shown in Figure 3.3. Our networks had five input nodes and one output node. We had a three-layer (of nodes) network for modeling a nonlinear relationship between the independent variables and the dependent variable. The number of hidden nodes was twice the number of input nodes + 1, which is a more common heuristic for smaller sample sizes. In the case of larger sample sizes, a higher number of hidden nodes are recommended. For our research, we tried two different sets of hidden nodes, 11 and 6, respectively.

5. *Input and Weight Noise.* One way to develop a robust neural network model is to add some noise in its input and weight nodes while the network is training. Adding a random input noise makes the ANN less sensitive to changes in input values. The weight noise shakes the network and sometimes causes it to jump out from a gradient direction that leads to a local minimum. In one of our experiments, we added input and weight noise during the network training phase.

The objective of our two experiments can be summarized as follows: in our first experiment, the objective was to identify the best ANN configuration and use this configuration in our second experiment. The objective of the second experiment was to use the proposed DEA-based training data selection to test if there are any performance differences when "efficient" versus "inefficient" training data are used.

Based on the design considerations, we conducted four different tests by varying the structural design and noise parameters in a two-layer network. Tables 3.2 and 3.3 illustrate the results of different structural designs for our four tests for network training on 100 cases and testing the trained network on 75 unseen cases, respectively. The first two tests represent the hidden number of nodes of 11 and 6 in a two-layer

TABLE 3.2 Performance of Four Different Structural Designs During the Training Phase

Number of Hidden Nodes	Input Noise	Weight Noise	RMS Error	Prediction Accuracy
11	0	0	0.037	97%
11	0.08	0.01	0.05	96%
6	0	0	0.044	97%
6	0.08	0.01	0.041	98%

TABLE 3.3 Performance of Four Different Networks During the Test Phase

Number of Hidden Nodes	Input Noise in Trained Network	Weight Noise in Trained Network	RMS Error	Prediction Accuracy
11	No	No	0.079	80%
11	Yes	Yes	0.20	37.3%
6	No	No	0.11	60%
6	Yes	Yes	0.098	64%

network. The numbers 11 and 6 were chosen to represent the $(2n + 1)$ and $(n + 1)$ heuristic, where n is the number of input nodes.

Tables 3.2 and 3.3 report two performance metrics—prediction accuracy and RMS error. Since our problem at hand is that of forecasting, prediction accuracy has a different meaning in our case. The prediction accuracy in our case implies the percentage of examples that satisfied the condition: |Actual output − Predicted output| ≤ 0.1. The prediction accuracy of less than 100% in training indicates that the network training was stopped at 5000 iterations. When evaluating the goodness of fit of the ANN models, RMS provides better information than prediction accuracy.

Although prediction accuracy numbers may make the reader believe that the ANNs are overtrained, we don't believe this is true. The reason for our belief is as follows: let's consider the zero input and weight noise training and test results of 11 hidden nodes and 6 hidden nodes simultaneously.

Focusing on RMS error only, we can see that RMS decreased for both training and test (for the no input and weight noise case) when the number of hidden nodes were changed from 6 to 11. The lowering of RMS with an increase in number of hidden nodes gives an indication that an even larger network (hidden nodes greater than 11) would have lowered the training RMS further. Also, in the case of overtraining, an expectation is that the RMS for training data would decrease and the RMS for the test data will either not change or would increase.

Clearly, the results did not indicate this situation. Overtraining in the ANN occurs when a very large size ANN is used. A large size ANN would have a tendency to memorize input cases and lower the training RMS significantly without lowering the test RMS. Our results indicate that the number of hidden nodes (for the no-noise case) not only decreased the training RMS but also decreased the test RMS. Thus we don't have any reason to believe that any significant overtraining problem was evident.

The reader can view Figures 3.9 and 3.10 to verify the lack of significant overfitting in both the training and test data. When random input and weight noise is considered, an expectation is that the smaller network will have a lower RMS as it has a lower degree of freedom. For example, in the case of 11 hidden nodes, 5 input + 1 threshold nodes, and 1 output node, random weight noise is added to 77 weights. In the case of 6 hidden nodes, random weight noise is added to 42 weights. Thus large size ANNs are more vulnerable to random weight noise and our results appear to confirm this hypothesis.

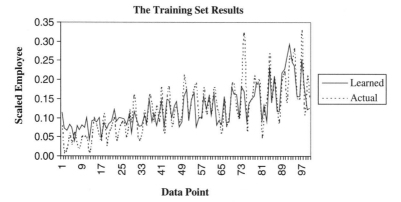

Figure 3.9 Performance comparison between the actual human resource requirements and the learned human resource requirements at convergence.

We plot the training and test results of our best performing first experiment with 11 hidden layers and no random input and weight noise in Figures 3.9 and 3.10, respectively. We use our best performing ANN in experiment #2.

As described earlier in the case, we divided the initial training set data into two sets of "efficient" and "inefficient" training data sets. The relative efficiency of the two sets is shown in Table 3.4. A two-layer, 11-hidden-node network was first trained on "efficient" training set examples and tested on the 75 test examples. A similar test was then conducted by training the ANN on "inefficient" training set examples and then testing on the 75 test cases. Figure 3.11 illustrates the forecasts generated for 75 test examples by training the two ANNs with "efficient" and "inefficient" cases.

Table 3.5 illustrates that ANN forecasting based on learning from "efficient" cases is higher than ANN forecasting based on learning from "inefficient" cases. A t-test on the difference of means between the ANN forecast values from learning from "efficient" and "inefficient" cases shows that the difference is significant at the 0.01 level of significance.

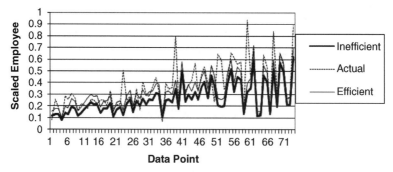

Figure 3.10 Performance comparison between the actual human resource requirements and ANN-predicted human resource requirements.

TABLE 3.4   Set of Efficient Hospitals (Efficiency > 53%)

Hidden Nodes	Type	Phase	RMS Error	Accuracy
11	Inefficient	Training	0.028	100%
11	Efficient	Training	0.032	98%
11	Inefficient	Test	0.11	54.7%
11	Efficient	Test	0.08	85.3%

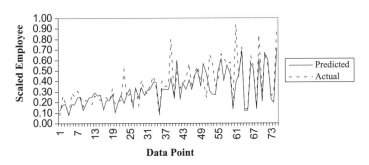

Figure 3.11   Performance comparison between the actual human resource requirements and efficient and inefficient ANN-predicted human resource requirements.

TABLE 3.5   The *t*-Test for Difference of Means for Predicting Unseen Cases Through Learning from "Inefficient" and "Efficient" Cases

Mean Difference	Correlation	*t*-Value	df	Two-Tail Significance
0.064	0.9976	−36.47	74	0.000[a]

[a] Significant at $\alpha = 0.01$.

The illustrated results in Figure 3.12 indicate that training ANNs on efficient cases increases the predictive accuracy by 5.3% when compared to training an ANN containing both "efficient" and "inefficient" cases (as shown in experiment 1). To determine if "inefficient" cases impair the predictive validity of the ANN, we plotted

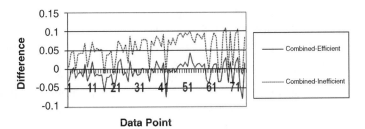

Figure 3.12   Plot of the difference between a combined efficient ANN forecast and a combined inefficient ANN forecast.

the differences in the forecast generated by the combined set (containing total of 100 examples) and the forecast generated by 50 "efficient" and 50 "inefficient" cases.

Based on the results plotted in Figure 3.12, it can be seen that the difference between the combined forecast and the forecast generated by learning from "efficient" cases is lower (RMS error $= 0.0245$) than the difference between the combined forecast and the "inefficient" forecast (RMS $= 0.0667$). This shows that for larger data sets, ANNs have a tendency to learn patterns from "efficient" cases. "Inefficient" cases in the training set lead to poor forecasting accuracy, hurt the monotonicity assumption, and foster inconsistent learning.

Among the key issues in the DEA assessment of technical efficiency are:

1. Comparison of sets of DMUs that perform uniform operations
2. Assumption that, for efficient DMUs, higher values of input should lead to higher values of output
3. No assumption made concerning the functional form that links the inputs and outputs

The following are similarities between ANNs and DEA models:

1. DEA and ANNs are nonparametric models.
2. Neither DEA nor ANNs make assumptions about the functional form that links its inputs to outputs.
3. DEA seeks a set of weights to maximize the technical efficiency, whereas ANNs seek a set of weights to derive the best possible fit through observations of the training data set.

In the case of using DEA for preprocessing data for ANNs, DEA eliminates some of the inconsistencies in the data that may hurt the monotonicity assumption. For example, if the training data set consists of two examples with the same inputs but different outputs, the example with a lower output receives a lower technical efficiency. The lower technical efficiency increases the likelihood of elimination of the case with the lower efficiency. For a slight difference in outputs, the difference in efficiency between the two examples may be low and both examples may be included in the training data.

However, for large differences in outputs, the case with a lower efficiency may clearly be considered as an outlier (as this may hurt the monotonicity assumption) and has a higher likelihood of elimination from the "efficient" training data. Whether or not the low efficiency case is finally eliminated from the "efficient" data set depends on its actual efficiency, number of total examples, and number of inputs, among other things. The possibility of eliminating a low efficiency case the from training data set is important when two examples show inconsistency; keeping both examples in the training data set would impair the ANN learning.

Because ANN learning is through "nonparametric regression-type" learning of a nonlinear curve, outliers and inconsistent data impair its learning. DEA models, on the other hand, are not as severe on the training data set where two examples may show the same input and slightly different output. In such a case, the DEA model would give a similar efficiency rating to both cases.

Unpredictability in an ANN occurs due to violation of the monotonicity property. It has been shown that the monotonicity neural network classification model can alleviate the overfitting problem. It has also been illustrated that ratio DEA models are consistent with the monotonicity property. For example, in an event where two DMUs have the same inputs and different outputs, the DMU with higher output will receive a higher efficiency score. Selecting the DMU with a higher efficiency score for training the ANN will reduce unpredictability as the selection of the DMU is consistent with the monotonicity property.

At first it may appear that selecting a DMU (hospital) from a different peer group can significantly alter the efficiency results for all the members in the group. Although this is true, the problem is not significant, as a change in efficiency scores doesn't alter the ranking/selection of the DMUs in the group significantly.

For example, if DMU_1 has a higher efficiency score than DMU_2 before DMU_3 was added, then even after the addition of DMU_3 the efficiency of DMU_1 will be higher than DMU_2. In other words, DEA models don't lead to *rank reversal* (i.e., an inefficient DMU will appear efficient due to addition of a new DMU) due to the addition of new DMUs. The individual efficiency scores may change reflecting the impact of each DMU on the ANN training. Since we select the top 50% of most efficient DMUs, an addition of an efficient DMU (even though it may change the efficiency scores of other DMUs) will not have a very big impact on the set of DMUs that would qualify as the top 50% "efficient."

For example, in our healthcare experiments, we selected 50 "efficient" DMUs as a set of "efficient" hospitals. If a new efficient DMU is added, although the efficiency scores will change, the new DMU will not change the relative ranking of the DMUs significantly. The new set of "efficient" hospitals (after the addition of the new DMU) will contain 49 hospitals from the previous set of "efficient" hospitals and the new efficient DMU. In effect, an efficient DMU will replace an inefficient DMU.

We received mixed results in our experiments on adding noise to neural networks. Although studies suggest that adding noise is beneficial and helps the ANN to avoid a local optimum, we found that the best performance was obtained by training ANNs on efficient test cases without noise. We believe that noise can sometimes increase the efficiency of inefficient cases by reducing the inputs randomly. Future research is needed to evaluate the impact of noise on network training performance.

Our case described a DEA-based training data screening approach to learn a monotonic nonlinear forecasting function. Using data from various hospitals throughout Pennsylvania and an ANN for the learning forecasting function, we have successfully tested our approach. Among the ANN model design and data preprocessing issues were technical efficiency of training data set, size, and random input and weight noise during the learning phase of our experiments.

Our experiments showed that selecting a training data set based on technically efficient cases is beneficial for the ANN performance on unseen cases. Our experiments on adding input and weight noise during the ANN training phase showed that noise helped the network with lesser nodes in the middle layer (6) predict well; however, for networks with a higher number of nodes in the middle layer (11), the addition of noise was detrimental for both its learning and predictive performance. Our conjecture is that a higher number of nodes in the middle layer helps the neural network store more sensitive patterns in a larger set of training examples. The addition of noise in a network with a higher number of layers distorts the learning of the sensitive patterns and hinders the predictive performance. In smaller networks (i.e., having few nodes in the hidden layer), a general pattern is learned, and the addition of noise reduces the overfitting of the network on the training set. Noise in smaller networks is more beneficial compared to noise in larger networks. Our conjectures on the impact of noise on the predictive ability of connectionist models are open for future tests.

In our experiments, we used a traditional back-propagation ANN algorithm without considering other algorithms. Several approaches have been tried in the literature that improve the performance of connectionist models over the traditional back-propagation. Feature construction can be used to improve learning and prediction of connectionist models and it has been shown that second-order gradient search and other gradient-free methods improve the performance of the back-propagation model that uses the steepest-descent search method. We believe that the performance of current ANNs can be improved further by considering other learning algorithms.

The successful implementation of our ANN model shows the promise of ANNs in forecasting the number of employees in the healthcare industry; however, it must be noted that the proposed model represents a subset of variables that impact employee requirements. Improved forecasts can be obtained by acquiring information in areas such as staff turnover rates and organizational reward structuring for employees involved in the facilities. Future research needs to be conducted in identifying other factors that impact the number of employees.

We used random sampling for dividing our sample into training and testing sets. A stratified sampling approach may be better suited as this approach will ensure the output values (number of employees) in both training and testing sets are similar. Furthermore, the current approach may be benchmarked against other data screening approaches described in the literature review. Sampling and benchmarking our approach with other approaches were considered out of scope of the current research. Future research may focus on addressing these issues.

In the course of our experiments, we examined evolutionary data mining techniques such as genetic algorithms and genetic programming. In our study, the use of genetic algorithms required that we prespecify the nonlinear relationships between our independent variables and the dependent variable. We concluded that this assumption was restrictive and thus removed the technique from further consideration. Genetic programming, however, did not require any prespecification of relationships between independent and dependent variables.

The design of a genetic programming model in studying employee requirement patterns required that we examine too many parameters (e.g., population size, mutation rate, initial tree size, crossover rate, function set, terminal set, etc.). With the lack of adequate knowledge regarding the exact impact these parameters had on our smaller sample, we did not consider this technique further. We do believe that both of these techniques may hold potential for learning about human resource requirement patterns. Future research may be focused on the performance comparisons of genetic programming and neural networks in learning and forecasting employees in the healthcare industry.

In healthcare as in other industries, it is important to have the capabilities to anticipate and predict. As we have stated previously, in order to optimally satisfy the needs of enterprise constituents and customers, executives must be able to anticipate emerging needs and requirements, to get ahead of the demand curve. Our monotonicity example illustrates this need for anticipating emerging needs for accurate data mining methods and algorithms.

In applying these data mining techniques, healthcare administrators must improve their ability to predict the timing of events needed to satisfy demand in a timely, accurate, and complete manner at the optimal rate, frequency, cost, and quality. These are nontrivial requirements and this is why executives are compensated handsomely for their performance, to recognize the importance of properties such as monotonicity.

Consideration of monotonicity is an example of improving data quality through awareness about anticipation, prediction and correct methods and algorithms.

3.4 BALANCE OF CONSEQUENCES AND REINFORCEMENT

Balance of consequences and reinforcement refers to the notion that positive implementation of smart data strategy should be rewarded and deviations should be dissuaded. Giving full disclosure to data uses, values, and asset protection reinforces positive behavior and exposes weaknesses for self-correction.

Deviations cost the enterprise, and therefore the source of deviation should be penalized in a manner that is sufficient to discourage noncompliance. Benefits that truly aggregate to the top of the enterprise should be shared equitably with participants in a manner that is significant to them.

3.5 COLLABORATION

Enterprise smart data strategy is an executive-led strategy promoting data responsiveness and qualified collaboration throughout the enterprise, and at successive levels of participation.

Since data are assets, access is subject to credentialing and privileging. The Princeton definition of credential is a "certificate: a document attesting to the truth of certain stated facts." IT professionals work with executives to establish certificates for people in the user community, granting qualified access to and use of enterprise data.

The notion of privileging comes from the medical community, in this example, the Army medical community: "privileging—(Clinical Privileges) the process of reviewing an individual's credentials through credentials committee channels to determine the authority and responsibility to be granted to a practitioner for making independent decisions to diagnose, initiate, alter, or terminate a regimen of medical or dental care."

We advocate applying the notion of privileging to enterprise data users such that credentialed users are permitted specific rights to manipulate, alter, or create and amend data.

3.6 ENTERPRISE PERFORMANCE OPTIMIZATION PROCESS

Enterprise performance optimization is a process that is unique to your enterprise. It is a performance differentiator. While it may have many common activities and characteristics with other organizations, some of which are competitors, what makes it most unique is how your enterprise attributes the process with enabling people and technology.

People attribute differentiators include:

- Skill
- Proficiency
- Knowledge
- Experience
- Currency
- Depth
- Breadth
- Ratio of senior experience to junior experience
- Associated costs
  - Recruiting
  - Retaining
  - Operating labor

Managing staffing and deployment of people in properly designed work environment as enablers to processes can be greatly improved through smart data strategy. People are a variable and so is the combination of people working together in organizations and applied to processes for the purpose of completing work as prescribed by the process design.

Technology attribute differentiators include:

- SOE paradigm
- Smart data engineering tools
- Open architecture
- Data exchange technology
- Smart application of open standards

3.7 ENTERPRISE PERFORMANCE OPTIMIZATION ARCHITECTURE

Enterprise performance optimization architecture depicts an array of disciplines required to optimize enterprise performance. A review of the architecture is given in Figure 3.13. The first column is labeled "Service-Oriented Enterprise & Smart Data Paradigm" that embraces three structural elements: (1) Smart data, (2) smart grid, and (3) smart services. Below that we list CMMI, six sigma, and balanced scorecard as applicable ideas and techniques. You will configure your own application of ideas and techniques from the body of expertise within your enterprise and augment with external contributions like ours.

The six-by-six array features disciplines that are employed by enterprises under executive management for the purpose of optimizing performance. What is the relationship among the elements?

The array is best read from top left and down the first column. Each column connotes an approximation of logical ordering, although the elements may be accessed as needed by the enterprise as one or combinations thereof to improve performance. The arrangement of elements by columns again connotes some logical ordering, although it is approximation for convenience.

Employing the process modeling technique, where would these disciplines reside?

Figure 3.14 illustrates the context diagram for optimizing enterprise performance, featuring primary inputs such as capital and materials being transformed under enterprise controls into outputs such as assets, costs, net gain and benefits, and outcomes.

Service-Oriented Enterprise & Smart Data Paradigm SOE Structural Elements 1. Smart Data 2. Smart Grid 3. Smart Services CMMI Six Sigma Balanced Scorecard	Performance Analysis	Performance Improvement Strategizing	Work Design	Performance Development & Training	Systems Engineering	Implementation
	Economic Analysis	Problem Solving	Process Engineering	Instructional Media Development	Interoperability Engineering	Operations Management
	Business Case Analysis	Decision Making	Knowledge Management	Performance Aids Development	Data & Information Engineering	Enterprise Services
	Requirements Analysis	Planning	Content Management	Tools & Equipment	Grid Engineering	Maintenance
	Project & Program Management	Sense Making & Predicting	Organization Development & Staffing	Balance of Consequences	Information Technology & Solution Development	Test, Evaluation & Certification
	Enterprise Resource Management	Financial Management	Configuration Management	Quality Assurance	Quantitative Research	Secrecy & Security

Figure 3.13 Enterprise performance optimization architecture.

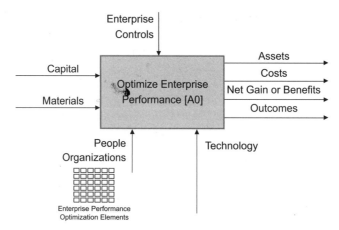

Figure 3.14   Employing the process modeling technique.

People and their organizations possess disciplines as arrayed in the "Enterprise Performance Optimization Architecture" as elements. The disciplines are attributes assigned to one or more people. This model says nothing about the number of individuals needed with these disciplines or about various combinations of disciplines residing among various people, as that is a function of enterprise size, scope, and complexity. Determining what is best for your enterprise is a part of proprietary differentiation. Furthermore, for each discipline type there are sources of best practices residing in academia and among technology vendors.

Terms

We prepared this glossary to explain what we mean by these terms. Many definitions are from Wikipedia, Princeton, and the Department of Defense; some are our own unique offerings.

Performance Analysis (PA).   This describes any analysis, spanning in scope from an enterprise view to process details and any combination in between. Where information technologists might apply the term to investigating a program's behavior, we use the term more for understanding process performance, that is, the sequence and timing of tasks and consumption of resources by enabling mechanisms to produce specific outcomes. Process modeling of the type we have used does not perform states and timing analysis, for instance. There are a host of techniques supporting detailed analysis and understanding of processes and activity performance by combinations of people and technologies performing work. Many techniques are embedded in drawing tools like Microsoft Visio™ and SmartDraw™, for example. More robust tools are needed for serious enterprise performance engineering and management.

Economic Analysis (EA).   This form of analysis is also scalable to a range in scope. One can use economic analysis to determine the cost effectiveness of a project or program by comparing benefits derived and the costs incurred. It is a way to estimate value expressed as worth of development in terms of impact on economy, that is, the impact of the stimulus package on the economy. On a more granular scale, it can be applied to determining completion of life cycle cost comparisons of project alternatives.

Business Case Analysis (BCA). This is a comparative analysis that presents facts and supporting details among competing business alternatives. For instance, the DoD application says that "the BCA should facilitate a determination whether to implement a Performance Based Logistics type arrangement by comparing the Government's cost of supporting the initiative versus the contractor's cost of supporting the items or material." There are two types—abbreviated (rough order magnitude, ROM) and final BCA.

Requirements Analysis (RA). In systems engineering and software engineering, this encompasses those tasks that go into determining the needs or conditions to meet for a new or altered product, taking account of the possibly conflicting requirements of the various stakeholders, such as beneficiaries or users. Requirements analysis is critical to the success of a development project. Requirements must be actionable, measurable, testable, related to identified business needs or opportunities, and defined to a level of detail sufficient for system design.

Project and Program Management (PPM). Project management is the discipline of planning, organizing, and managing resources to bring about the successful completion of specific project goals and objectives. Program management is the same discipline applied to a broader scope and scale, multiple projects, or multiple tasks.

Enterprise Resource Management (ERM). This term is used to describe comprehensive resource management throughout the enterprise. It includes planning and organizing and controlling resource application and various rates of consumption. It is a subset of enterprise performance optimization (EPO).

Performance Improve Strategizing (PIS). The act of strategizing how to improve performance. We used to call this performance improvement engineering (PIE), but to emphasize that this is an executive and management responsibility and that it is strategic, we changed the term. It is a systematic plan of action.

Problem Solving (PS). The Wikipedia definition [14] is "a part of thinking considered the most complex of all intellectual functions; problem solving has been defined as [a] higher-order cognitive process that requires the modulation and control of more routine or fundamental skills.

It occurs if an organism or an artificial intelligence system does not know how to proceed from a given state to a desired goal state. It is part of the larger problem process that includes problem finding and problem shaping.

Problem solving is of crucial importance in engineering when products or processes fail, so corrective action can be taken to prevent further failures. Forensic engineering is an important technique of failure analysis which involves tracing product defects and flaws. Corrective action can then be taken to prevent further failures."

We suggest that problem solving begins with performance problem analysis (PPA) to produce a problem definition on which to apply creative and innovative skills to produce a solution.

Decision Making (DM).   The Princeton definition [15] is "the cognitive process of reaching a decision; 'a good executive must be good at decision making.'"

Planning (P).   The Princeton definition [15] is "an act of formulating a program for a definite course of action; 'the planning was more fun than the trip itself.'"

Sensemaking and Predicting (SP).   "Sensemaking is the ability or attempt to make sense of an ambiguous situation. More exactly, sensemaking is the process of creating situational awareness and understanding in situations of high complexity or uncertainty in order to make decisions. It is 'a motivated, continuous effort to understand connections (which can be among people, places, and events) in order to anticipate their trajectories and act effectively'" [16].

Predicting is the act of reasoning about the future.

Financial Management (FM).   This is management related to the financial structure of the company and therefore to the decisions of source and use of the financial resources, which is reflected in the size of the financial income and/or charges [17].

Work Design (WD).   One definition has it that "in organizational development (OD), work design is the application of Socio-Technical Systems principles and techniques to the humanization of work." We have a more specific definition to add in context with our enterprise performance optimization process.

Work is performed by people, by technology, and by combinations such that people may use technology or operate technology to complete a task as evidenced by specific outcomes. Using the process modeling lexicon, a process is made up of activities and an activity may be equated to a task or subtask.

Work design combines process modeling and the attribution of people and technology to the model such that (1) individuals may be assigned who possess certain skills, knowledge, and proficiencies, (2) under constraints of budget and time, (3) to produce superior results.

Job models may be the products of work design that become inputs to organization development that is akin to the process of organization design.

Process Engineering (PE).   This term is customarily employed in chemical and biological process industries to define activities that produce an end result. That is really the same way that we use the term, although we apply the IDEF0 definition to define processes as a relationship of activities with associated inputs, controls, outputs, and mechanisms. Our definition applies to all processes.

Knowledge Management (KM).   The Wikipedia definition [18] "comprises a range of practices used in an organization to identify, create, represent, distribute and enable adoption of insights and experiences. Such insights and experiences comprise knowledge, either embodied in individuals or embedded in organizational processes or practice."

Content Management (CM$_1$). This "is a set of processes and technologies that support the evolutionary life cycle of digital information. This digital information is often referred to as content or, to be precise, digital content. Digital content may take the form of text, such as documents, multimedia files, such as audio or video files, or any other file type which follows a content lifecycle which requires management" [19].

Organization Development and Staffing (ODS). To paraphrase Richard Beckhard, organization development (OD) is a planned, top–down, organization-wide effort to increase the organization's effectiveness and health.

George worked under Beckhard's supervision in a program at Sherwin Williams Company. From experience at AT&T, Bendix, GTE, and other clients, George developed a refined use that is not intended to eclipse the OD profession, but to position some application of the discipline.

ODS as used in our context is a collaboration, facilitated by OD, including executive leadership, financial management, process design, and work design to identify the optimal organization of workers/employees under supervision to produce desired outcomes.

Modern organization must be highly adaptive, and a smart data environment supports adaptive performance and agility. This must be reinforced through a balance of consequences.

Configuration Management (CM$_2$). This "is a field of management that focuses on establishing and maintaining consistency of a product's performance and its functional and physical attributes with its requirements, design, and operational information throughout its life" according to Wikipedia [20]. We would extend the definition to apply to all enterprise artifacts that are the assets describing how the enterprise works and how it performs.

To our surprise, Wikipedia features an IDEF0 model for configuration management as shown in Figure 3.15.

Performance Development and Training (PDT). This is the discipline of improving worker performance through a combination of actions that begin with (1) validating work design and job design, (2) ensuring that workers have the right tools and equipment, (3) ensuring that work is performed with a proper balance of consequences and business rules, (4) checking for errors of omission and commission, (5) identifying knowledge and skill deficiencies, and ultimately (5) training if necessary.

Instructional Media Development (IMD). In today's world, the standard for instructional media must be as high as commercial standards for entertainment because that is both the competition for attention and the standard expected by

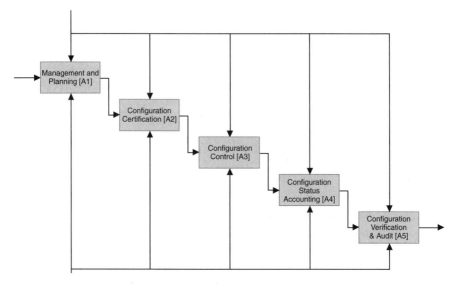

Figure 3.15   Configuration management.

people. Media development includes everything from still visual presentations to audio and motion visual media, including interactive programming, and integrated delivered via computer systems and online media distribution.

Performance Aids Development (PAD).   This includes things like checklists and information linking that guides workers through troubleshooting and problem solving activities or prompt workers in communication at Help Desks, for instance.

Tools and Equipment (TE).   These are defined as hardware, software, and other equipment that support work performance.

Balance of Consequences (BOC).   Workers of all kinds need to experience the difference between good work and deficient performance by enterprises, ensuring that desired performance receives proper consideration and reward and that deficient performance is dissuaded.

Quality Assurance (QA).   Wikipedia [21] says that QA "refers to planned and systematic production processes that provide confidence in a product's suitability for its intended purpose. It is a set of activities intended to ensure that products (goods and/or services) satisfy customer requirements in a systematic, reliable fashion."

Systems Engineering (SE).   Systems engineering "is an interdisciplinary field of engineering that focuses on how complex engineering projects should be designed and managed. Issues such as logistics, the coordination of different teams, and automatic control of machinery become more difficult when dealing with large, complex projects. Systems engineering deals with work-processes and tools to handle such projects, and it overlaps with both technical and human-centered disciplines such as control engineering and project management [22].

Interoperability Engineering (IE). Interoperability engineering is "the ability to exchange and use information usually in a large heterogeneous network made up of several local area networks" [23].

Data/Information Engineering (DIE). "In software engineering DIE is an approach to designing and developing information systems" [24].

Smart data engineering embraces metadata management, open systems, interoperable data exchange, semantic interoperability, and other advanced technologies to leap beyond the grasp of legacy systems and proprietary encumbrances.

Grid Engineering (GE). Wikipedia [25] describes a smart grid as "representing a vision for a digital upgrade of distribution and long distance transmission grids to both optimize current operations, as well as open up new markets for alternative energy production." Though inspiring our ideas, the definition of grid engineering refers to the net-centric infrastructure needed to support smart data environments.

Information Technology Solutions and Services (ITSS). These are any solution elements developed, delivered, or otherwise provided by information technologists as mechanisms for process enablement and work performance.

Quantitative Research (QR). "Quantitative research is the systematic scientific investigation of quantitative properties and phenomena and their relationships. The objective of quantitative research is to develop and employ mathematical models, theories and/or hypotheses pertaining to natural phenomena. The process of measurement is central to quantitative research because it provides the fundamental connection between empirical observation and mathematical expression of quantitative relationships" [26].

Implementation (I). "Execution: the act of accomplishing some aim or executing some order; the agency was created for the implementation of the policy [15].

Operations Management (OM). OM "is an area of business that is concerned with the production of good quality goods and services, and involves the responsibility of ensuring that business operations are efficient and effective. It is the management of resources, the distribution of goods and services to customers" [27].

Enterprise Services (ES). These are typically information technology services, although they can be any common or shared services that support vertical functions horizontally across the enterprise. In the smart data paradigm, these are specific services that support smart data implementation such as a data exchange utility that facilitates mapping from legacy domains to a neutral standard for exchange.

Maintenance (M). Maintenance is activity involved in maintaining something in good working order.

Test, Evaluation, and Certification (TEC). TEC is the independent or objective process of testing, evaluating, and certifying performance whether it relates to products or people.

Secrecy and Security (SS). Secrecy implies limited access and security consists of measures taken as a precaution against theft, espionage, or sabotage.

Prepare for Smart Data and Smart Data Strategy

Notice that the description of activities or tasks for preparing for smart data strategy is in the same syntax as the IDEF0 model, although it is expressed in words instead. Some executives may like to read the works, but notice that missing from the activity list is the relationship of inputs, controls, outputs, and mechanisms.

Activities

1. Model enterprise processes: top–down.
   1.1. List primary processes.
   1.2. Identify primary inputs.
       1.2.1. Identify and attribute data inputs.
       1.2.2. Identify and attribute data outputs.
       1.2.3. Classify data assets.
   1.3. Identify primary controls.
   1.4. Identify primary outputs.
   1.5. Identify enabling mechanisms.
2. Make an inventory of enterprise performance optimization data and requirements.
   2.1. Make a list of key questions needed to optimize performance in the enterprise.
   2.2. Assess how well plans, decision making, problem solving, sense making, and predicting are supported today.
   2.3. Assess data interoperability and exchangeability with attention to openness, seamlessness, and actionability.
3. Assess methods and algorithms associated with using data.
   3.1. Identify methods used to analyze, interpret, and otherwise operate data.
   3.2. Identify best methods and algorithms.
   3.3. Determine how to point data to best methods or otherwise align data with best methods.
4. Determine modeling, simulation, and prediction requirements.
   4.1. Identify essential models for dynamic display.
   4.2. Identify simulation requirements (i.e., what-if scenarios).
   4.3. Identify prediction and forecasting requirements.
5. Assess smart data readiness: as-is/to-be/gap analysis.
   5.1. Describe as-is performance.
   5.2. Describe to-be performance.
   5.3. Describe the gap and how to close it.
6. Assess executive user satisfaction.

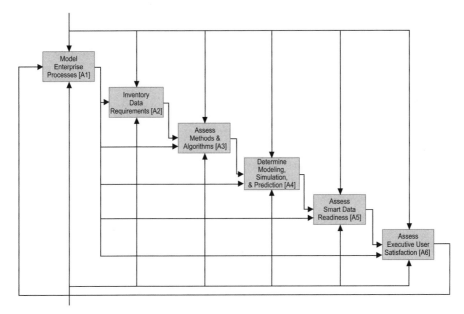

Figure 3.16 Prepare for smart data and smart data strategy.

Figure 3.16 shows an IDEF0 process model at the first level of decomposition and, without being fully attributed, provides a context for preparing for smart data and smart Data strategy as we discussed.

What might be the controls on this process? Mission, strategy (including smart data strategy), values, business rules, laws, and regulations are all possible sources.

What are the mechanisms performing the work? Herein there would be a hierarchy for each activity: domain expert supervisor and domain experts. Individuals would be organized into departments or units as the ordering or work has logical affinity.

When approaching this or any enterprise performance optimization activity, we suggest that you resist starting with assumptions about the organization. Instead, start by assigning talent to the effort, and then evaluate how to best organize the talent. The reason we say this is because existing organizations may often be a part of the deficiency and executives must preserve a degree of objectivity when establishing this effort.

Model Enterprise Processes

Just as the list of tasks is made to identify activity needed to prepare for smart data and smart data strategy, the first step is to identify primary business processes needed to fulfill your enterprise mission and to produce primary outcomes and results. This of course assumes that you know your primary audience of customers, prospects, and constituents to be served by your enterprise in a superior manner.

Input for this activity would surely include your balanced scorecard if that is available. It is worthwhile constructing a *balanced scorecard*, which is, of course, another complementary process. (Balanced scorecard [28] is a framework that translates a company's vision and strategy into a coherent set of performance measures, developed by Robert Kaplan and David Norton and first published in the *Harvard Business Review* in 1992.)

It is important not to start with an organization chart in identifying processes because, as you know now, people and organizations are enabling mechanisms that *attribute* processes but *are not* the processes. Keeping this straight is a means of auditing the relevance of current organization structures, which may or may not be correctly assigned or deployed.

This underscores that the modeling process must be objective, reflecting the executive team's best thinking.

For each process, identify primary controls that we also call sources of *leadership and integration*, such as laws and regulations, business plans and strategies, rules that are self-imposed, and rules that are conveyed or imposed.

You may want to assign specific responsibilities for leading each of these tasks, such as assigning identification of operations processes to the chief operating officer and assigning identification of controls to the compliance officer or contracts and legal department or a combination thereof, depending on the complexity of the enterprise.

Daniel S. Appleton differentiated enterprises by organization forms: unitary (U), multiple organizations (M), and partnership (P), for instance. The essence is to recognize that enterprises are variable in organization size and scope and this affects the complexity of enterprise modeling and performance engineering.

Before Michael Hammer introduced business process reengineering (BPR) in 1990, Appleton and George were engaged in developing the concept of business engineering with attention to two primary processes: (1) new business engineering and (2) business reengineering. These processes and associated constructs for enterprise performance optimization are as valid today as they were a decade ago for a couple of reasons: (1) turnover among executives is high (knowledge and experience is lost), (2) survivability rate of enterprises is low, and (3) people have difficulty building on wisdom as a result.

It is true that enabling technology has advanced and, with its application, new catalysts have emerged, such as smart data.

New business engineering is a process developed by George to support the GTE venture program and is roughly modeled in Figure 3.17. Four primary activities comprise the process: (1) ideation, (2) conceptualization, (3) validation, and (4) deployment.

The rough model for business reengineering is shown in Figure 3.18, [29] as defined by Daniel S. Appleton.

Appleton and George operated on the principle that one cannot reengineer a business that was not engineered to begin with. Therefore the discipline is called *business engineering*. At the time, Appleton proposed a formula for optimizing business performance that is applicable to commercial business:

Yield = Increasing revenue + Increasing quality/Decreasing cost + Decreasing time

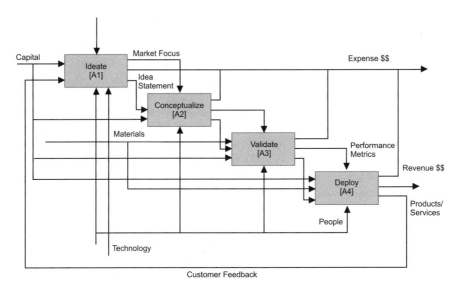

Figure 3.17   New business engineering sketch.

One familiar with differential calculus would observe that the dimension of time must be added to the formula such that

$$Y = [(R_2 + Q_2/C_2 + T_2) - (R_1 + Q_1/C_1 + T_1)]$$

Of course, you may observe that you really can't add currency and quality, or currency and time, but you can quantify these factors such that they do produce a ratio with a larger number desirable as the numerator and the smaller as the denominator.

These are simple examples of how to address a need to increase revenue and improve performance from a process view. Sketches like these beg for more detailed decomposition, as it is called, to identify, describe, and explain the details.

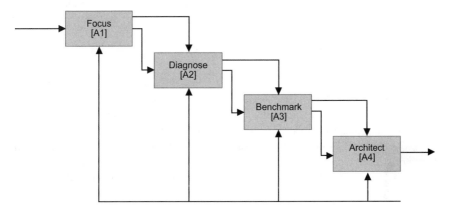

Figure 3.18   Business reengineering sketch.

As-Is Situation

Typically, when IT professionals address data strategy, they do so with attention to certain topics, such as data integration and quality, metadata, and data modeling. They suggest certain roles and responsibilities within the IT structure. They address performance and measurements, most likely in an IT context, such as "Does data strategy support IT strategy, presuming that IT strategy supports corporate strategy?" Herein is a problem because data should be addressed closer to the executive, we believe. Corporate strategy is supported by smart data strategy that is implemented by IT strategy.

Security and privacy are issues that we address as credentialing and privileging, which are complementary but with a different focus from cybersecurity and fire-walling the infrastructure, for instance. We think smart data strategy and the associated technology hold potential for reducing threats by gating access and partitioning users.

Such things as DBMS selection is out of the scope of this book, as that is an IT responsibility, although we should note that DBMS providers often supply the data engineering tools, as they are embedded.

Dealing with unstructured data is often a topic of priority for data management—as it is for smart data strategy—and we advocate a common system of data exchange that is independent from legacy applications and seeks to eliminate maintenance of brittle application interfaces.

Business intelligence, a topic that is in the scope of smart data strategy, is most aligned with executive needs and priorities. The Wikipedia definition of business intelligence is as follows [30]: "Business intelligence (BI) refers to skills, knowledge, technologies, applications, quality, risks, security issues and practices used to help a business to acquire a better understanding of market behavior and commercial context. For this purpose it undertakes the collection, integration, analysis, interpretation and presentation of business information. By extension, 'business intelligence' may refer to the collected information itself or the explicit knowledge developed from the information." The trouble with this definition is that it appears to be another label for encapsulation that isn't particularly useful because it lacks discrete explanation (nice for marketing, but not practical in implementation).

As discussed previously, value of data and ROI are upfront topics that we have already addressed.

Model the current enterprise, attributing the as-is situation with cost, time, and quality metrics. This takes a special effort, so develop a strategy to prioritize where to begin. Start with the requirements for the executive's enterprise performance optimization concept of operations and operational architecture.

"A Concept of Operations (abbreviated CONOPS) is a document describing the characteristics of a proposed system from the viewpoint of an individual who will use that system. It is used to communicate the quantitative and qualitative system characteristics to all stakeholders. CONOPS are widely used in the military or in government services, as well as other fields" [30].

"Operational Architecture is an information management system represented in terms of the business processes it supports, and how information related to conduct of the business processes passes through the system's components" [31].

How is the enterprise performing now?

To-Be Performance

How should the enterprise be performing in the future?

In the past we used the term *toolkit* to describe what we thought executives required to achieve optimal performance. Then we applied such labels as *business intelligence* to describe the domain in which we wanted to provide improved IT support. Then we used *portfolio* to describe a set of dashboards and software to support management.

Today, we advocate considering the enterprise executive and management responsibility as a complete system. As it branches to successive layers throughout the enterprise, one might say that it is a *system of systems*. We advocate stronger enterprise branding such that the enterprise has a high state of continuity and understanding that promote optimal data exchange and seamless transactions.

Model and attribute the to-be performance with cost, time, and quality metrics, reflecting what is expected from changed processes.

Gap Analysis

How will the outcomes change? What are the cost, time, and resources to change?

When performing a gap analysis using the methods introduced here, there are a host of possibilities to consider:

- Comparison of actual planned outcomes versus actual outcomes—specific results, product and assets produced, costs expended, revenues and other benefits gained or lost.
- Comparison of planned and actual enabling mechanism resources—people by various granular analyses, technologies and their actual versus planned performance, all costs associated, and all time consumed.
- Evaluation of the cost of compliance with controls levied and those that are self-imposed.
- Cost of capital and cost of materials, planned versus actual.

3.8 SCOPING, SCHEDULING, BUDGETING, AND PROJECT AND PROGRAM MANAGEMENT

Enterprises often embrace iterative and incremental development, a cyclic software development process created in response to the weaknesses of the waterfall model. It starts with an initial planning and ends with deployment with the cyclic interaction in between.

Executive Specification

Executives' Enterprise Performance Optimization: Concept of Operations and Operational Architecture

CONOPS   The viewpoint is that of the executive user. Adopt a smart data strategy as a lead element of optimizing enterprise performance.

Evidence of the smart data strategy will manifest in the following ways:

1. Enterprise strategy and value for data assets.
2. Executive performance questions—routine and unique and an associated continuous process.
3. Management by the numbers—smart data serves executives with actionable data accompanied by best methods and algorithms.
4. Enterprise process models—fully attributed.
5. Data asset inventory—life cycle attribution and variable valuation.
6. Enterprise performance evaluation process—real time and continuous.
7. Enterprise predictive management—improved accuracy of anticipation, near-term and longer views.
8. Enterprise data exchange utility based on standard exchange model, interoperability, and openness.
9. Increased agility and adaptability and reduction of brittle and costly interfaces.
10. Application of semantic mediation technologies.
11. Enterprise staffing includes support for smart data strategy with strategic performance development and additions or augmentation of needed skills to implement and to support analysis.
12. Enterprise performance optimization system.

OA

1. The operational architecture will include embracing the service-oriented enterprise (SOE) paradigm.
2. Create a model-driven data exchange utility or subscribe to model-driven exchange services (i.e., cloud computing).
3. Shed allegiance to brittle standards.
4. Add or strengthen an executive's performance optimization analytical service utility.

Enterprise Performance Optimization System   The combination of CONOPS, OA, and implementation produces the enterprise performance optimization system. This discussion begs for examples of possibilities. We have stayed clear of specific vendor discussions, although we advocate that such a system should be vendor neutral or else the enterprise will be drawn into a proprietary platform. A central vendor requirement is to respect your value for openness and interoperability and for model-driven data exchange.

REFERENCES

1. *Federal CIO Council Strategic Plan 2007–2009*, http://www.cio.gov/documents/CIOCouncilStrategicPlan2007.

2. *The Small Business Economy*, an annual chapter in *The Small Business Economy: A Report to the President*, by Brian Headd, with accompanying appendix data tables by Brian Headd and Victoria Williams (http://www.sba.gov/advo/research/sbe.html).

3. P. C. Pendharkar and J. A. Rodger, "Maximum Entropy Density Estimation Using a Genetic Algorithm," *2006 Proceedings INFORMS*, 2006.

4. http://en.wikipedia.org/wiki/Machine_learning.

5. http:/www.mccip.org.uk/arc/2006/glossary3.htm.

6. Department of Computer Science, University of California–Santa Cruz.

7. "A Tutorial on Clustering Algorithms," http://home.dei.polimi.it/matteucc/Clustering/tutorial_html/.

8. David E. Goldberg, *Genetic Algorithms in Search of Optimization and Machine Learning*, Addison-Wesley, Boston, 1989.

9. J. A. Rodger, "Book Review of Encyclopedia of Information Science and Technology," *Information Resources Management Journal*, **18**(3), 92–93, 2005.

10. P. C. Pendharkar and J. A. Rodger, "Response to Lin Zhao's Comments on Technical Efficiency-Based Selection of Learning Cases to Improve Forecasting of Neural Networks Under Monotonicity Assumption," *Journal of Forecasting*, 739–740, 2004.

11. P. C. Pendharkar and J. A. Rodger, "Technical Efficiency-Based Selection of Learning Cases to Improve Forecasting Accuracy of Neural Networks Under Monotonicity Assumption," *Decision Support Systems*, 117–136, 2003.

12. P. C. Pendharkar, S. Nanda, J. A. Rodger, and R. Bhaskar, "An Evolutionary Misclassification Cost Minimization Approach for Medical Diagnosis." in Parag C. Pendharkar (Ed.), *Managing Data Mining Technologies in Organizations: Techniques and Applications*, Chap. 3, pp. 32–44. Idea Group, Inc., Hershey, PA, 2003.

13. P. C. Pendharkar, J. A. Rodger, and K. Wibowo, "Artificial Neural Network Based Data Mining Application for Learning, Discovering and Forecasting Human Resources Requirement in Pennsylvania Health Care Industry," *Pennsylvania Journal of Business and Economics*, **7**(1), 91–111, 2000.

14. http://en.wikipedia.org/wiki/Problem_solving.

15. wordnet.princeton.edu/perl/webwn.

16. http://en.wikipedia.org/wiki/Sensemaking.

17. Heva On-Demand Glossary.

18. en.wikipedia.org/wiki/Knowledge_Management.

19. en.wikipedia.org/wiki/Content_management.

20. en.wikipedia.org/wiki/Configuration_management.

21. en.wikipedia.org/wiki/Quality_assurance.

22. en.wikipedia.org/wiki/Systems_engineering.

23. www.audioenglish.net/dictionary/interoperability.htm.

24. Professor Robert Glushko, "Information System and Service Design: Strategy, Models, and Methods," University of California at Berkeley, 2008.

25. en.wikipedia.org/wiki/Smart_Grid.

26. en.wikipedia.org/wiki/Quantitative_research.

27. en.wikipedia.org/wiki/Operations_management.

28. Robert Kaplan and David Norton, "Balanced Scorecard," *Harvard Business Review*, 1993. http://hbr.harvardbusiness.org/2005/07/the-balanced-scorecard/ar/1.

29. Daniel S. Appleton, business engineering, http://www.dacom.com/interaction/bio.html.

30. en.wikipedia.org/wiki/Business_intelligence.

31. en.wikipedia.org/wiki/CONOPS.

32. www.wtec.org/loyola/digilibs/d_01.htm.

Chapter **4**

Visionary Ideas: Technical Enablement

Be a yardstick of quality. Some people aren't used to an environment where excellence is expected.

—Steve Jobs

4.1 TODAY'S POSSIBILITIES

Through a series of experiences serving various clients, we have been able to test elements of our ideas. We observe in the contemporary world around us the technologies in action that, when guided by smart data strategy, can produce higher impact results. Our aim is to motivate you to carry out the vision because you are the leaders of government and commercial enterprises where only you can make it happen, where the needs are greatest, and the requirements most important.

In Chapter 1 we discussed the context and value of data and presented the service-oriented enterprise strategic paradigm. We offered a process model for enterprise performance optimization. In Chapter 2 we discussed elements and positioned data as being at the heart of creative and innovative management, including a robust technical case study. Chapter 3 presented barriers with opportunities and possibilities for overcoming them. Now, in this chapter, we present ideas for applying smart data accompanied by examples of technical enablement.

A crucial difference is leading with executive authority over data and enterprise performance optimization instead of following technology. Technology is essential, but without proper leadership and emphasis, technology may go adrift and can produce less than desirable results, including a culture that inhibits optimal performance.

Smart Data: Enterprise Performance Optimization Strategy, by James A. George and James A. Rodger
Copyright © 2010 John Wiley & Sons, Inc.

We spoke with Linda Carr, president and CEO of TMR, Inc., a woman-owned business where she is a subject matter expert in enterprise governance. Her clients include CIOs of significant federal government departments and agencies. She says that "CIOs face several predicaments: How do they determine their organization's target enterprise architecture (EA) that document the future IT business architecture in terms of where the agency wants to be in two or three years? How do CIOs align technology with their organizations' business and mission goals? How do CIOs determine users' business requirements, performance measures, and acceptable levels of risk? How do CIOs determine how IT dollars should be invested?" She then emphasized the need for collaboration in determining answers to these critical questions.

Carr says that "governance is political" and goes on to describe it as something that some view as a "web of bureaucratic, time-consuming processes that seek to impose rules without reasons." To offset this, she says the governance process must be consensus based and results oriented, and we believe, most importantly, that it must be supported at the highest levels of the organization.

In fact, we believe that enterprise governance must be addressed in concert with data governance beginning with top executives, as they have a direct stake in the outcome as we have said consistently. Furthermore, enterprise governance is what we identify more completely as the leadership and integration architecture for the enterprise for which there is no option and for which much of the content is imposed through laws and regulations. Organizations that don't pay attention to this might end up like General Motors, off track and in the hands of someone else.

Smart data ideas are tested in a variety of ways that appear as standalone projects and proofs-of-concept. This does not mean that smart data is necessarily on the bleeding edge which usually connotes great risk. It just means that priorities have not been properly aligned with the idea that enterprises should exploit the potential advantages.

Technology solutions are sufficient for aggressive implementation, and developments toward the semantic web are surely supportive of smart data ideas.

Employ Accounting for Data as a Catalyst for Change and Improvement

Applying smart data is accomplished in three dimensions beginning with executives embracing the idea to employ accounting for data as a catalyst for change and improvement. Follow the data through processes and to outcomes to identify problems and to realize opportunities. Do this in the most direct manner since a focus on data means a focus on the primary business.

Smart data has three dimensions:

1. Performance optimization in an enterprise context
2. Interoperability technology
3. Data-aligned methods and algorithms

Problems and opportunities will be solved and pursued through application of improved data engineering, improved data exchange strategies, and technical implementations including decision support. Improvements are made in conjunction with contemporary initiatives such as service-oriented architecture (SOA) (tactical) and related technical dimensions, as part of the service-oriented enterprise (SOE) paradigm (strategic).

DoD Architecture Framework (DoDAF)

Perhaps the closest to addressing the opportunity comprehensively is the DoD Architecture Framework (DoDAF) initiative, which is a prerequisite step toward modeling the defense enterprise. However, missing is the smart data strategy that is closer to the end game of optimizing enterprise performance. Enterprise modeling and architecture initiatives need executive-driven and data-focused strategy to produce results.

The models must be brought to life immediately and deployed with analytical executive support in contrast to waiting for thoroughness or completeness, as we must manage with the models we have, not the ones we ultimately want. Bringing them to life requires attributing them with data, linking to distributed data sources with configuration management over distributed repositories or the best architecture for supporting customers.

Gaps in functionality can be closed, although models without the ability to support executives in real time have no actionable value. That is why they must be employed in a live support system.

Envision the defense secretary being able to query his enterprise performance optimization support system for up-to-the-minute answers supporting his planning, decision making, sense making, and predicting. Data from this system also feeds the president's system as defense is one department node of many on the federal executive system.

Modeling the enterprise is not an end game. It is a vital step that is made relevant in context with enterprise engineering discipline and the SOE paradigm that accounts for dynamic relationships between governance (control architecture), processes, data, and enablement.

In the U.S. government enterprise, much of the control architecture's governance is the product of Congress whereby laws and regulations place constraints on executives and on processes. Reconciling discrepancies between executive intent and associated outcomes and congressional constraint is sorted out in the legal system.

When discussing complex enterprise models, Linda Carr expressed the need to refine subjects into simple diagrams that make sense to executives. She shared experience that executives glaze over when they see complex models, even though they are accurate depictions.

To that, we conclude that the executive interface should provide the most concise and simplest depictions. Based on a common and simple modeling technique, such as we have discussed, executives will become familiar with the drill-down process as it manages layers and levels of information with an accompanying enterprise road map so no one gets lost. The more automated the system, the better it is in this regard. What has happened in the past is much modeling without transformation into an executive support system, and that can be confusing if not confounding.

Of concern from an executive perspective is that applying engineering discipline to an enterprise is a big effort that never ends. In fact, this is the new management paradigm. Enterprise executives who are equipped with the disciplines and strategy for optimizing performance must parse the days such that they are making the right tactical decisions while guiding the ship on the right course. We want executives to embrace the idea that they need an enterprise performance optimization support system that serves them both tactically and strategically. Such a system is the manifestation of the enabling technology for smart data strategy implementation.

Getting the enterprise performance optimization support system in place requires rapid execution such that its daily service to executives justifies its continuous development and refinement. We want it to be something they can point to and to say this or that is serving me well or not so well, such that IT professionals can address improvements with precision and efficiency.

While what goes on in the IT background is necessarily complex, we cannot let the complexity interfere with the necessity for executive clarity in executing their complex responsibilities.

At the rollout of some of the modeling artifacts at a defense and industry manufacturing conference in Las Vegas in December 2007, we asked the DoD panel a question: "What are you doing about data? What role will ISO 10303 Product Life Cycle Support (PLCS) play in your approach, for instance?" Note that this is a robust international standard supporting all manufactured products and yet the leadership did not recognize it.

The panel members looked puzzled and then the moderator replied: "We were not anticipating questions about data, though perhaps we should have." This instance underscores what we have said about the need for executive attention and comprehensive strategy that begins with data. There is a net-centric data strategy at the DoD and we are certain everyone was aware. Yet it was not foremost in the discussion, in part, because "net-centric data" is more closely associated with enabling infrastructure and not the data itself. In addition, net-centric data strategy "belongs to" the National Information Infrastructure (NII), another initiative assigned to the DoD CIO and this illustrates that strategy is stovepiped.

Pose the same question about the standard to some American manufacturing companies and you will get a more proactive response because some have had to use the standard in order to comply with their global customers' requirements. A standard that was largely born from the efforts of the National Institute of Standards and Technology (NIST) and the Department of Defense is attended by a few manufacturers in America, by many more manufacturers in Europe, and nearly ignored by the DoD at home. What is wrong with this picture? To the credit of the Aerospace Industries Association (AIA), ISO 10303 AP 239 was adopted as a standard for aerospace maintenance data in April 2008.

Missing still is a DoD smart data strategy that will comprehensively apply the standard to all weapons systems, incorporated with the proposed method of data exchange that will break out from legacy environments and catalyze a host of improvements, but only with smart executive leadership.

Federal Enterprise Architecture

The Federal Enterprise Architecture (FEA) initiative is intended to support calibration of information technology investments in the federal government. The FEA may be thought of as the parent to the DoDAF, for instance. These two initiatives are a source of control—rules and guidance for enterprise performance optimization in our way of thinking. The Clinger–Cohen Act is the governing law.

The Act states "Federal Enterprise Architecture (FEA) provides a common methodology for information technology (IT) acquisition." While this is useful it does not say that the "FEA is the centerpiece for enterprise performance optimization."

Our view is to shift the emphasis from IT to executive responsibility for performance optimization employing smart data strategy.

Figure 4.1 shows the structure of the U.S. Federal Enterprise Architecture Framework (FEAF) components as presented in 2001. The FEA is built with a collection of reference models that develop a common taxonomy and ontology for describing IT resources:

- Performance reference model
- Business reference model
- Service component reference model
- Data reference model
- Technical reference model

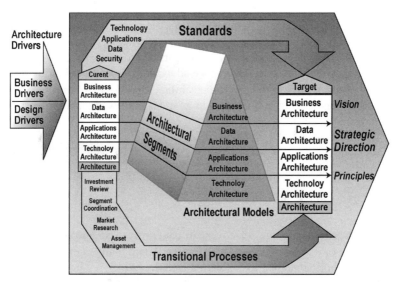

Figure 4.1 Structure of the U.S. Federal Enterprise Architecture Framework (FEAF) components.

To be accurate, the federal government says that enterprise architecture (EA) is a management practice for aligning resources to improve business performance and to assist agencies better to fulfill their core missions. "An EA describes the current and future state of the agency, and lays out a plan for transitioning from the current state to the desired future state. Federal Enterprise Architecture is a work in progress to achieve these goals."

The U.S. Federal Enterprise Architecture (FEA) is an initiative of the U.S. Office of Management and Budget that aims to comply with the Clinger–Cohen Act and provide a common methodology for information technology (IT) acquisition in the U.S. government. It is designed to ease sharing of information and resources across federal agencies, reduce costs, and improve citizen services.

Sometimes a graphic artifact such as Figure 4.1 that was introduced in 2001 and appears in Wikipedia says much in retrospect, as in fact, it says very little. Reading left to right it connotes the following:

- There are business and design drivers for enterprise architecture—begging the question, "What design drivers would ever trump business drivers?"
- Standards emanate from the current architecture, which manifests in future architecture—begging for explanation, "To what extent do legacy standards impact the future?"
- Business architecture apparently includes data architecture, application architecture, and technology architecture. What do they mean? Do they mean that data is modeled in a neutral view? Are applications proprietary or unique to specific software? Does technology provide enablement?
- Somehow, the future contains principles and vision and strategic direction.

This all seems deficient if not completely backwards. Yet in our efforts to date, we operate under this guidance and it gives motivation to wanting to turn this situation on its head. While the government's idea to share its data openly as part of "data.gov" initiative is a good idea, the priority that we advocate is executives employing smart data to accomplish their primary responsibility, that is, to optimize enterprise performance. Data sharing is fine, as long as it does not distract from the primary responsibility and focus.

The data reference model (DRM) is described as follows. Structurally, the DRM provides a standard means by which data may be described, categorized, and shared. These are reflected within each of the DRM's three standardization areas:

- *Data Description.* Provides a means to uniformly describe data, thereby supporting its discovery and sharing.
- *Data Context.* Facilitates discovery of data through an approach to the categorization of data according to taxonomies. Additionally, it enables the definition of authoritative data assets within a COI.
- *Data Sharing.* Supports the access and exchange of data where access consists of ad hoc requests (such as a query of a data asset), and exchange

consists of fixed, recurring transactions between parties. It is enabled by capabilities provided by both the data context and data description standardization areas.

There is a specific international standard governing the description of data, ISO/IEC 11179 (formally known as the ISO/IEC 11179 Metadata Registry (MDR) standard); it is the international standard for representing metadata for an organization in a Metadata Registry. Our intent in this book is not to go into the technical details of data engineering, but to inform executives about the environment and context for this activity.

Now, we want to review core ideas about smart data and to demonstrate some examples of how they have been tested and implemented. Following that, we will discuss a way ahead.

Smart Data Definition Expanded

To review, our ideas about smart data include three different dimensions that are expanded in the description here:

1. Performance optimization in an enterprise context
   Not traditional systems engineering by stovepipes or verticals
   Enterprise wide scope
   Executive user leadership
   Outcome focused
2. Interoperability technology
   Data engineering
   Model-driven data exchange
   Semantic mediation
   Metadata management
   Automated mapping tools
   Service-oriented enterprise paradigm that includes smart data, smart grid, and smart services
   Credentialing and privileging
3. Data-aligned methods and algorithms
   Data with pointers for best practices
   Data with built-in intelligence
   Autonomics
   Automated regulatory environment (ARE) and automated contracting environment (ACE)

We have covered the first element of smart data performance optimization in an enterprise context, although some additional case studies will provide additional examples. The second element, interoperability technology, deserves a little more explanation.

"Data engineering" is a discipline that involves enterprise modeling and data in concert. It involves developing an enterprise concept of operations and operational architecture, in collaboration with specialists who have these skills.

"Model-driven data exchange" is something for which we will provide case examples, although basically it means that enterprise components develop a data model in a neutral form that can be used in developing a translator to which data from legacy systems and from external sources are mapped to a neutral state before being mapped again to another user environment. The mapping and exchange process is accomplished through servers for this purpose.

"Semantic mediation" identifies the service and process of accommodating different users with different words and sometimes the same words with different meanings. By employing the technology of tagging and mapping, it is possible to support natural language processing, for instance.

Other terms describing interoperability technology have been discussed, and our cases will provide more detailed examples. Frustrating is that we still need an enterprise executive client so that we can demonstrate how these ideas are applied from the very top down.

The third element, "data-aligned methods and algorithms," is something for which we have robust academic and scholarly examples. They are domain specific and technical, so we ask you to work with us to transfer the concepts to your own needs and environment.

Case Studies

Forensic Science Research

There are two types of case studies presented in this section: (1) scientific/academic and (2) anecdotal/programs and projects. For both types, we apply our smart data labels to help you identify and understand our thinking in practice. The scientific/ academic examples are more rigorously detailed, while the anecdotal programs and projects may be easier to relate. Both demonstrate the effort we have taken to test our ideas in different layers of the enterprise and for different types of enterprises. While the case studies are specific to certain domains, the lessons are transferable horizontally across domains, departments, and enterprises.

Case Study: Medical Automation   Here is a specific case study that demonstrates application of our thinking and is applied in the area of healthcare (military medical environment). This example shows more scientific rigor than other examples that are reflective of a government customer environment [1].

Actual Name.   Adapting the "Task-Technology-Fit (TTF) Model to the Development and Testing of a Voice Activated Medical Tracking Application." The task-technology fit (TTF) theory holds that IT is more likely to have a positive impact on individual performance and be used if the capabilities of the IT match the tasks that the user must perform [2]. Goodhue and Thompson developed a measure of task-technology fit that consists of eight factors: quality, locatability, authorization, compatibility, ease of use/training, production timeliness, systems reliability, and relationship with users. Each factor is measured using between two and ten questions with responses on a seven-point scale ranging from strongly disagree to strongly agree.

Background. The customer is the Joint Military Medical Command of the U.S. Department of Defense.

Goals. Validate that voice activated medical technology application produces reliable and accurate results while affording cost and time savings and convenience advantages.

Decision. Should the voice activated medical tracking be applied throughout the military medical community?

IT Support. Provide a methodology and algorithms for validation. Help standardize data capture, recording, and processing.

This example illustrates three level views of smart data with the following case: adapting the TTF model to the development and testing of a voice activated medical tracking application (VAMTA). The TTF asserts that for an information technology to have a positive impact on individual performance, the technology (1) must be utilized and (2) must be a good fit with the tasks it supports.

"Enterprise context" suggests that we are not using traditional systems engineering by stovepipes or verticals and that the systems are enterprise-wide in their scope and utilize executive user leadership.

VAMTA is an enterprise-wide application that has been championed and endorsed by top management in the military. To accomplish this task of adopting novel technology, it was necessary to develop a valid instrument for obtaining user evaluations of the VAMTA.

The case was initiated in order to study the application with medical end-users. This phase of the case provided face validity. The case demonstrated that the perceptions of end-users can be measured and evaluation of the system from a conceptual viewpoint can be documented, in order to determine the scope of this nontraditional application.

The case survey results were analyzed using the Statistical Package for the Social Sciences (SPSS) data analysis tool to determine whether TTF, along with individual characteristics, will have an impact on user evaluations of VAMTA.

The case modified the original TTF model for adequate domain coverage of medical patient-care applications. This case provides the underpinnings for a subsequent, higher level study of nationwide medical personnel. Follow-on studies will be conducted to investigate performance and user perceptions of the VAMTA system under actual medical field conditions.

As the medical community continues its quest for more efficient and effective methods of gathering patient data and meeting patient needs, increased demands are placed on IT to facilitate this process. The situation is further complicated by the segmented nature of healthcare data systems. Healthcare information is often encapsulated in incompatible systems with uncoordinated definitions of formats and terms. It is essential that the different parts of this organization, which have different data systems, find ways to work together in order to improve quality performance. The VAMTA has been developed to contribute to that effort by enhancing the electronic management of patient data.

Much has been written about end-user perceptions of IT, but few studies address user evaluations of IT as it applies to voice activated healthcare data collection. This case focuses on the development and testing of an instrument to measure the performance and end-user evaluation of the VAMTA in preventive healthcare delivery. It is not inconceivable that, in the future, a significant amount of healthcare data will be collected via applications such as VAMTA.

The TTF model (Figure 4.2) is a popular model for assessing user evaluations of information systems. The central premise for the TTF model is that "users will give evaluations based on the extent to which systems meet their needs and abilities." For the purpose of our study, we define user evaluations as the user perceptions of the fit of systems and services they use, based on their personal task needs.

The TTF model represented in Figure 4.2 is very general, and using it for a particular setting requires special consideration. Among the three factors that determine user evaluations of information systems, technology is the most complex factor to measure in healthcare. Technology in healthcare is used primarily for reporting, electronic information sharing and connectivity, and staff and equipment scheduling.

Reporting is important in a healthcare setting because patient lives depend on accurate and timely information. Functional departments within the healthcare facility must be able to access and report new information in order to respond properly to changes in the healthcare environment.

Our ideas about smart data include the dimension of the interoperability of the technology. Our case shows that there are four types of smart data that are reported in a healthcare facility. These include scientific and technical smart data, patient-care smart data, customer satisfaction smart data, and administrative smart data.

These four types of smart data embrace data engineering, model-driven data exchange, semantic mediation, metadata management, automated mapping tools, and the service-oriented enterprise paradigm that includes smart data, smart grid, smart services, and credentialing and privileging.

Scientific and technical smart data requires metadata management in order to provide the knowledge base for identifying, organizing, retrieving, analyzing, delivering, and reporting clinical and managerial journal literature, reference smart

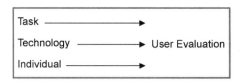

Figure 4.2 Task-technology fit (TFF) model.

data, and research data for use in designing, managing, and improving patient-specific and departmental processes. Semantic mediation is required to manage the volume of content that confronts healthcare personnel.

Patient-care information is specific data and information on patients that is essential for maintaining accurate medical records of the patients' medical histories and physical examinations. This process requires automated mapping tools because patient-specific smart data are critical to tracking all diagnostic and therapeutic procedures and tests.

Maintaining model-driven data exchange of accurate smart data about patient-care results and discharges is imperative to delivering quality healthcare. Customer satisfaction smart data is gathered from external customers, such as a patient and his or her family and friends. Customer satisfaction smart data is gathered from surveys and takes into account sociodemographic characteristics, physical and psychological status, attitudes, and expectations concerning medical care, the outcome of treatment, and the healthcare setting. The adoption of a service-oriented enterprise paradigm that includes smart data, smart grid, and smart services are essential to maintaining customer satisfaction.

Credentialing and privileging are necessary for managing the administrative smart data that is reported in a healthcare facility. It is essential for formulating and implementing effective policies at both the organizational and departmental levels. Administrative information is necessary to determine the degree of risk involved in financing expansion of services.

As the technological infrastructure of organizations becomes increasingly complex, IT and the adoption of the smart data paradigm is increasingly being used to improve coordination of activities within and across organizations. Computers and video networks provide long-distance healthcare through medical connectivity. Doctors interact with each other and ancillary medical personnel through e-mail, video, and audio means. A difficult patient case in a rural area or on board a ship can be given expert specialist attention simply by using "distance" medicine. Not only can patient records, text, and documents be transmitted instantaneously via electronic means, but live video, X-rays, and other diagnostic parameters can be discussed in an interactive manner with live discussions.

As the availability of external consultative services increases, sharing of smart data through connectivity has become increasingly important. Connectivity allows diagnoses to be made in remote locations using electronic means, and information sharing decreases the chances that mistakes will be made in a healthcare setting. Connectivity leads to shared smart data and care that comprises the continued, coordinated, and integrated activities of different people from different institutions applying different methods in different time frames, all in a combined effort to aid patients medically, psychologically, and socially in the most beneficial ways.

In addition to reporting and electronic information sharing and connectivity, smart data is used hand in hand with IT to extensively improve staff and equipment scheduling in healthcare settings. IT-based scheduling, utilizing smart data, can lower healthcare costs and improve the utilization of physical and human resources. Scheduling using statistical, time series, and regression analysis is conducted to achieve lower costs through rationing assets (e.g., ambulatory service and real-time forecasting of resources).

The purpose of this case was to develop and test a valid survey instrument for measuring user evaluations of VAMTA, as a novel means to expedite smart data delivery, in preventive healthcare [3]. The findings of the case study were used to test the instrument that was used in a preliminary assessment of the effectiveness of the VAMTA system and the applicability of smart data in the TTF model to the VAMTA application. The development of the instrument was carried out in two stages.

The first stage was item creation. The objective of this first stage was to ensure face and content validity of the instrument. An item pool was generated by interviewing two end-users of IT, obtained from a pool of medical technicians. The end-users were given training on the module for 2 days and invited to participate in the study. These subjects were selected for reasons of geographical proximity of the sample and, in many cases, the existence of personal contacts onboard ship.

An interview was also conducted with one of the authors of this study, who has approximately 10 years experience as an IT end-user. In addition, the domain coverage of the developed pool of items was assessed by three other end-users from three different ship environments covered in the survey.

None of the end-users, who were a part of the scale development, completed the final survey instrument. All the items were measured on a five-point Likert scale ranging from "strongly agree" to "strongly disagree." Next, the survey instrument was utilized in a pilot study in which end-users tested the VAMTA. This pilot study provided face validity, in that the perceptions of end-users could be measured, and evaluation of the system from a conceptual viewpoint could be documented. A total of 33 end-users were used in this phase to test the VAMTA.

They reported their perceptions of the VAMTA in the survey instrument, which was provided after their training and testing. The pilot study results were analyzed using SPSS and Microsoft Excel to determine whether TTF, along with individual characteristics, had an impact on user evaluations of the VAMTA. For the case, the original TTF model was modified to ensure adequate domain coverage of medical and preventive healthcare applications.

The third smart data dimension revolves around data-aligned methods and algorithms that utilize data with pointers for best practices, data with built-in intelligence, autonomics, an automated regulatory environment (ARE), and an automated contracting environment (ACE).

Our case used data-aligned methods for the 33 individuals who participated in the testing of the VAMTA, including 11 females and 22 males, with various levels of experience and education. Experience ranged from test subjects who had never seen the application before they tested it, to developers who built the application. A number of test subjects had limited medical backgrounds. Education ranged from high school graduates to people with doctorate degrees. Data with pointers for best practices was utilized so that each test subject was shown a demonstration of the VAMTA prior to testing.

Test subjects were then required to build a new user account and speech profile. Subjects ran through the application once using a test script to become familiar with the application. Next, the test subjects went through the application again while being videotaped. No corrections were made to dictated text during the videotaping. This allowed the tracking of voice dictation accuracy for each user with a new speech profile.

Test subjects completed the entire process in an average of 2 hours. Afterward, each test subject completed a questionnaire to determine user demographics, utility and quality performance, system interface, hardware, information management, and overall system satisfaction and importance. This survey instrument also allowed the test subjects to record any problems and suggest improvements to the system. Data with built-in intelligence techniques recorded any problems that the test subjects encountered in the survey instrument.

Autonomics were employed by the VAMTA development team, to document bugs in the tracking system, and targeted by the test architect for the appropriate problem solving action. The performance of VAMTA during testing was measured in terms of voice accuracy, voice accuracy total errors, duration with data entry by voice, and duration with data entry by keyboard and mouse. Viewed together, these statistics provide a snapshot of the accuracy, speed, and overall effectiveness of the VAMTA system.

Each test subject's printout was compared with a test script printout for accuracy. When discrepancies occurred between the subject's printout and the test script, the printouts were compared with the automated video recordings to determine whether the test subjects said the words properly, stuttered or mumbled words, and/or followed the test script properly.

Misrecognitions occurred when the test subject said a word properly and the speech program recorded the wrong word. This automated regulatory environment provided a solid baseline for the lab experiments and led to the following results.

The accuracy of voice recognition, confirmed by videotaped records of test sessions, averaged 97.6%, with 6 misrecognitions (Table 4.1). The minimum average voice recognition was 85%, with 37 misrecognitions. The maximum average voice recognition was 99.6% with 1 misrecognition. Median voice recognition was 98.4%, with 4 misrecognitions.

TABLE 4.1   VAMTA: Voice Accuracy During Testing

	Misrecognitions with Video	Number Correct with Punctuation and Video	Percentage Accurate with Video
Average	6	241	97.6%
Minimum	1	210	85.0%
Maximum	37	246	99.6%
Median	4	243	98.4%
Males	22		
Females	11		
Count	33		

TABLE 4.2   VAMTA: Voice Accuracy Total Errors During Testing

Total Errors	
Average accurate	95.4%
Minimum accurate	85.0%
Maximum accurate	99.2%
Median	96.0%
Count	33

Total errors include both misrecognitions and human errors. Human errors occurred when a test subject mispronounced a word, stuttered, or mumbled. The total accuracy rate of the VAMTA was 95.4% (Table 4.2). Human error accounted for 2.2% of the total errors within the application.

The duration of each test subject's voice dictation was recorded to determine the average length of time required to complete a medical encounter while using the VAMTA. The average time required to complete a medical encounter in which data entry was conducted by voice was 8 minutes and 31 seconds (Table 4.3). The shortest time was 4 minutes and 45 seconds, and the longest time was 23 minutes and 51 seconds.

While the majority of test subjects entered medical encounter information into the VAMTA only by voice, several test subjects entered the same medical encounter information using a keyboard and mouse. The average time required to complete a medical encounter in which data entry was conducted with keyboard and mouse was 15 minutes and 54 seconds (Table 4.4). The shortest time was 7 minutes and 52 seconds, and the longest time was 24 minutes and 42 seconds.

The average duration of sessions in which data entry was performed by voice dictation was compared to the average duration of sessions in which data entry was performed with a keyboard and mouse. On average, less time was required to complete the documentation of a medical encounter using VAMTA when data entry was performed by voice instead of with a keyboard and mouse.

TABLE 4.3   Medical Encounter Duration with Data Entry by Voice

Average time—Voice	0:08:31
Minimum time—voice	0:04:54
Maximum time—voice	0:32:51
Median	0:06:59

TABLE 4.4   Medical Encounter Duration with Data Entry by Keyboard and Mouse

Average time with keyboard	0:15:54
Minimum time with keyboard	0:07:52
Maximum time with keyboard	0:24:42
Median	0:15:31

The average time saved using voice versus a keyboard and mouse was 7 minutes and 52 seconds per medical encounter. The duration of each medical encounter included the dictation and printing of the entire Chronological Record of Medical Care form, a Poly Prescription form, and a Radiologic Consultation Request/Report form. (See Tables 4.5–4.7.)

TABLE 4.5 VAMTA Test and Evaluation (T&E) Data Human Errors

Subject Number	Total with Punctuation	Human Errors	Number Right Minus Human Errors	Percentage Accurate Human Errors	Sex
1	247	0	247	100.0%	F
2	247	12	235	95.1%	F
3	247	12	235	95.1%	F
4	247	3	244	98.8%	M
5	247	0	247	100.0%	F
6	247	0	247	100.0%	F
7	247	2	245	99.2%	M
8	247	2	245	99.2%	M
9	247	9	238	96.4%	M
10	247	16	231	93.5%	M
11	247	15	232	93.9%	M
12	247	1	246	99.6%	F
13	247	0	247	100.0%	F
14	247	10	237	96.0%	F
15	247	13	234	94.7%	M
16	247	7	240	97.2%	F
17	247	6	241	97.6%	M
18	247	14	233	94.3%	M
19	247	1	246	99.6%	M
20	247	3	244	98.8%	F
21	247	9	238	96.4%	M
22	247	7	240	.97.2%	M
23	247	6	241	97.6%	M
24	247	3	244	98.8%	M
25	247	0	247	100.0%	M
26	247	4	243	98.4%	F
27	247	1	246	99.6%	M
28	247	2	245	99.2%	M
29	247	11	236	95.5%	M
30	247	3	244	98.8%	M
31	247	1	246	99.6%	M
33	247	1	246	99.6%	M

Average Accurate	Minimum Accurate	Maximum Accurate	Median	Count
97.8%	93.5%	100.0%	98.8%	33

TABLE 4.6   VAMTA T&E Data Total Errors

Subject Number	Total	Total Errors	Number Right	Percentage Accurate	Sex
1	247	13	234	94.7%	F
2	247	13	234	94.7%	F
3	247	14	233	94.3%	F
4	247	6	241	97.6%	M
5	247	4	243	98.4%	F
6	247	4	243	98.4%	F
7	247	8	239	96.8%	M
8	247	4	243	98.4%	M
9	247	13	234	94.7%	M
10	247	25	222	89.9%	M
11	247	22	225	91.1%	M
12	247	2	245	99.2%	F
13	247	37	210	85.0%	F
14	247	26	221	89.5%	F
15	247	17	230	93.1%	M
16	247	12	235	95.1%	F
17	247	8	239	96.8%	M
18	247	19	228	92.3%	M
19	247	6	241	97.6%	M
20	247	9	238	96.4%	F
21	247	20	227	91.9%	M
22	247	12	235	95.1%	M
23	247	11	236	95.5%	M
24	247	10	237	96.0%	M
25	247	3	244	98.8%	M
26	247	8	239	96.8%	F
27	247	8	239	96.8%	M
28	247	4	243	98.4%	M
29	247	14	233	94.3%	M
30	247	10	237	96.0%	M
31	247	5	242	98.0%	M
33	247	3	244	98.8%	M

Average Accurate	Minimum Accurate	Maximum Accurate	Median	Count
95.4%	85.0%	99.2%	96.0%	33

While the effectiveness of the VAMTA during the pilot study testing can be estimated from statistics on voice accuracy and duration of medical encounters, the applicability of the TTF model to the VAMTA system can be determined by an analysis of the end-user survey instrument responses. Multiple regression analysis reveals the effects of VAMTA utility, quality performance, task characteristics, and individual characteristics on user evaluations of VAMTA (Table 4.8).

TABLE 4.7 VAMTA T&E Data Female

Subject Number	Total with Punctuation	Miss Recognitions with Video	Number Right with Punctuation and Video	Percentage Accurate with Video	Sex	Time Started Voice	Time Stopped Voice	Time to Complete Voice
1	247	13	234	94.7%	F	13:35:43	14:08:34	0:32:51
2	247	1	246	99.6%	F	14:26:53	14:32:18	0:05:25
3	247	2	245	99.2%	F	10:49:42	11:01:15	0:11:33
5	247	4	243	98.4%	F	11:18:13	11:23:27	0:05:14
6	247	4	243	98.4%	F	13:35:02	13:42:16	0:07:14
12	247	1	246	99.6%	F	14:28:28	14:33:37	0:05:09
13	247	37	210	85.0%	F	10:49:39	11:01:28	0:11:49
14	247	16	231	93.5%	F	9:45:02	10:07:50	0:22:48
16	247	5	242	98.0%	F	10:41:12	10:47:18	0:06:06
20	247	6	241	97.6%	F	10:45:40	10:51:11	0:05:31
26	247	4	243	98.4%	F	11:06:50	11:12:58	0:06:08

Average Accurate		Minimum Accurate		Maximum Accurate		Median		Total Females
96.6%		85.0%		99.6%		98.4%		11

Average Time Voice		Minimum Time Voice		Maximum Time Voice				Median
0:10:53		0:05:09		0:32:51				0:06:08

TABLE 4.8   One-Way ANOVA for Regression Analysis

Degrees of Freedom	Source	Sum of Squares	Mean Squares	F Value	$p > F$
4	Regression	2.664	0.666	6.735	0.001[a]
29	Residual	2.867	0.099		
33	Total	5.531			

[a]Significant at $P = 0.001$.

The results indicate that overall end-user evaluations of VAMTA are consistent with the TTF model. The F value was 6.735 and the model was significant at the $p = 0.001$ level of significance. The R-square for the model was 0.410. This indicates that model-independent variables explain 41% of the variance in the dependent variable. The individual contributions of each independent variable factor are shown in Table 4.9.

While the Table 4.8 data reveals the suitability of the TTF model, Table 4.9 reveals another finding. Based on the data shown in Table 4.9, according to the case study user evaluations of VAMTA, utility and quality performance are the major factors that affect the management of smart data by VAMTA. The VAMTA case revealed findings related to the effectiveness of the survey instrument and the VAMTA itself. As a result of the case study, the VAMTA follow-up questionnaire has been proved to be a valid survey instrument.

By examining end-user responses from completed surveys, analysts were able to measure multiple variables and determine if the TTF model was applicable to the VAMTA system. The survey's effectiveness should extend to its use in future studies of VAMTA's performance in providing smart data for preventive healthcare in a national setting.

Analysis of the actual end-user responses supplied during the case study confirmed that the TTF model of smart data does apply to VAMTA. In the case survey responses, the VAMTA system received high ratings in perceived usefulness and perceived ease of use. This suggests that the VAMTA shows promise in medical applications for providing smart data. The survey responses also revealed that utility and quality

TABLE 4.9   Individual Contribution of the Study Variables to the Dependent Variable

Source	Degrees of Freedom	Sum of Squares	Mean Square	F Value	$p > F$ 0.022[a]
Ease of use	1	1.577	1.577	5.834	0.007[a]
Navigation	1	2.122	2.122	8.459	0.0000[b]
Application	1	0.211	3.80	4.59	0.0000[b]
Operation	1	0.912	0.912	3.102	0.008[a]
Understandable	1	0.965	0.965	3.196	0.0085[a]

[a]Significant at $p = 0.05$.
[b]Significant at $p = 0.001$.

performance are the major factors affecting the management of information by VAMTA.

In the future, end-users who want to improve the management of healthcare information through use of VAMTA will need to focus on utility and quality performance as measured by perceived usefulness and perceived ease of use of the VAMTA system.

In addition to providing findings related to TTF and the utility and quality performance of VAMTA, the case study demonstrated ways in which the VAMTA system itself can be improved. For example, additional training with the application and corrections of misrecognitions will improve the overall accuracy rate of this product. The combined findings resulting from the case study have laid the groundwork for further testing of VAMTA. This additional testing is necessary to determine the system's performance in an actual medical setting and to define more clearly the variables affecting the TTF model when applied to delivering smart data in that setting.

In the case study, efforts were focused on defining the smart data technology construct for preventive healthcare applications [4]. It is our contention that applying smart data is accomplished in three dimensions beginning with healthcare executives embracing the idea to employ accounting for patient and other data as a catalyst for change and improvement in the healthcare delivery process.

In this case, we have followed the data through processes and arrived at outcomes that identify problems and help us to realize opportunities for improving collection of healthcare data. We have also demonstrated that this data collection should be done in the most direct manner, by focusing on smart data that in turn focuses on the primary business principles.

This case also demonstrates that smart data has three dimensions. These dimensions include optimization performance in an enterprise context, interoperability of technology, and data-aligned methods and algorithms. The case also illustrates that problems and opportunities will be solved and pursued through application of improved data engineering applications, such as VAMTA, which improve data exchange strategies and technical implementations, including decision support.

While our case accomplished many insights, limited work was done to define tasks and individual characteristics of the end-users of the application. To complete this work, future research should focus on defining the task and individual characteristics constructs for the TTF model for measuring user evaluations of IT delivering smart data, in preventive healthcare.

Case Study: Enterprise Logistics Information Trading Exchange (ELITE) Now, we will share a case study experience that further supports some of these parameters and lessons learned. We have said that smart data is a product of data engineering, so we will define the technical skills and position descriptions that comprise data engineering disciplines derived from our experience.

When starting and staffing the ELITE program, we had to plan for work to be done and to assign labor categories to the work. Work was categorized into four main activities and with subordinate activities:

1. Requirements Planning
    1.1. Meet with client management, stakeholders, and technical points of contact.
    1.2. Analyze current system architecture.
    1.3. Based on the performance work statement, update and create business models while researching and analyzing system enhancements.
    1.4. Evaluate and select products required to support the to-be architecture—including risks and dependencies.
    1.5. Finalize system architecture.
    1.6. Document interoperability requirements and specifications and publish and solicit feedback.
2. Discover, Design, and Develop
    2.1. Identify relevant elements, content, and usage within the subject business procedures.
    2.2. Obtain schema and sample databases from each stakeholder.
    2.3. Import application models and metadata into COTS tools and identify sources of information.
    2.4. Analyze data and process models with respect to mapping ISO 10303-AP239 PLCS standard or comparable standard to the customer domain.
    2.5. Develop new use case business process models as required.
    2.6. Generate Express models to extend the PLCS information model.
    2.7. Develop and update ontology files as applications are incorporated into the enterprise model.
    2.8. Incorporate vendor neutral software architecture.
    2.9. Develop transformation software to facilitate interoperability rules and processes.
    2.10. Develop updated test plans.
    2.11. Perform first level programmer testing and debugging.
    2.12. Perform integrated functional testing and debugging.
    2.13. Execute interoperability demonstration.
    2.14. Create version controlled baseline for independent verification, validation, and acceptance.
    2.15. Schedule and implement release of complete product version.
3. Install and Test
    3.1. Develop a plan for site implementation, installation, and initial operating capability (IOC).
    3.2. Conduct installation training survey (user information roles and responsibilities).
    3.3. Schedule and conduct training.
    3.4. Develop implementation schedule.

3.5. Configure and install site unique information.

3.6. Conduct pre-IOC.

3.7. Perform site IOC.

4. Accredit and Transition to Sustainment

4.1. Develop system security authorization agreement to obtain DoD Information Technology Security Certification and Accreditation Process (DITSCAP).

4.2. Develop plan for system migration.

4.3. Attain DITSCAP.

4.4. Implement Help Desk.

4.5. Schedule change control meeting.

4.6. Publish results.

This level of effort was performed by 13 individuals as follows:

- Program manager
- Project manager
- Integration and interoperability architect
- Transformation mapping specialist/data architect
- Process manager
- Process modeler
- Adapter developer
- Technical infrastructure services administrator
- SME/functional expert
- Knowledge manager
- Content editor
- Repository specialist
- Business case, cost, economic analyst

The effort lasted approximately one year.

The performance of this work produced the technical solution to facilitate interoperable data exchange among disparate users without changing legacy systems. The technical solution is an artifact that is leverageable and transferable to a host of Department of Defense needs under an executive strategy and direction to do so.

Model-Driven Data Exchange

As part of the ELITE program, we invented a better way to exchange data. Generally, observe the following:

- A product producer of a highly engineered product (helicopter) is responsible for total product life cycle support.
- The customer is the U.S. Department of Defense and each military service.

- Military services have different roles and responsibilities as users and product repair and maintainers. (For example, Army provides central repair and maintenance for all military services for the H60 helicopter.)
- All participants have different and disparate information technologies.
- All participants have different and disparate processes.
- Common are overarching contracts, laws, and regulations.
- Common is the DoD net-centric data strategy.
- A shared outcome is to maximize weapon system availability for warfighters while minimizing repair costs and cycle time.
- Fluid is the developing Federal Enterprise Architecture (FEA) and its influence as guidance over department initiatives such as the DoDAF.

The Department of Defense Architecture Framework (DoDAF) is a reference model to organize the enterprise architecture (EA) and systems architecture into complementary and consistent views inspired by the Zachman framework: operational view, systems/services view, and technical/standards view. In our next example, we employed the Zachman framework as if we were performing in an enterprise context and found it useful. The dynamics in this case are likely common among many other weapon systems programs, and generally representative of systems and circumstances throughout defense and military services operations. Therefore the resulting solution should be transferable and applicable throughout the defense enterprise community.

At the outset, missing is executive leadership and sponsorship at the Joint Command level since multiple military services are involved. The project is characterized as an information technology pilot demonstration that is intended to be a spiral development. Spiral development is intended to be an iterative process, advancing on merit from phase to phase (Figure 4.3).

The spiral process in this circumstance, and similar to many others, is out of context of the mainstream program funding and budgeting process, which we will discuss as another case study analysis. What this means is that in the absence of executive sponsorship and being out of context of mainstream funding, even a successful demonstration is likely to reach a dead end.

This circumstance is symptomatic of a problem, not just in the DoD, but in government enterprise in general. It underscores why a smart data strategy is needed as a catalyst to change the way government does business with a top–down focus on data and resulting performance outcomes that is a product of how work gets done. This is too big a problem to bite off in this simple case instance, although the symptoms are clearly evident everywhere one turns.

We developed and learned what is necessary through two principal programs: Enterprise Logistics Information Trading Exchange (ELITE), and the Joint Strike Fighter Single Point of Entry (JSF SPOE). ELITE has no security restriction and the results are in the public domain, whereas the JSF SPOE program is a secure program and under restrictions. Therefore we will focus on ELITE and will identify other open programs that illustrate experience similar to JSF SPOE.

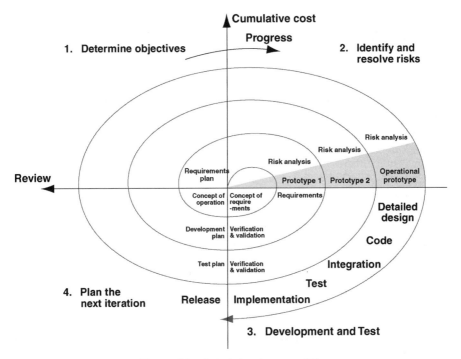

Figure 4.3   Spiral development [5].

In addition, we have performed an analysis of industry technology providers from which we will crystallize our definition. For instance, DataFlux, a SAS Company, ran an advertisement with the headline "Make Your Data Behave." The ad copy describes how data arrives into an organization in a variety of forms and conditions that may well be described as chaotic and lacking discipline. The DataFlux product somehow transforms this into actionable data that you can use.

The fact that this company, among others, is promoting this message is validation of our notion that data quality can be improved and that data can be made to behave as you wish. The question is how to accomplish this. How is data exchange executed? Our goal is to accomplish this by leveraging open interoperability and leapfrogging rigid standards and brittle interfaces.

The ELITE case study readily extends to needs and applications throughout any enterprise and that is why we share it. We describe transferability in the Chapter 5.

Shared Data Exchange

Speaking with people supporting the U.S. Immigration Customs Enforcement (ICE) Agency of the Department of Homeland Security (DHS), we know that their shared repository was developed nearly 10 years ago, and five divisions that use this capability put the highest trust in the information they create and manage themselves versus what they share from others. There is no sense of value for a common data model belonging to the department, and congressional direction and regulations are also stovepiped. This is a typical situation that manifests in a variety of ways. Problems arise from losing opportunities for enforcement and protection that could come from synergy among divisions. The following instance is as applicable to the DHS as it is to any other department, and any commercial enterprise facing similar dynamics. Some try to solve this problem through improved data governance, although changing the technical approach can also catalyze change and improvement as demonstrated here. In the end, an effective solution must be comprehensive and top–down.

When supporting the ELITE program for the Department of Defense, we were presented with the following problem:

Situation   Sikorsky is the prime contractor and producer of the H60 helicopter in its various models and delivered to each of the military services that are customers. Each military service deploys the helicopters on different types of missions and in different operating environments. The DoD uses the Army as the central repair facility for all H60s.

Sikorsky had a performance-based logistics contract and sought end-to-end visibility about its products throughout their life cycle such that the company could optimize support at the lowest cost. Accomplishing this is a win–win for the government customer and producer-supplier.

Visibility was constrained because each military service recorded information differently in various flight data and maintenance log books. The data was captured as represented on forms, with each service having different form designs and recording different data in different ways and in different locations.

The Army had difficulty tracking the use and maintenance of Navy helicopters. When Navy helicopters arrive at the Army depot for repair, history and configuration management are suspect. In many cases, not having reliable data about the Navy's helicopter would lead to repair and parts replacement that may have been redundant or otherwise unnecessary, adding excess cost and time.

We even heard an instance when a helicopter was repaired on one side of the runway, moved to the other, and had the same repair made on the other side because the repair history did not move as fast as the helicopter.

When confronted by the circumstance and a genuine lack of configuration management, all parties agreed that the situation could be improved. Sikorsky was highly motivated. The Army was motivated because one of its repair customers, the Navy, was a willing participant. All were wary because none had money for a big change and were dependent on a small pool of pilot funding to pursue possibilities.

Requirements   Constrained by resources to address the problem, the government customer could have adopted a top–down strategy and attempted to direct all of the services to adopt a single system or to standardize. That would require a command decision from the Joint Chiefs as it affected all military services using the same weapon system. Wars in Iraq and Afghanistan consumed executive bandwidth.

A directed strategy would likely fail because resources required for massive change are unavailable, and dictates are rarely successful. As important, no one in a position of executive power was engaged to impose such a directive. Thus this case must be characterized as an information technology-driven initiative by default.

As such, the program inherited strategic direction from the DoD data strategy [6]:

> The Data Strategy provides the basis for implementing and sharing data in a net-centric environment. It describes the requirements for inputting and sharing data, metadata, and forming dynamic communities to share data. Program managers and Sponsors/Domain Owners should comply with the explicit requirements and the intent of this strategy, which is to share data as widely and as rapidly as possible, consistent with security requirements. Additional requirements and details on implementing the DoD Data Strategy are found in another section . . .. Specific architecture attributes associated with this strategy that should be demonstrated by the program manager include:
>
> - Data Centric—Data separate from applications; applications talk to each other by posting data. Focus on metadata registered in DoD Metadata Repository.
>
> - Only Handle Information Once—Data is posted by authoritative sources and made visible, available, and usable (including the ability to re-purpose) to accelerate decision-making. Focus on re-use of existing data repositories.
>
> - Smart Pull (Versus Smart Push)—Applications encourage discovery; users can pull data directly from the net or use value added discovery services. Focus on data sharing, with data stored in accessible shared space and advertised (tagged) for discovery.
>
> - Post in Parallel—Process owners make their data available on the net as soon as it is created. Focus on data being tagged and posted before processing.
>
> - Application (Community of Interest (COI) Service) Diversity—Users can pull multiple applications (COI Services) to access same data or choose same applications (Core and COI Services) for collaboration. Focus on applications (COI service) posting and tagging for discovery.

The presence of this strategy was powerful because it had been adopted by defense executives and was already on the books to guide our type of project.

The government customer placed certain requirements on solution development beginning with "no big bang" solution—meaning make it something readily implementable with limited resources. First, it was recognized that for the Army, Navy, and Sikorsky to replace all of their disparate legacy systems was out of the question. The solution would have to be as noninvasive as possible. Second, the solution would be a system of data exchange among all parties that assured accessibility to actionable data.

The creative vision for a solution was derived from working with the customer lead technical representative. We envisioned creating an exchange server that would host a neutral exchange mechanism to which disparate users could map data for exchange to any other qualified user in the community.

Needed was a basis for neutral exchange. We characterized the data as being aircraft maintenance data that is a subset of weapon system data that is a subset of manufactured product data. We were aware of an international standard for product data that has been under development for many years and that is generally characterized as applicable to this situation.

The ISO standard was identified as an asset worth leveraging since considerable DoD and Department of Commerce National Institute of Standards and Technology (NIST) investment had already been sunk in the international initiative. Europeans, principally the Norwegian Navy, have applied the standard, for instance, although no one had applied it to aircraft data or anything else in the DoD.

The International Standards Organization (ISO) addresses product data in the Product Life Cycle Support Standard, ISO 10303. ISO describes itself as "a non-governmental organization that forms a bridge between the public and private sectors. On the one hand, many of its member institutes are part of the governmental structure of their countries, or are mandated by their government. On the other hand, other members have their roots uniquely in the private sector, having been set up by national partnerships of industry associations" [7].

An application set within the standard addresses product maintenance data [8].

ISO/TS 10303-1287:2005 specifies the application module AP239 activity recording. The following are within the scope of this part of ISO 10303:

- recording the usage of a product;
- recording the maintenance activities on a product;
- recording the use of resources during usage of a product;
- recording the use of resources during the maintenance of a product;
- recording observations against a product or its support solution;
- relating the record of an activity to the work order, work request and work definition from which it arose.

Sikorsky, the Army, and the Navy agreed that the ISO 10303 Product Life Cycle Support Standard AP 239 could serve as the basis for neutral exchange.

Breakthrough Discovery

The breakthrough discovery from this effort is using common models as a medium for data exchange that is applicable to all defense enterprise integration problems centered on exchanging information based on rigid standards and brittle interfaces.

The Office of the Secretary of Defense stakeholder for this effort was the Unique Identifier (UID) Program Office. UID is an effort by the DoD to put a unique identifier on parts of a certain value to increase asset visibility and to comply with congressional directives to improve asset control and management. The stakeholder was interested in populating the UID repository with current configuration management data that would be generated from the data sharing solution.

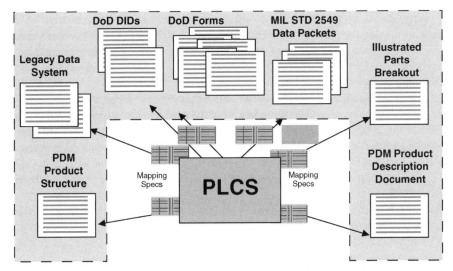

Figure 4.4   PLCS integrates transactions, providing extensibility.

Given this agreement, the next step would be to examine the primary forms and data collected on maintenance logs by military services users. When talking with military services representatives who work in the environment under consideration, we spoke in their language and with examples to which they could relate. All sorts of data are used by the defense and military systems as shown in Figure 4.4. All can be mapped to a common PLCS model for data exchange.

Aircraft maintenance records come from the Air Force Logistics Command, Navy Air, Army Command, Suppliers, Depots, and Program Management Office (Figure 4.5).

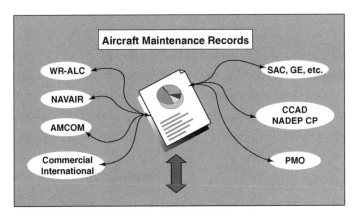

Figure 4.5   A shared data process.

Why are Army and Navy forms different for the same weapon system? The basic difference has to do with the environment in which they operate—water versus sand storm—and different types of missions. Since the Army must repair Navy helicopters, they must accommodate the operating characteristics of the Navy environment.

Why does the Sikorsky producer have a different system? A better question is why do the military services and DoD not key off of the producer? Answers include: (1) the DoD customer did not want to be hostage to the prime contractor, and (2) military services are autonomous cultures and have difficulty embracing common processes and common needs. This is changing because the culture of "jointness" is maturing. The notion that the DoD is hostage to a prime contractor needs to be explored further, but not here.

The data from these forms were mapped to the standard around which the exchange utility would be developed.

Project Objectives

1. Improve acquisition visibility through implementing PLCS standards in the defense industrial enterprise.
2. Convert defense industrial business information objects into PLCS data exchange objects (DEX).
3. Integrate PLCS/DEX with COTS information integration tools.

Improvement. The effort is to improve data exchange between dependent organizations. ELITE tasks include the following:
- Establish data exchange capability between the Army and Sikorsky.
- Establish data exchange capability between the Navy and Army.
- Establish data exchange capability between the Navy and Sikorsky.
- Establish passive ability to update the UID registry.

Scope. The H60 Helicopter Program and Aviation Maintenance actions define the scope. UID is a small subset of the larger requirement.

Architectural Strategy. The architectural strategy is to facilitate data exchange through the use of a common information model as the medium for exchange.

Consequences. Problems manifest in a host of consequences:
- Repairing and replacing parts unnecessarily at great expense
- Inadequate visibility about products and materials leading to oversupply and undersupply and errors in deployment
- Excessive system sustainment expense
- Deficient utilization of scarce human resources
- Underexploitation of legacy systems
- Inability to support contemporary best practices

In the course of this pilot project, there was no attempt to pursue a host of improvement opportunities that became apparent in the course of analysis as that was out of scope. It should be noted, however, that the military services processes were excessively labor intensive and that there was much duplication of effort associated with them. The fact that the "IT-oriented" effort was so narrowly focused and was performed in the absence of executive leadership to guide smart data strategy resulted in much lost opportunity to improve performance.

The solution produced by ELITE included the creation of data exchange servers at the Army, Navy, and Sikorsky, whereby the participants could share data in a standard manner without changing respective applications. For the pilot demonstration, Microsoft BizTalk Servers hosted the PLCS data translator for exchange. Other brands of servers could be used to implement the solutions as vendor independence is a part of smart data strategy. It was critical to have expertise in the ISO standard and application development process using ISO. For this, we had to contract with European consultants due to the scarcity of experience in the United States.

So, how is this making the data smart?

The Army, Navy, and Sikorsky each had proprietary ways in which to represent helicopter maintenance data, and these differences precluded seamless data sharing. Adopting a standard data exchange utility enabled seamless exchange without disrupting the respective applications environments (Figure 4.6).

In the course of this project and other DoD activities, we observed that the legacy systems span technologies that are as old as 40 years. Technologies that are associated with state-of-the-art solution development include:

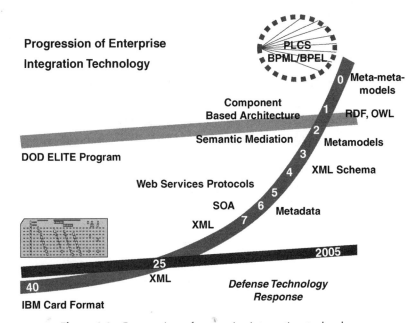

Figure 4.6 Progression of enterprise integration technology.

- ISO 10303 PLCS
- BPM/BPEL
- Metadata management
- Component-based architecture
- RDF/OWL
- Semantic mediation
- XML schema

We recommend that organizations keep a visible record like that shown in Figure 4.7 to track which standards are being used by the organization and to understand where they are on their maturity life cycle.

The solution development activity involved modeling and mapping from the respective business environments of participants to the ISO PLCS standard domain for neutral exchange. A characteristic of the standard is that it is made extensible to users by the use of reference data.

The construct of the international standard used for model-driven data exchange is applicable and transferable to all domains and this is a most powerful finding.

Business Case for Product Life Cycle Support (PLCS) Standard as Enabling Technology for Smart Data Strategy

Resulting from the ELITE experience, we discovered the power of the PLCS standard. Since it is designed to support any manufactured product—the standard is used to describe geometry, physical characteristics, materials, and manufacturing processes, as well as virtually anything needed to support the product throughout its life cycle— our solution for data exchange can be applied to all manufactured products. Therefore

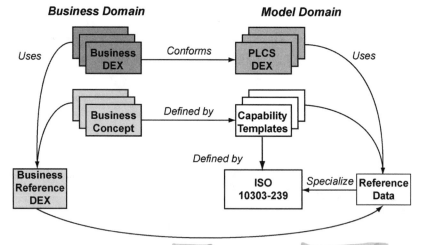

Figure 4.7   Data exchange (DEX) business objects as PLCS submodels.

we advocate consideration of ISO standards as a part of enterprise data exchange solutions aligned with smart data strategy. We were asked to provide a business case for PLCS as a part of the project conclusion and the major points of the case are included in our discussion.

When promoting change and improvement that is characteristically incremental, without a "big bang," it is a challenge to get executive attention to what are credibly "big bang" results and potential. Now hear this! Attention! Admirals, generals, and enterprise executives, you are on a critical path to capitalizing on this important transformational technology.

Your information technologists deep inside your systems development and application zone may insert the technology, but dots must be connected to reach your priority needs. The business case is multidimensional: it begins by addressing tactical requirements but expands to become a strategic asset. The following argument will demonstrate this and why efforts to apply PLCS technology deserve your approval.

Action is needed to kill off the enemy—resource hogging legacy systems and their sustainment, as well as deficient processes. You may not be able to replace all of your legacy systems, but you can leap over their deficiencies. Needed is better technology to enable the exchange of logistics and other technical information among trading partners (e.g., defense customers, prime contractors, and suppliers). Without it, efforts to automate exchange of information in legacy environments are grossly constrained.

Continuing current practices is costly, as evidenced by expensive interface maintenance, high code rework costs, and performance deficiencies in sense making, decision making, problem solving, and planning. Executives and managers simply do not have timely visibility of information describing products and materiel that are subject to deployment, sustainment, and consumption.

In the warfighting environment, improvements are overdue. The technology discussed here is a breakthrough in preventing legacy systems from consuming scarce resources at the expense of immediate performance improvement. Stagnant systems and standing still are not options.

At the top of the chain of command, you are presented with a large number of demands and policy mandates. Your capacity for change and improvement is limited. You must prioritize, assess risks, and make your best resource bets on opportunities that produce timely results with the greatest assurance for success. While CIOs do their best to advise in the information technology domain, operational demands take precedence and may bury your best opportunity.

Our argument for PLCS presents the following:

1. Identify and describe performance improvement opportunities enabled by PLCS.
2. Describe the work and investment needed to improve performance applying this technology.
3. Describe expected break-even and payback characteristics.
4. Identify target problems suitable for change and improvement.

5. Describe specific roles and responsibilities.
6. Describe implementation barriers and how to overcome them.

These accomplishments are supported by lessons learned from the subject experience.

Breakthrough Discovery During Requirements Definition   Problems with current information exchange include the following:

1. There is no cross-military component data integration; that is, communication about aviation maintenance data is via telecommunications and manual processes.
2. Information is exchanged in periodic batches with much manual data entry and file uploading.
3. Information is out of date or nonsynchronous with other databases reporting the same type of information.
4. Interfaces used to link systems are brittle, hard to change, and costly to maintain.
5. Planning, problem solving, sense making, and decision making about products and material are deficient, negatively affecting warfighter wait time.
6. Prime contractors cannot optimally support performance-based maintenance contracting without adequate information from the DoD components about their products and usage characteristics.

Observe that these "problems" and associated "opportunities" from fixing them will be viewed with varying relevance to where you are in the hierarchy of your enterprise, be it defense enterprise, service components, or defense supply chain. The champion for PLCS enabling technology needs to be sufficiently high in the organization to promote optimal benefits through aggregation. Also observe that different functions will have varying levels of affinity with the specific opportunities.

In a real-world instance, a Navy vice admiral needs to reduce the work force by 20%. How can PLCS transformational technology make a difference? The ELITE Program demonstrated that using a model-driven approach to enterprise integration with PLCS, as for the ontology for aviation maintenance data exchange, the work performed by five people can be done as well by two.

In addition, the benefits from speed in automating the associated process resulted in work being done in minutes that formerly took days and weeks. The challenge is getting on the admiral's radar screen with this type of information and to make the case that enterprise integration strategy using PLCS can make a difference. A standards-based approach to information interoperability provides the warfighter with data electronically that does not require human manipulation. Consequently, the OEM community receives timely reporting of parts usage and dependability in support of performance-based logistics. This is a win–win situation for all stakeholders.

Focusing only on the problem within scope of the ELITE Project and contract caused customers and stakeholders to miss even greater opportunity. This is an instance for our advocating higher order smart data strategy and executive attention to areas that have significant positive impact that is enterprise-wide.

We pushed the strategy and enabling technology to a higher level for attention, and to explain and promote understanding for action.

- The data exchange project represented a strategy for defense enterprise integration that is much broader and more encompassing than the subject application.
- PLCS represented enabling technology as a component for improving defense enterprise integration.

Identify and Describe Performance Improvement Opportunities Enabled by PLCS Assistant Deputy Secretary of Defense Materiel Supply & Service Management was the first of our executive customers to connect the dots from PLCS to cutting warfighter wait time through improving materiel visibility. In addition, the UID director represented the DoD weapon system life cycle management pursuit for information interoperability that is the crux for improving visibility into as-built and as-maintained information about weapon systems.

An IT director at Sikorsky was a pioneer in seeing how the enabling technology combined with the model-driven exchange strategy could aid implementation of performance-based maintenance contracts. The Army's IT specialists were pioneers with intense focus on automating operations and improving operational performance from the Army to Sikorsky and from the Army to the Navy.

The Deputy Chief of Staff U.S. Army G4 said in addressing related requirements: "Life Cycle Management Commands (LCMs) and Program Executive Officers (PEOs) must act in concert to promote data interoperability and equipment standardization. A major challenge is to define the best use of UID data to support transformation of Army logistics processes. Marking items is not enough" [9].

Benefits are realized only when all of the technical connections are made with introduction of the common information model called PLCS DEX for aviation maintenance.

To simplify a part of the business case, we developed a summary view as shown in using the "greatest thing since sliced bread" metaphor. For executive management, we backed up the overview with the best facts and information available to support it.

Essentials for making new technology and associated benefits operational include:

- Understanding and adopting a model-driven integration strategy
- Subscribing to open standard solutions (not allowing proprietary impediments)
- Making information interoperability a critical success factor
- Promoting the value and requirement for visibility in whatever functional domain where applied

Stay focused on why we are doing these things:

- Lower operating and sustainment costs
- Less code rework
- Lower interface costs
- Increased agility
- Leveraged legacy systems and assets
- Enabled to performance-based contracting—maintenance and logistics

Observe desirable characteristics:

- Small investment
- Quick payback
- Long-term continuous returns
- Improved joint planning, problem solving, sense making, and decision making

Press for evidence to support these claims. Bear in mind that because transformational technology is new and continuously advancing, it is imperative to frequently verify with evidence of progress. The ELITE Program employed spiral development as the means for accomplishing this. The first spirals of the ELITE Program demonstrated that (1) model-driven data exchange works as a replacement for point-to-point and hub-and-spoke connectivity, and (2) using an international standards-based model for exchange (PLCS) is useful as an enterprise ontology or framework for achieving open interoperability. Open interoperability has its own benefits that are already proved elsewhere.

The Army and Sikorsky offered testimonials attesting to significant improvements that included labor saving and increases in productivity. Additional development spirals included gathering more performance data in cooperation with customer-users.

Describe the Work and Investment Needed to Improve Performance Applying This Technology Implementing a new technology strategy around the use of an emerging technology like PLCS will become easier as new tools to support implementation become available. The ELITE Program provided an opportunity to interact with tool vendors to guide their progress. Today, the degree of difficulty is no greater than any significant effort to produce higher performance. Living with current shortcomings means accepting cost buildup over the long haul. Implementing PLCS holds promise for significantly reducing total IT life cycle support costs with a target reduction of at least 50%.

The PLCS methodology that is a part of the standard addressed the relationship between processes and data. The effort to improve performance using this technology requires higher-level change management akin to business process reengineering or simply business engineering. When conducting pilot projects such as ELITE, we are conscious that we are operating below the radar of the much needed enterprise oversight and context. We addressed the enterprise context in the service-oriented enterprise description that described the paradigm for achieving higher performance from enterprise integration addressing smart data, smart grid, and smart services as core elements.

Organizations approach change from technology insertions in different ways. The ELITE Program revealed that data format clauses in contracts must be considered and sometimes amended to introduce new technology, even when the benefit is readily apparent. Attending agreements may be critical on the path to paying or getting paid. While the Army customers for ELITE were tactically adaptive, the Navy required a more formal mechanism, including memoranda of agreements among participants. Policies and guidelines may govern technology insertion, which must be considered in implementation planning.

Because core PLCS technology is available in the form of a standard aviation maintenance DEX, new implementers will have a foundation on which to develop. While aviation maintenance PLCS DEX is still evolving, it is prudent to follow the following steps:

1. Requirements planning
2. Design/discovery/demonstration
3. Installation and test
4. Accreditation and transition to sustainment

The magnitude of the effort depends on the scope that you determine based on business concepts or process domains where you want to apply it. For example, some propose that if your target is code rework reduction, for every hour devoted to preparing information for automated exchange using this approach, you will save 10 hours of code development and a corresponding 100 hours of rework. Others point to being able to manage information automatically in a timely, accurate, and complete manner that was not previously possible. More will find that reducing interface costs and interface maintenance difficulty is a priority target. Developing the hypotheses for proving it is the responsibility of implementers, as their needs and priorities are business-specific and unique. Metrics like these are meaningful only when actions are taken to realign resources to higher yield and best uses, or eliminate them and their unnecessary cost and time.

Work Requirements

1. *Work to Develop Enabling Technology.* When introducing new technology, you will hear about the effort that was devoted to produce it, and about the effort needed to sustain it. There is no "free lunch." Today, the program developed the first application of PLCS to aviation maintenance that may be applied to all DoD aviation programs from mature programs like the Sikorsky H60 helicopter to the Lockheed Joint Strike Fighter. As the community of participation expands, so will contributions from changes and improvements. Enabling technologies must be maintained,

especially when rooted in international standards. Government and industry must actively maintain PLCS.

2. *Work to Implement Enabling Technology.* PLCS becomes a catalyst for change and improvement in the domain of aviation maintenance. PLCS can be applied to the maintenance of other defense products. In all cases, implementers will (1) examine current processes, (2) examine data exchange requirements, and (3) plan for change and improvement that will affect how work is done with improved automation. Organizations must engineer change and improvement to produce defined outcomes. PLCS is intended to be relatively easy to adopt, but it must be recognized that people and systems will perform differently as a result. Expect to discover significant new things that data users will be able to do that they could not imagine before.

3. *Work to Sustain Enabling Technologies.* As with any standards approach, the community of participants must maintain the standard. In the past several years, new standards and technologies have emerged. The new work force is more likely to be learning the new technologies in place of the old one. Determining the optimal mix of skill, knowledge, and contemporary experience in the work force presents new challenges, although with substantial pay-off for those who get it right.

Describe Expected Break-even and Payback Characteristics   PLCS is a catalyst for change and improvement. To achieve results, it must be accompanied by a strategy, and the technology must be directed to a need or problem area. Aviation maintenance is the target we chose, although we also know that others considered targeting Army land vehicles. Since PLCS is designed with broad applicability and adaptability, it is possible to migrate from one target domain to another.

The way to approach consideration of the business case begins by addressing the following:

1. How we do things now (list the areas of pain—major cost, schedule and risk impacts)
2. Future state vision

Identify Target Problems Suitable for Change and Improvement   The target for implementing a PLCS solution is high-dollar assets (products) that have a long useful lifespan.

Describe Specific Roles and Responsibilities   A prerequisite is including ISO 10303 PLCS DEX as a requirement in contracting language so that industry partners and services must comply.

Describe Implementation Barriers and How to Overcome Them   Gather metrics that substantiate the cost associated with implementing information exchange currently and in the new environment employing PLCS. With continuing demonstration of success, it will be increasingly easier to justify the investment to achieve information interoperability.

Future Considerations

New initiatives like the Joint Strike Fighter Single Point of Entry Program could incorporate lessons learned from the ELITE Program and include application of PLCS. (And they did.)

Case Study: Defense Logistics Backorder Smart Data Application

An example of current problems and deficiencies from poor data strategy can be drawn from the Defense Logistics Agency (DLA) failures to predict backorders. We suggest that the adoption of a smart data philosophy along with an SOE paradigm would eliminate the *bullwhip effect* that has been observed in supply chain management and documented in many industries. It is generally accepted that variability increases as one moves up the supply chain [10].

> *Actual Name.* "Application of Forecasting Techniques and a Naive Bayesian Network to Determine the Feasibility of Predicting Backorder Age, Unfilled Backorders and Customer Wait Time."
>
> *Background.* The customer is the Defense Logistics Agency.
>
> *Goals.* Validate that backorder age, unfilled backorders, and customer wait time can be predicted utilizing correct methods and algorithms such as a naive Bayesian network.
>
> *Decision.* Is there a bullwhip effect that interferes with decision making within the DLA supply chain?
>
> *IT Support.* Provide a naive Bayesian network methodology and algorithm for validation. Help standardize data capture, recording, and processing to prevent propagating these errors in the future.

In the case "Application of Forecasting Techniques and a Naive Bayesian Network to Determine the Feasibility of Predicting Backorder Age, Unfilled Backorders and Customer Wait Time," we divide the variability into behavioral and operational causes and quantify the effect in a multi-echelon inventory system [11].

We propose that the failure of the DLA to adopt SOE principles led to their misunderstanding of backorder inventories. These deficiencies could have been overcome by utilizing smart data wrapped in an SOE paradigm.

Our model demonstrated how behavioral misperceptions of feedback due to information distortion and distorted demand signal processing led to operational time delays and lead-time variability involving forecast errors during replenishment lead time.

We applied these results to the multistage supply chain and demonstrated that there is a reverse bullwhip effect that can be reduced by reducing information distortion, improving collaborative demand planning, using an effective ERP system, and the instituting kanban countermeasures. Implementation of smart data and artificial intelligence methods, within the SOE paradigm, could solve many of these perceived backorder delays in customer wait time.

This case demonstrates that modeling an enterprise, such as the DLA, is not an end game. We show that modeling is a vital step that is made relevant in context with the enterprise engineering discipline and the SOE paradigm that accounts for dynamic relationships between governance, processes, data, and enablement.

The case further illustrates our concern, from an executive perspective, that applying the engineering discipline to an enterprise is a big effort that never ends. Enterprise executives, such as those found in the DLA, are equipped with the disciplines and strategy for optimizing performance. However, they must parse the days such that they are making the right tactical decisions while guiding the ship on the right course. We want executives to embrace the idea that they need an enterprise performance optimization support system that serves them both tactically and strategically.

Our case gives an example of how behavioral perceptions can cloud decision making that would be better served by applying the correct methods and algorithms. We show that a perceptual system is the manifestation of a need for enabling technology that embraces a smart data strategy implementation.

Getting the enterprise performance optimization support system in place requires rapid execution such that its daily service to executives justifies its continuous development and refinement. Our case study illustrates how improvements are needed by the DLA and how adoption of the three legs of the smart data paradigm will help them to improve their processes.

We show that a "war room" mentality is not a substitute for a smart data strategy. We want the DLA to adopt our smart data strategy to encompass a method or algorithm to which they can point and say "this or that is serving me well or not so well," so that IT professionals can address improvements with precision and efficiency.

Algorithms and methods are needed to accomplish these tasks, not a bullwhip effect. While what goes on in the IT background is necessarily complex, we cannot let the complexity interfere with the necessity for executive clarity in executing complex responsibilities. There has been much interest in coordinating these complexities among the manufacturers, distributors, wholesalers, and retailers in the supply chain management.

In this case, we studied the behavioral misperceptions of feedback due to information distortion and distorted demand signal processing from a war room and compared unfilled backorders from operational time delays and lead-time variability involving forecast error, during replenishment lead time. A Bayesian network was employed to show that items perceived to be on backorder were in fact no more likely than any other item to increase customer wait time.

This phenomenon is called the "bullwhip effect" because of the rippling panic that can occur in distribution channels when there is a real or perceived shortage of a product. In industry, such distribution quirks happen frequently, but they rarely trickle down to the consumer level.

These largely psychological panics typically unwind when consumers regain confidence that supplies remain plentiful. However, in a "war room" environment, perceptions of a time delay in unfilled backorder, which do not exist at the operational level, give false feedback that shortages exist, when in fact the operational supply chain is robust.

Because the war rooms are dominated by feedback from high ranking decision makers, these misperceptions, due to distorted signal processing, are elevated to a level that sends a reverse ripple effect through the supply chain. This reverse bullwhip effect leads to "a bull in a china shop" syndrome, where a high-level decision-making consumer sends reverse ripples, upstream through the supply chain. Left unabated, the supply chain will react in a manner that is not indicative of the true picture and show variations between orders and sales that do not exist.

This case study illustrated possible problem types within the DLA. The case illustrated a deficiency in leadership and integration by relying on a war room mentality, whenever an algorithm mechanism shows contrary results.

There was also a deficiency in mission, values, policy, regulations, and rules. The hardware, software, databases, and communications systems of the system were working well; it was the failure of the people, following policies and procedures of the system, to listen to the smart data, which led to poor decision making.

A deficiency in strategy, framework, planning, sense making, systems, skill, proficiency, execution, balance of consequences, tools, equipment, processes, infrastructure, and organization were evident as sources of errors in determining backorder aging. These customer wait times would have been decreased through the various opportunities of better use of capital, better use of people, better use of technology, better use of materials, and better competitive advantage.

The importance of supply chain management (SCM) cannot be overemphasized. SCM is the tracking and coordination of the flow of material, information, and finance in the supplier–customer network. We present an empirical study of the determinants of backorder aging for the Battlefield Breakout Backorder Initiative (B3I), which studies the effects of the flow of material, information, and finance through the main supply chain agents: supplier, manufacturer, assembler, distributor, retailer, and customer. Since supply chains are complex systems prone to uncertainty, statistical analysis is a useful tool to capture their dynamics.

By using statistical analysis of acquisition history data and case study reports, regression analysis was employed to predict backorder aging using National Item Identification Numbers (NIINs) as unique identifiers. Over 56,000 NIINs were identified and utilized for our analysis. The results of this research led us to conclude that it is statistically feasible to predict whether an individual NIIN has a propensity to become a backordered item.

There are two types of applications in an SCM system: planning and execution. Planning applications use statistics and advanced algorithms to determine the best way to fill an order. Analysis of overhead times obtained from each of the suppliers can be used to find the fastest resource for their needs.

On the other hand, execution applications implement the day-to-day activities of the supply chain flows. They track the physical status of goods, the management of materials, and financial information involving all parties. In this case, we will focus on the planning application.

In the B3I case, the DLA needed to promote change and improvement that is characteristically incremental, without a "big bang," however, it is a challenge to get executives' attention unless the results have "big bang" potential. Now hear this, Gump is going to be a shrimp boat captain! Attention! Admirals, generals, and enterprise executives, if you are committed to capitalizing on this important transformational technology that predicts backorders, then it is imperative that you drop your behavioral mental map and pay attention to the smart data in front of you.

Your information technologists deep inside your systems development and application zone may insert the technology, but dots must be connected to reach your priority needs. If you do not heed this advice, then a bullwhip effect will be perpetrated and all reality of the truth on determining causes of backorders will be lost. The business case is multidimensional and begins by addressing tactical requirements for part replacements, but expands to become a strategic asset for predicting backorders.

The following argument will demonstrate why efforts to apply smart data and technologies, such as the PLCS format discussed in the Sikorsky case, deserve your approval for a system that will effectively predict backorders.

Action is needed to kill off the enemy—resource hogging legacy systems and their sustainment, as well as deficient processes. You may not be able to replace all of your legacy systems, but you can leap over their deficiencies. Needed is better technology to enable the exchange of logistics and other technical information among trading partners (e.g., defense customers, prime contractors, and suppliers). Without it, efforts to automate exchange of information in legacy environments are grossly constrained and the causes of backorders will never be determined.

Continuing current DLA practices is costly, as evidenced by expensive interface maintenance, high code rework costs, and performance deficiencies in sense making, decision making, problem solving, and planning. Executives and managers simply do not have timely visibility of information describing products and materiel that are subject to deployment, sustainment, and consumption, in order to predict backorders.

In the warfighting environment, improvements are overdue. The technology discussed here is a breakthrough in preventing legacy systems from consuming scarce resources at the expense of immediate performance improvement. Stagnant systems and standing still are not options. Bullwhip effects will not be tolerated.

These changes can only occur at the top of the chain of command, and these executives are presented with a large number of demands and policy mandates for which their capacity for change and improvement is limited. These decision makers must prioritize, assess risks, and make their best resource bets on smart data opportunities that produce timely results, with the greatest assurance for success, without relying on behavioral war room hysterics. While these decision makers do their best to advise the supply chain information technology domain, operational demands take precedence and may bury their best intentions.

A supply chain is a complex system that is prone to uncertainty. It is natural that statistics will play a major role in its study. Statistical analysis is also a handy tool that captures the SCM dynamics. Many researchers have used empirical studies and algorithms to unravel the different distribution, planning, and control aspects of SCM.

These approaches include stochastic, fuzzy models, multi-agent models, and subsystem models. Optimal inventory modeling of systems using multi-echelon techniques have also been employed to study backorders.

These studies have provided useful insights with different sensitivity and reliability. Intelligent multi-agent systems, using distributed artificial intelligence, could be built into the planning application to provide an intuitive and exploratory tool that can be enhanced as needed. Decision makers can also use it for a quick overview of the supply chain process and to investigate supply chain issues through independently constructed modules and scenarios.

Ideally, the topology of a supply chain is defined by nodes and arcs. The nodes or entities interact with each other through the arcs or connectors as an order is fulfilled in a supply chain. Each node performs five main actions with regard to the order life cycle: creation, placement, processing, shipping, and receiving. The order has to be created by an inventory policy or customer request. This creation is made known to the supply chain by placement.

There may be processing delays with the order transport. The order has to be shipped to the origination node, and, when it is received, the order is consummated. Statistical analysis can help to model inventory and demand management. It will integrate such actions by investigating the inventory and demand processes owned by the nodes that stock items.

This design will provide much flexibility in understanding component stocking through the supply chain. The ultimate planning system may be run on a server or a PC, since web-based planning applications have become a prominent source of software deployments. The planning application domain is shown in Figure 4.8.

The importance of supply chain management (SCM) cannot be overemphasized. SCM is the tracking and coordination of the flow of material, information, and finance in the supplier–customer network. We present an empirical study of the determinants of backorder aging for the Battlefield Breakout Backorder Initiative (B3I) that studies the effects of the flow of material, information and finance through the main supply chain agents: supplier, manufacturer, assembler, distributor, retailer, and customer.

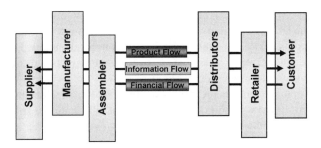

Figure 4.8   Planning application domain.

The Defense Logistics Agency (DLA) seeks to resolve a continually growing list of problem parts for the war effort and military readiness. Despite the many reasons a part may have problems, a streamlined method of resolving these issues must be quickly developed and implemented in order to provide needed parts to our fighting men and women. A proactive approach needs to be developed to reduce the long-term costs of maintaining readiness of DLA supported weapons systems.

The DLA and the military services employ a multi-echelon supply system to provide items to the operational community. When this supply system is unable to satisfy a demand, a backorder is created, indefinitely extending customer wait time (CWT) for the item.

Backorder age is affected by many factors associated with the item's demand and procurement history and by past support funding and the economics of suppliers. To improve the responsiveness of the supply system, DLA is seeking innovative, nontraditional ways of more quickly resolving backorders (thus reducing CWT).

The DLA implemented a "war room" concept at its defense supply centers and created the Battlefield Backorder Breakout Initiative (B3I) to research, develop, test, and evaluate policies, procedures, processes, tools, and methodologies to reduce the backorder aging problem for critical warfighter supply items.

Additionally, the services and DLA seek ways to analyze supply chains as well as factors affecting supply and demand to predict and address potential backorder problems before they occur. As part of the DLA initiative to analyze the supply chain to address the creation of supply backorders, we developed a mechanism that can be embedded in the requisition process to identify items of supply with a potential or propensity to become backordered, even before orders are placed.

With this type of predictive tool available to them, DLA personnel would be able to move more efficiently and effectively to address the problem of backorder creation, thus freeing up critical items of supply to the warfighter.

The DLA supports the military services by providing a wide range of food (subsistence), medicines and medical equipment (medical), construction and industrial materials, fuels and lubricants, clothing and other fabric-based items (clothing and textile), and repair parts for the various weapon systems and associated equipment. When a military organization needs items of supply, it requests (requisitions) the items from the DLA.

The DLA supports such requests by procuring and storing items in defense depots (stocking) and then shipping items from the depots (issuing) to the organization requesting them (DLA direct) or by letting contracts with vendors who, when instructed by the DLA, ship directly to the organization requesting the item (Customer direct). In doing its job, it manages over 5.2 million items of supply. Daily, the DLA responds to over 54,000 requisitions and lets over 8200 contracts and contract actions, including delivery orders and modifications [12].

In order to manage the procurement, stocking, and issuance of these 5.2 million items, the DLA employs 21,000 people and uses an information technology system known as its Business System Modernization (BSM) system. BSM handles the full range of DLA business functions from procurement, to inventory management, to the financial aspects of its operation. The DLA is not able to stock all items, and those items it does stock (or items for which it sets up contracts for direct delivery) may not be fully stocked at levels that will satisfy all requisitions within the time frame requested by the customers.

This situation may be due to funding constraints, but it may also stem from the inability to accurately forecast/predict the demand for items. When the DLA experiences stock outages—that is, when a customer requisitions items from the DLA and the DLA is unable to fulfill all or even part of the requisition—the requisition is labeled as being "on backorder" or in backorder status, increasing customer wait time (CWT).

The number of backorders is one of the key metrics used by the DLA to measure the performance of the order fulfillment aspect of its system and is monitored by the highest levels of management. Backorders and backorder reduction have been and still are a focus of attention by DLA management.

The total time necessary to procure and receive/ship items can be broken down into two component times: the time needed to procure the item (administrative lead time—ALT), and the time required to produce and deliver the item (production lead time—PLT). Accurately knowing/predicting total lead time (the sum of ALT and PLT) is as important to successfully maintaining inventory as predicting demand. This is a pilot outcome study, involving exploratory research.

Therefore the PHDM/RHDM data should provide an adequate sample size and sufficient power, for a multivariate regression test, on backorder aging. Here, we will identify a set of factors that may be used to forecast backorder aging root causes as measured in days. Using field data on backorders from various sources and the multiple regression model, we will empirically test the impact of possible variables such as the ALT variable that involves time to award the contract and the PLT variable that leads to delays in vendor delivery.

Possible independent variables that may be investigated include sole source and limited competition, type of material, delinquent and terminated contracts, never bought before materials, tech data availability, last time bought/number of days ratio, proportion of replenishable to nonreplenishable items, organization by quantity ordered, price, purchase request/buyer ratio, planned backorder, delay in issuing items, the vitals priority score, and micropurchases.

Drawing on the supply chain management (SCM) literature, we deduced that these independent variables may be used to predict backorder aging in days, as a dependent variable. We expect our results to indicate that a relationship exists between the application of the independent variables and the age of the backorder.

Our approach to the ongoing process of data analysis and development of a backorder and a CWT predictive mechanism may best be described as employing the spiral development process (SDP) model. The SDP model is a development process combining elements of both design and prototyping-in-stages, in an effort to combine advantages of top–down and bottom–up concepts.

It employs an iterative, incremental change approach. Also known as the spiral life cycle model, it is a systems development method (SDM) used in information technology (IT). This model of development combines the features of the prototyping model and the waterfall model. The SDP model is intended for large, expensive, and complicated projects. We believe that development of a predictive mechanism is technically feasible, and in fact we have identified more than one possible mechanism to accomplish the project goal.

The complexity of the SCRI tasking demanded a superior analytical approach, first to investigate the feasibility of a predictive model, and later, to explore avenues that would lead to the development of that model. Throughout the duration of the project, we followed a rigorous analytical approach in a number of ways. First, we utilized a relevant literature review and used structured interviews and brainstorming with DLA experts to hypothesize the major causes of backorder aging. Data was gathered, scrubbed, and analyzed using appropriate statistical techniques generally accepted in the industry, such as multiple regression analysis, to determine probabilities.

Using sensitivity analysis and Ishikawa diagrams, we investigated, conceptualized, and brainstormed in order to determine the component elements of ALT and PLT. These components include controlled inventory, Acquisition Advice stock codes (AAC), Acquisition Method Suffix Code (AMSC) Procurable Codes, Precious Metal Indicator Codes (PMIC), Tech Documentation Codes, Item Standardization Code (ISC) Categories, Water Commodity Categories, Controlled Inventory Item Code (CIIC) Categories, and Criticality Categories.

By training with that data, and investigating some baseline descriptive statistics, we discovered retrospective patterns in the data that support the hypotheses that long ALT and long PLT lead to increased birth to ship dates. Once the feasibility of integrating data to develop a predictive mechanism was established, more sophisticated methods were required to provide actual prospective forecasting of backorder aging, employing B3I data.

In order to accomplish this task, it was necessary for the subject matter experts to develop discrete numeric variables from the qualitative string characters present in the data set. First, logistic regression analysis was employed to find these prospective relationships. Second, in a later iteration of the SDP, Bayesian networks utilized the discrete variables to determine the probabilities of these factors impacting birth to last ship date, customer wait times, and logistic response times. Another iteration of the SDP may allow for the triangulation of the Bayesian probabilities and qualitative data in order to develop an expert system for B3I and backorder aging prediction.

The SDP of planning, analyzing, designing, and implementing a solution can be seen in our mechanism development methodology. We followed an iterative, incremental method of literature review, conceptualization, hypothesis development, descriptive statistical analysis, retrospective regression analysis, utilization of discrete variables to run logit, and determination of prospective Bayesian network probabilities.

The next step in the SDP is to utilize this information in order to develop an expert system. These methods were chosen for the reasons described above. Even with improved demand forecasting as a result of CDP and the concurrent ability to reduce backorder age caused by inaccurate forecasting, the process could be enhanced even further with the development of a process/mechanism for identifying which items have a probability of causing significant backorder age. Several successes can be drawn from the analysis of the PHDM and RHDM databases.

Descriptive statistics provide several interesting results. For example, the combined RHDM and PHDM datasets contain 56,052 birth date to last shipment (BLS) entries. There are 1833 BLS entries that are over 360 days and 48,136 BLS entries that are less than 90 days. It can be seen that there is a statistically significant difference between the means for less than 90 days and greater than 360 days for quantity ordered, administrative lead time, production lead time, and unit price of the NIIN.

These conclusions suggest that the profiles of the less than 90 day sample and the greater than 360 day backorders are distinctly different. A majority of the AAC groups fall into the D, H, J, and Z categories for both <90 and >360 days. For both categories, groups 1 and 3 account for a majority of the AMC groups, whereas C, G, H, and P are the major AMSC groups.

Discriminant analysis also provides some interesting insights. For example, ALT, PLT, unit price, and quantity ordered correctly classify ACC groups D and H approximately 77% of the time. The overall model is significant at $p = 0.000$. Therefore if the ALT, PLT, unit price, and quantity ordered are known, the buyer can use this model to predict if the item is stocked or nonstocked and order accordingly. ANOVA for the model is significant at $p = 0.000$. Regression analysis supports our hypotheses that there is a possible relationship between ALT, PLT, and ACC code with birth to last shipment date ($p = 0.00$). There is also support for the hypothesis that suggests there is a possible correlation between unit price and birth to last shipment date ($p = 0.005$).

However, no relationship was found between quantity ordered and birth to last shipment date ($p = 0.957$). Since the slope of the curve is positive for PLT, ALT, and unit price, it can be concluded that an increase in any of these three variables will lead to increased birth to last shipment time, and vice versa. AAC codes D and H (stocked) have an inverse relationship with birth to last ship date. Therefore any D or H item that is in stock takes less time to ship than those that are not stocked.

SOE Paradigm Is Main Contention

One of the main contentions of our book is that the SOE paradigm was advanced as a best-practice approach to achieving results from enterprise integration, placing enterprise context at the forefront and endpoint of all enterprise integration initiatives. SOE is presented as a replacement for traditional systems integration approaches, such as the one employed currently by the DLA, that are too often narrowly focused, and from which results are functionally siloed, rarely producing the magnitude of improvement planned, and costing far more than expected.

The SOE initiative addresses the causes and the remedies and identifies what actions and elements can improve optimizing enterprise performance with computer automation. In our example, the war room mentality of the decision makers led to a bullwhip effect that hid the real issues affecting the supply chain.

SOE was developed in concert with a subset of U.S. Department of Defense (DoD) customers who manage the most complex enterprise in the world. Because optimizing performance in the defense enterprise is dependent on supply chains and complex relationships between government customers, contractors, subcontractors, and suppliers, this may be the source of best practices in addressing associated problems and needs. One limitation of this study was the sample size.

While the original set of 11,000 real-world backorders appears to be a fairly large sample size, it only represents a subset of all the possible combinations of known variables that may be observed in the real world. Increasing the sample size to include more real-world projects will leave fewer missing values in the conditional probability tables.

As more projects become available in future, researchers may be able to obtain more complete probability distributions of the birth to last ship date variable. Principally, the concept of backorder, as currently used by the defense logistics community, is vague and provides for no differentiation between temporal backorder and B3I war room criticalparts.

Under the B3I initiative, if the lack of a critically needed part is hindering military operations, the defense logistics community is mobilized quickly to procure, manufacture, and deliver the part. However, many of these B3I parts are not temporally backordered—that is, they do not have long lead times. It is not that the procurement system and industrial base are functioning abnormally; rather, in the judgment of officers involved in ordering parts identified as B3I parts, the parts were simply critical enough that they required additional expediting and due diligence.

Temporal backorders, on the contrary, are parts that have long administrative and/or production lead times. This project addressed a problem of estimating birth to last ship date distribution for backorder aging. Future studies may endeavor to add qualitative data for richness. Gathering more recent data may also prove useful to understanding the backorder phenomenon.

The use of nonparametric methods such as neural networks, tabu search, genetic algorithms, Bayesian networks, Markov processes, simulation, ant colony optimization and particle swarm applications may also provide insights when investigating the B3I phenomenon.

4.2 CALIBRATING EXECUTIVE EXPECTATIONS

Messages to executives from this experience include the following:

- Executives must oversee technology-driven improvements to maximize benefit, since there is no auto pilot.
- The presence of data strategy can provide needed detail direction in the absence of executive hands-on involvement, but this is not a substitute as both are needed.
- Activities and projects that are narrow in focus may harbor benefits and solutions that are broadly applicable.
- Exploiting benefits from local and specific applications requires executive insight, foresight, and leadership so they do not become latent, underutilized, or lost opportunities.
- Formalize strategy to guide and direct implementation activities, as this focuses improvement behavior.

- Fund improvements commensurate with proofs-of-concept and follow-through.
- Technical data exchange can be accomplished via model-driven capability with expected and advertised benefits.
- Special contemporary skill, knowledge, and experience are essential to achieving technical solutions.
- Technology is a catalyst and complementary with executive and managerial leadership to optimize performance.

Thinking About Federal Procurement Through the Lens of Smart Data Strategy

The U.S. government "procurement problem" has caught the current president's attention as he and the administration believe more competition and accountability are needed. From the view of the president, one aspect of the problem is too much outsourcing to contractors whereby they are performing work that the president believes should be done by government workers as contractors are too expensive and perform without sufficient integrity, oversight, and authority.

One question the executive might ask is, What work should be performed by government workers and what should be outsourced to contractors? What are the business rules or policies governing this decision?

Does the workload support career staffing, or does workload rise and fall such that contractors provide a more flexible means of getting work done? What is the government's capacity to employ workers in a deep economic recession?

In our terms, does the government have sufficient capacity to process and manage the volume of procurements with government staff, given the policy direction that government professionals are more desirable than contractors to perform this work?

Can smart data be employed in context with automated contracting technology to improve reviews and focus more efficiently on prospective problems? Can contracting problems be predicted based on an analysis of past performance? What procurement attributes might indicate potential problems such that they can be prevented or mitigated in a timelier manner?

There are other aspects of the problem. According to a *Washington Post* article [13], Frank Cramm, a senior economist at the Rand Corporation, says that "most major weapon systems tend to be 10 to 15 years old" by the time they make it through the cycle and into deployment. When we hear this, we don't know if this is good, bad, or indifferent because there is insufficient data to form a conclusion.

In our approach, we are at square one in the six by six matrixes, problem analysis. What is the problem definition? What data is needed to support the findings bounded by the problem definition?

By the time our book is in print, the administration will have launched an investigation into the procurement activity, as has every previous administration. Would starting this activity with a smart data orientation make any difference?

We believe that the focus would be more precise and the pursuit would be more objective, at least. Getting the problem definition right is very important to avoid the investigations being endless, actionless, and otherwise fruitless.

Defense Acquisition Subset

One of the most perplexing challenges and problems in the U.S. government is called defense acquisition. Every federal department in the government has an acquisition process, and all have common regulations imposed in the Federal Acquisition Regulations (FAR) and in various directions from the Office of Management and Budget as well as from the congressional process for planning, funding, and authorization.

Because the defense budget is the largest, it is a very big target for discussion and example.

The Government Accountability Office (GAO) is the audit, evaluation, and investigation arm of the U.S. Congress and is part of the legislative branch of government (formerly known as the General/Government Accounting Office).

The GAO is headed by the Comptroller General of the United States, a professional and non-partisan position in the U.S. government. The Comptroller General is appointed by the President, by and with the advice and consent of the Senate, for a 15-year, non-renewable term. The President selects a nominee from a list of at least three individuals recommended by an eight member bipartisan, bicameral commission of congressional leaders. The Comptroller General may not be removed by the President, but only by Congress through impeachment or joint resolution for specific reasons (*See Bowsher v. Synar*, 478 U.S. 714 (1986)). Since 1921, there have been only seven Comptroller Generals and no formal attempt has ever been made to remove a Comptroller General. The long tenure of the Comptroller General and the manner of appointment and removal gives GAO a continuity of leadership and independence that is rare within government [14].

On March 3, 2009, we read a story entitled "On Pentagon Waste: Just Give It to Us Straight—Often, Bureaucracy-Speak Masks the Real Story," by Anna Mulrine. [15] Her story was interesting because she focused on the phraseology of Washington bureaucrats who are simply not straight talkers. Sometimes it can be hard to get the point through the Washington bureaucratese. She tells the story wonderfully and we quote her.

Take, for instance, a new Government Accountability Office report on Pentagon overspending presented by Michael Sullivan, the GAO's director of acquisition and sourcing management, in testimony before the Senate Armed Services Committee on Tuesday.

The GAO study of defense weapons acquisitions finds that the Department of Defense "commits to more programs than resources can support."

Translation: It doesn't say "no" often enough.

What's more, the report notes that the DoD "fails to balance the competing needs of the services with those of the joint warfighter."

Translation: The services do not coordinate their needs, so they often duplicate systems and run up costs.

There is a body—with the fantastically bureaucratic name of Joint Capabilities and Integration Development System (JCIDS)—that is supposed to prevent such duplication, but it reportedly approves nearly all of the proposals it receives from the services rather than prioritizing them. The services are supposed to coordinate with each other, but, according to the GAO, nearly 70 percent of the time they don't.

The study, released today, points to the Pentagon's "acceptance of unreliable cost estimates based on overly optimistic assumptions."

Translation: Defense officials wink at unrealistic price tags for complex weapons systems, leaving their successors—and taxpayers—to pay the full costs sometime in the future.

Moreover, the study adds, "DoD officials are rarely held accountable for poor decisions or poor program outcomes."

Mulrine reports further:

The GAO study concluded that the Pentagon must more effectively resist "the urge to achieve revolutionary but unachievable capabilities," instead "allowing technologies to mature in the technology base before bringing [them] onto programs, ensuring requirements are well defined and doable, and instituting shorter development cycles."

In other words, they need more realistic goals with shorter timelines.

This in turn would make it easier to estimate costs accurately and, better still, keep them from skyrocketing out of control.

We are talking about the DoD's program being $295 billion over the original cost estimates, with programs running almost two years behind schedule. The DoD gets less than what it contracted for.

Mulrine continues:

Jacques Gansler, the chairman of the Defense Science Board's Task Force on Industrial Structure for Transformation, testified that the Defense Department needs more acquisition officials who are qualified to review defense contracts. Currently, he said, they are short-staffed and their importance to the Pentagon's workforce is undervalued.

This, Gansler added, introduces opportunities for fraud, waste, and abuse. He testified that there are currently 90 acquisition fraud cases under review from the war zones.

From this it sounds like the answer is to throw more analysts and investigators at the problem. While the work load has surely increased and may justify adding to the investigative resources, we are concerned more about predicting problems and preventing them before they materialize. This can be accomplished by (1) process reengineering and (2) smart data strategy.

Integrated Defense Acquisition, Technology, and Logistics Life Cycle Management Framework

We are looking at an 11in. × 17in. color chart with the same title as this section. It is too much to include in the book, and the subject is too broad to address comprehensively as it would take another book dedicated to it. However, for the purpose of applying smart data thinking as we have discussed, let's consider it as a potential application.

Say President Obama calls up with Secretary Gates by his side and says: "We want to do something significant with Defense Acquisition because the process is killing us, eating up precious resources, and not delivering support to warfighters as we want. How would you address the problem and our need to produce meaningful results in two years?"

We might reply: "Thank you, Mr. President, and we are glad you are giving us 24 months because Phase 1 will be problem analysis and definition and Phase 2 will be strategic action, each of which are 12 month segments."

Using the chart before us, a 7 column by 3 column matrix with granular subdivisions and block diagram enhancements (sounds like Arlo Guthrie in Alice's Restaurant), here is our approach for the purpose of highlighting value added difference from smart data thinking. We are referring to a chart used by the Defense Acquisition University to explain the Acquisition, Technology, and Logistics process.

Executive-Driven Request   We have established that the request is from the Chief Executive and that is progress.

What does the Chief Executive really want to know?

The planning, programming, budgeting, and execution process is described as biennial calendar-driven. This means that there is an "on-year" and "off-year," a planning and budgeting process and a tweaking process, respectively.

It starts with the White House National Security Strategy. When the new president arrived, he inherited the situation and his first year is an "off-year." He provides fiscal guidance based on his strategy and based on events on the ground—wars and economic meltdown.

Nothing works without capital and since the capital system is broken, fixing it takes priority. However, certain assumptions must be made about the probability that fixes in motion will succeed. Contingencies must be planned in the event that they don't fix the problem.

Calculations of this nature are made by the Office of Management and Budget (OMB) and by the Congress and must be reconciled to produce a more certain prediction. This is a fluid process, although estimates must be frozen with some certitude for the process to continue with credibility.

The Office of Secretary of Defense and the Joint Chiefs of Staff develop a National Military Strategy and Joint Planning Document as input into the Strategic Planning Guidance which is used to formulate the budget (POM) and all of this begins in January and concludes by August.

Between September and February, there is a reviewing and budgeting process that results in the DoD budget being submitted to the OMB in January and becoming the President's budget for submission to Congress on the first Monday in February.

Once in Congress, the budget rolls from the Budget Committee to the Authorization Committee to the Appropriations Committee with various testimony and appeals and resulting in appropriations and allocation.

That describes the legal process, though underlying this is the Defense Acquisition System. Herein is a complex system for which the model before us identifies inputs and outputs as we prescribe, and with accompanying detailed activities. As a control on the Defense Acquisition System is the Joint Capabilities Integration and Development System (JCIDS). As presented in Mulrine's account, JCIDS is criticized for (1) giving everything a pass and (2) not achieving its mission to prevent duplicate programs among the military services. From a cursory review, these things happen because (1) there is simply insufficient time for JCIDS to perform its duty, and (2) military services have already determined how to game the system. Most important, however, is that the senior executive must insist on a different outcome, and that is mission critical.

Since the process is already modeled, the path to optimizing performance of the Defense Acquisition System must include determining performance metrics for the current process. From appearances, it seems that there is an abundance of base-touching in a short period of time with insufficient and deficient resources to perform required work. That observation is substantiated by Dr. Gansler as Mulrine reported.

Applying the smart data strategy would minimally ensure that all of the performance data is accessible, addressable, and transparent. Full attribution of the current system will likely reveal opportunities for improvement that will first appear as trouble symptoms, which must be validated and refined into specific performance problem descriptions.

Apply Professional Disciplines   The proper disciplines must be deployed and applied in the correct combination to determine such things as the following:

1. Are the processes being performed in a timely manner such that outputs from one activity are entering dependent activities with sufficient time to process? (Note that the IDEF0 modeling technique does not answer states and timing questions, and therefore a specific modeling technique is applied for that analysis.)
2. What is the human resource profile and footprint supporting activities? Answering this requires job descriptions and staffing profiles including labor costs.

3. What is the human resource profile of executives and managers overseeing the work? Is it sufficient?

Joining the top–down planning, there is a culture of bottom–up, warfighter-driven requirements planning that must be factored as an essential component. The process is also described as evolutionary and dynamic as technology continuously improves along the end-to-end acquisition process. There is a strong predictive element in this planning based on global threat scenarios as well as economic and technical scenarios.

Performance Optimization Process For the DoD, much of what we suggest as the performance optimization process is in motion such as processes being modeled and information inputs and outputs being accounted for. It appears that the addition of a formalized smart data strategy can help organize and advance the initiative. Moving the data issues closer to executives and their decision support is an improvement.

Needed is a technical review to weed out data exchange and management technologies that are not the best technologies available for today's environment.

1. *What are the questions and what are the data?* Describing the Defense Acquisition System with full performance metric attribution, and displaying it as a real-time system for management use, including modeling and simulating scenarios, would be a major improvement.

2. *Can data be engineered to make a difference?* Data can be engineered such that it is fully exchangeable throughout the system, leveraging ISO model-driven data exchange. Such improvement can not only reduce performance costs but can greatly improve management control and responsiveness in an exceedingly difficult to manage global environment.

4.3 FIVE YEARS FROM NOW

There is a struggle underway at present between technology vendors that want to control the customer environment with proprietary solutions and customers that need and want independence via interoperable and open systems. Because of the depressed global economy and consolidation among leading technology vendors and systems integrators over the past 10 years, this leaves enterprises with fewer alternatives. Perhaps a resurging economy will turn this circumstance around with enterprises leading the way with greater control over their destinies.

Globalization forces enterprises to accommodate greater diversity in languages, rules, and cultures such that technologies like semantic engineering become most relevant. Needed are better tools for data engineering—better metadata management, better mapping, better data exchange utilities.

In the simplest terms, we have discussed that having a model of all of the major processes in the enterprise is essential to precise management. Having an enterprise data model depicting all of the data that fuels processes and that are outputs from processes is essential to precise management. Being able to associate specific data to specific processes is essential to precise enterprise performance management. Understanding the impact of business rules on processes and data is essential to precise enterprise management.

We reviewed our progress with Dan Appleton and he suggested that we consider a paper written by Mark Fox and Michael Gruninger called "Enterprise Modeling" that was published by the American Association for Artificial Intelligence nearly 10 years ago [16]. "Why a 10 year-old paper," you might ask? Sometimes work advances in spurts, and we think we have been marking time for 10 years when it comes to executives attending the subject. So we will review what was said with the intent to see what needs to be updated and to share wisdom that should be preserved and applied.

Enterprise Modeling

Here is an analysis of the subject paper as we were in communication with Mark Fox, Professor of Industrial Engineering at the Department of Computer Science at the University of Toronto.

Validating some of our ideas, the paper begins by acknowledging the need for competitive enterprise to be exceedingly agile and integrated. Our difference on this view might be with the term "integrated" as we have discussed because it sometimes implies a degree of rigidity that conflicts with the notion of agility. The article intends to promote the need for enterprise models and introduces the ideas that models may be generic and deductive.

A generic enterprise model implies that businesses of a certain type, such as manufacturing, for instance, may appear to have a common construct shared among others in the same industry classification. A deductive model is one that is derived from analysis and reasoning about how the enterprise operates. We will consider these ideas further in this discussion as does the paper.

Something striking occurred to us as we considered the notion of generic models: enterprise resource planning (ERP) software packages that have been popular for the past 10 years are based on the idea that industries of common types basically perform the same way, or furthermore, that they should perform the same way as software vendors develop solutions that are based on best practices. Then again, one must ask: Does not all of that "sameness" undermine the notion of competitive advantage? Wherein does competitive difference appear?

To answer these questions, one must (1) have a model of the enterprise processes fully attributed with inputs, outputs, control, and mechanisms; (2) have an enterprise data model; and (3) have additional information about work flow and timing states. Herein, we are in agreement with Fox and Gruninger.

According to the authors, "an enterprise model is a computational representation of the structure, activities, processes, information, resources, people, behavior, goals, and constraints of a business, government, or other enterprise." We surely agree with that definition but are attracted to the term "computational."

Tom Gerber wrote an entry for *Encyclopedia of Database Systems*, [17] called "Ontology," in which he offered synonyms: "computational ontology, semantic data model, ontological engineering" [18].

> Ontologies are typically specified in languages that allow abstraction away from data structures and implementation strategies; in practice, the languages of ontologies are closer in expressive power to first-order logic than languages used to model databases. For this reason, ontologies are said to be at the "semantic" level, whereas database schema are models of data at the "logical" or "physical" level. Due to their independence from lower level data models, ontologies are used for integrating heterogeneous databases, enabling interoperability among disparate systems, and specifying interfaces to independent, knowledge-based services. In the technology stack of the Semantic Web standards, ontologies are called out as an explicit layer. There are now standard languages and a variety of commercial and open source tools for creating and working with ontologies [17].

Inserting this reference illustrates how the same adjectives are applied to describe different aspects of the subject where the first instance describes enterprise models and the second instance dips into database and data processing constructs, and both subjects are relevant to the discussion, although more to the information technologist than to the executive.

Returning to the question about enterprise models and their commonality, Fox and Gruninger pose the following questions: "Can a process be performed in a different way, or can we achieve some goal in a different way? Can we relax the constraint in an enterprise such that we can improve performance to achieve new goals?" Such questions illustrate the need for enterprise management to be constantly thinking and challenging the models before them in pursuit of continuous improvement.

From this, of course, one must assess the impact of changes, and as we have illustrated changes ripple through the model and eventually are reflected in data. Operationally, management can perform "what-if" analysis and present different scenarios.

Fox and Gruninger say that a principal motivator is to integrate the enterprise for the purpose of achieving agility. They cite Nagal and Dovel's four principles of agility: (1) Use an entrepreneurial organization strategy, (2) invest to increase strategic impact of people and information on the bottom line, (3) use the organization strategy as a dynamic structure both inside and outside the enterprise, and (4) adopt a value-based strategy to configure products and services into solutions for your customers. They follow this up with a dose of business process reengineering (BPR) talk from Hammer and Champy.

We're not jaded in this review as we have been there and so have many of our readers.

Federal Government Information Technology and Smart Data Alignment

Federal government practices combine with commercial technology vendors and commercial enterprise to produce a composite view about the alignment of current practices and trends with smart data. To determine the current view of federal government information technology and alignment with smart data, we surveyed the federal CIO and DoD CIO websites. We also searched our three major subjects and subordinate subjects to produce a composite view from that of the CIO with the assumption that information technologists exist for the purpose of supporting the executive in optimizing enterprise performance.

Much of the federal CIOs' focus is lost in activities that are remote from the primary mission, or too deep in the technology layer to extrapolate significance. It is not the aim of this book to address this finding, however. The circumstance begs for executives to assert leadership over the technology community that focuses attention on smart data as a place to begin and not get lost in the infrastructure.

Technology push comes from advanced technologists working with things like the semantic web and vendors pushing middleware, applications, and hardware infrastructure. It is difficult at times for enterprises to escape vendors' technology push. Executives must take an active role in setting priorities, to stay focused on primary operational outcomes.

Our view is that pursuing the semantic web and related visions is a peg for the next 5 years, and that immediate focus should be on streamlining the enterprise to address primary requirements to achieve and sustain competitive superiority. Our smart data strategy provides laser focus.

Federal CIO Topics

We are examining federal CIO topics from the enterprise executive viewpoint. As an executive needing to optimize performance in today's environment, cash is king because capital is scarce. Taking care of customers and anticipating and attending customer needs in a superior manner is essential to keeping them, and sustaining the enterprise in hard times. In addition, accounting for enterprise assets and protecting them are essential. Reducing IT expenditures and ensuring high service levels is a priority challenge.

Using the approach that we recommend for executives in assessing enterprise performance, consider that the topics addressed by the CIOs are enabling technologies. Technologies support business processes and determine how data is processed and exchanged in various ways.

Omitted from the federal CIO website's "featured content" is (1) explicit attention to enterprise data and (2) explicit discussion about enterprise executives user needs. Those are pretty serious omissions.

Cloud Computing

Federal CIOs care about "cloud computing" as that is the featured news on the Federal CIO Council website. Therefore we will investigate key ideas to determine if and how they relate to smart data thinking. For the executive, cloud computing is an infrastructure and enterprise services alternative akin to outsourcing, but more like subscribing to IT capabilities as a utility and paying for what you use.

From a smart data perspective, we are evaluating how the cloud computing alternative can deliver data to the executive more effectively and efficiently, how the alternative supports enterprise interoperability such that data exchange is frictionless, and how enterprise assets are protected in the process.

The "cloud computing" briefing on the website is presented by Hewlett Packard (HP), a vendor, with Rick Fleming being the HP cloud computing practice lead. Fleming implies "the cloud" is the next generation beyond "the Web," and that we should have our heads in the cloud about now as we transition from "information and e-commerce" to "virtualized services."

The smart data paradigm identifies smart services as an element of the service-oriented enterprise architecture and we are talking about the same thing. The briefing says cloud computing is "a pool of abstracted, highly scalable, and managed infrastructure capable of hosting end-customer applications and billed by consumption," citing Forrester Research for this definition. As with all paradigms, cloud computing will have different meanings to different members of the user community.

The presentation says that technology over the Internet will shift from existing applications and infrastructure to contract-based consumption. That is another way to say outsourcing.

In our smart data description, we suggested that certain services might best be provided by a utility, such as interoperable data exchange.

In contrast, we suggest that the user community, probably led by the principal head of the value chain, collaborate with representative members to determine a smart way to exchange data. Our model suggests that (1) the user community will determine an open and interoperable standard that will serve as a neutral model for exchange, (2) data exchange servers either centralized or dispersed among the community will enable data exchange that is independent from legacy applications and infrastructure, (3) the user community will employ semantic mediation technology to accept diverse lexicons, (4) a common system of configuration management and collaboration will be employed to support the user community, and (5) based on a value set by the head of an enterprise, there will be governance control (all or shared) over the technology enabler.

We don't suggest a particular set of vendors to support the concept. It is possible that government customer-led virtual enterprises might determine that a commercial and government utility is most desirable. A commercial services utility might emerge to sell data exchange services to various enterprises, commercial and government. Most important is that the solution should evolve from customers' requirements versus vendor push.

What distinguishes the application of utility services are the decisions that users make about the international standards that are most applicable to their type of enterprise.

The shift, as envisioned by HP, is from an enterprise services model to a global services model, whereby the bandwidth and scale of the global provider would dwarf that of an enterprise, which would be one of many customers. From a vendor viewpoint, HP suggests the shift is from a cost center to a value/revenue center concept. This thinking is consistent with our notion that data are assets and data products are value-added assets.

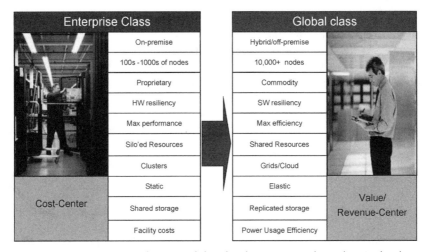

Figure 4.9   Massive scale-out and the cloud. (Courtesy of Hewlett Packard.)

Note in Figure 4.9 the HW to SW resiliency, siloed to shared resources, and clusters to grids/clouds (suggesting that cloud may replace the grid notion).

When researching, we discovered the following comparison that negates equating the cloud with the grid. Grid computing is a cluster of networked, loosely coupled computers working to perform very large tasks. Utility computing is more the packaging of computing resources as described in the HP briefing. Some suggest that the cloud is an evolution from the grid computing model.

> Indeed many cloud computing deployments as of 2009 depend on grids, have autonomic characteristics and bill like utilities—but cloud computing can be seen as a natural next step from the grid-utility model. Some successful cloud architectures have little or no centralized infrastructure or billing systems whatsoever, including peer-to-peer networks like BitTorrent and Skype and volunteer computing like SETI@home.

Addressing the federal CIOs, HP uses the terms "adaptive infrastructure, business technology optimization, and automated services environment." The smart data paradigm aims higher with emphasis on enterprise performance optimization. We appreciate automated services and we equate that with autonomics, a common descriptive.

Then the presentation shifts to emphasizing how the cloud is made secure. That is nontrivial, although out of the scope of our discussion. Essential from our viewpoint is credentialing and privileging users.

Review of Other Cloud Computing References

In addition to the one featured on the Federal CIO Council website, which is likely there as a matter of happenstance, we searched for others and here is what we discovered as it relates to smart data. Omitted from the federal CIOs discussion is a data focus and any attention to data engineering or open standards and technology that will make data smart. Furthermore, the most serious deficiency is the absence of focus on the executive user as a principal customer.

The basic premise of Cloud Computing is that users are not required to have knowledge or expertise or central control over enabling technology. We return executive focus to enterprise data and suggest that leaping to the cloud must be a more deliberate and thoughtful evolution.

While cloud service providers like Amazon, Microsoft, Google, Sun, and Yahoo may push forward, and early adopters like GE, L'Oreal, and P &G gain experience, enterprises should place much higher priority on understanding and managing their enterprises. Computation organized as a public utility may emerge from enterprise-led requirements, and not by technology push.

Cloud computing is taking hold in software development by providing dynamic, flexible tools and processes that can be shared across teams and have traceability across projects. Such a flexible environment can be accessed from a virtual private cloud within the corporate data center or from public clouds. The pooled physical and virtual machine capacity of a cloud can be used for more flexible automation and reuse across projects. This concept of looking at servers as clouds or as pools of virtual resources allows developers to manage and store software from one interface and the visibility to charge back resources per project.

The purpose of the following case study is to demonstrate, from the Microsoft side, how a cloud can be utilized to set up and run a BizTalk® server that enables interoperable and optimized smart data transmissions. BizTalk is a business process management server, and therefore a holistic management approach. It provides business effectiveness by focusing on aligning organizations with the wants and needs of clients. The purpose is to optimize business processes [20].

Actual Name. "Construction and Adaptation of a Cloud in the Development and Testing of a BizTalk Server."

Background. The customer is the DLA in this example but it can be applied to any business.

Goals. Utilize cloud computing concepts to construct and validate an integrated BizTalk server for facilitating supply chain management.

Decision. Should the BizTalk cloud be applied throughout the government and private supply chain communities?

IT Support. Provide a methodology to help standardize data capture, recording, and processing in a supply chain environment and integrate with existing ERPs to form a SOE in the future that will save costs and improve efficiencies.

BizTalk enables the user to communicate through the use of adapters with different software systems. Business processes can be automated and integrated through it. In the following case, those applications are described which are necessary to set up BizTalk for a project. These applications include Microsoft Virtual PC, Orchestration, SharePoint, SQL, and SOAP/Web.

This case helps to reiterate our smart data paradigm that, for the executive, cloud computing is an infrastructure and enterprise services alternative akin to outsourcing.

However, more like subscribing to IT capabilities as a utility and paying for what you use, we used Microsoft Virtual PC to set up both operating systems, such as Server 2008, and many of the applications necessary for BizTalk to perform its functions. In the orchestration application we use a UML sequence diagram to illustrate how parts and messages may be relayed through the cloud, via a network of various vendors that point back to the purchaser. In our example, we used the Defense Logistics Agency as the focal point.

From a smart data perspective, we are using this case for evaluating how the cloud computing alternative can deliver data to the executive more effectively and efficiently. The physical schema describes how the cloud alternative supports enterprise interoperability such that data exchange is frictionless, how enterprise performance is optimized, and how enterprise assets are protected in the process.

In our cloud, the first thing that occurs is a request from the military that it needs something such as bulletproof glass, armor, bullets, or screws. The request is sent to the Defense Logistics Agency, which basically is a warehouse for the military. If the DLA does not have what was requested, the DLA will talk to the specific vendor who can provide the order, such as Lockheed Martin or Boeing. Now the message is with the vendor.

The vendor will send out an order to the private smaller companies for the item that was requested. The company will manufacture what was requested and send it on to the vendor. As soon as the smaller company sends it to the vendor, the DLA will be informed of the order. The vendor will then ship it to the DLA and from there it will be shipped to whoever requested the product in the beginning of the process.

Figure 4.10 illustrates how an order is processed, from the DLA to a vendor. The DLA sends the order to the vendor, together with information about the specification and the quantity. The vendor responds with information about delivery time and other pertinent smart data.

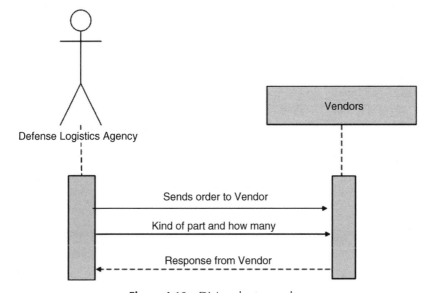

Figure 4.10 DLA order to vendor.

Microsoft SharePoint® products and technologies is a collection of applications and software elements that include, among a growing selection of components, browser-based collaboration functions, process management modules, search modules, and a document-management platform.

The SharePoint component of the cloud can be used to host websites that access shared workspaces, store information and documents, as well as host defined applications such as wikis and blogs. All users can manipulate proprietary controls called "web parts" or interact with pieces of content such as lists and document libraries.

Although SharePoint is a Microsoft product, either Internet Explorer or an open source browser can be used to access SharePoint. Using Mozilla Firefox, for example, SharePoint could be accessed through a IP-Address, together with a user name and password. Groups can be created and individual users added to these groups. This leads to the capability to share documents among users by uploading them to SharePoint, while at the same time having the capability to set different properties for each user, such as read/write capabilities.

Figure 4.11 shows the Central Administration screen of Microsoft SharePoint. The Administrator can customize and enable and disable different services dependent on the needs of the specific project.

SQL Server is a database server that is a major component of our theoretical cloud. It provides different services and is capable of implementing Third Party Applications into its structure. In this example, Microsoft SharePoint is tied to SQL Server as illustrated in Figure 4.12.

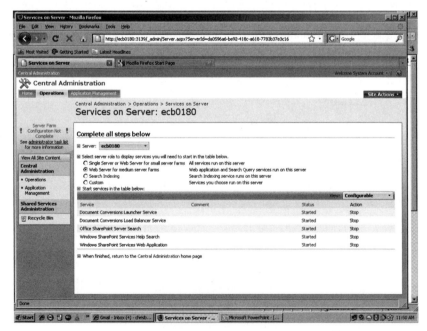

Figure 4.11   Screenshot of Microsoft SharePoint®.

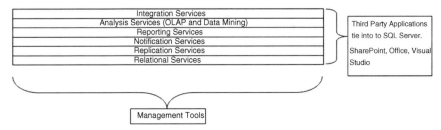

Figure 4.12   Services of SQL Server.

Figure 4.12 shows the different available services of SQL Server. The Integration Service is used to solve complex business problems by copying or downloading files, sending e-mail messages in response to events, updating data warehouses, cleaning and mining data, and managing SQL Server objects and data.

For data mining applications, Analysis Services let the user design, create, and visualize data mining models [21]. Reporting Services help to create, deploy, and manage reports for organizations, as well as programming features that enable the user to extend and customize the reporting functionality. Notification applications can be run on a server engine using the Notification Services. The Replication Services allow creation of backups of any database assigned to the service.

The SOAP/Web portion of our cloud invokes a Web service via orchestration such as the BizTalk orchestration. With the help of Visual Studio, the SOAP/Web cloud components can be configured. By choosing a web reference, Visual Studio automatically creates an orchestration port. In addition, it creates a schema for Web service request and response message, retrieved from a Web service WSDL document. Via a port configuration wizard, a logical port on the orchestration map is created which can bind to the Web Service endpoint reference, as demonstrated in Figure 4.13.

Figure 4.13 shows the flow of messages from the Source on the left to the Destination on the right side of the screen. New messages can be created through the orchestration view pane and mapped by appropriate request and respond schemas.

Setting up all of the above applications makes it possible to run a functioning BizTalk server. Understanding how the individual components of the cloud function is necessary for executives to better understand the complexity and interactions occurring in a business process management server, and the consequent flow of smart data. Of course, the Microsoft components of the cloud can be connected to other open source and proprietary applications such as Oracle and SAP for an even more extensive cloud to enable the smart data transmissions and for executive decision making and problem solving.

In BizTalk, as in any application, security and information assurance is a major concern. Wikipedia states that [22]:

Information assurance (IA) is closely related to information security and the terms are sometimes used interchangeably. However, IA's broader connotation also includes

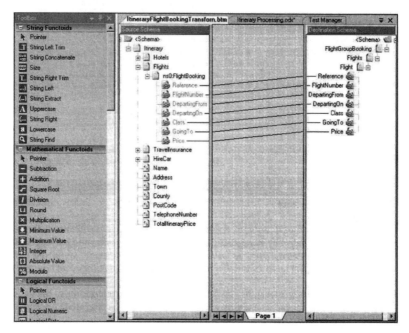

Figure 4.13   Screenshot of the Toolbox of Visual Studio.

reliability and emphasizes strategic risk management over tools and tactics. In addition to defending against malicious hackers and code (e.g., viruses), IA includes other corporate governance issues (that embrace the smart data paradigm), such as privacy, compliance, audits, business continuity, and disaster recovery.

Further, while information security draws primarily from computer science, IA is interdisciplinary and draws from multiple fields, including accounting, fraud examination, forensic science, management science, systems engineering, security engineering, and criminology, in addition to computer science. Therefore, IA is best thought of as a superset of information security. Information assurance is not just Computer Security because it includes security issues that do not involve computers. The U.S. Government's National Information Assurance Glossary defines IA as "measures that protect and defend information and information systems by ensuring their availability, integrity, authentication, confidentiality, and non-repudiation. These measures include providing for restoration of information systems by incorporating protection, detection, and reaction capabilities."

A smart data paradigm would not be complete without consideration of security and information assurance. Information operations protect and defend information and information systems by ensuring confidentiality, integrity, availability, authentication and nonrepudiation. IA regulations apply to the following items: Department of Central Intelligence Directive (DCID)/Joint Air Force Army and Navy (JAFAN)/DoD Information Assurance Certification and Accreditation Program (DIACAP), and the DoD Information Technology Security Certification and

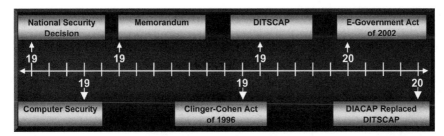

Figure 4.14 Evolution of IA.

Accreditation Process (DITSCAP). This includes enabling restoration of information systems by incorporating protection, detection, and reaction to invaders who would harm our smart data.

Many regulations require IA such as the 1998 Presidential Directive 63. Poor security endangers careers, businesses, agencies, and countries. If the system does not have authority to operate (ATO) because of IA noncompliance, the system will not be allowed to connect to another Department of Defense (DoD) network. The evolution of IA can be seen in Figure 4.14.

There are several types of security threats that our smart data paradigm needs to recognize. Internal threats consist of 90% of all attacks and involve disgruntled employees, spies, terrorists, and unintentional damage. External threats include hackers, terrorist groups, foreign countries, and protesters. For example, the United States Strategic Command (US STRATCOM) is attacked an average 80 times per day. SIPRNet is not completely secure and it was compromised in 2006. TACLANE and encryption do not guarantee security.

A crucial difference in the way that top management looks at IA, from a smart data approach, is demonstrated by leading with executive authority over data and enterprise performance optimization instead of only following technology, to ensure security. Prevention not reaction is the future of IA.

Smart data ideas on IA are tested in a variety ways that appear as standalone projects and proofs-of-concept. This does not mean that the smart data approach to IA is necessarily on the bleeding edge, which usually connotes great risk. It just means that priorities have not been properly aligned with the idea such that enterprises are exploiting the potential advantages. While technology solutions are sufficient for aggressive implementation, the following preventative directives are surely supportive of smart data ideas.

The DoD has mandated that all Information Assurance Vulnerability Management (IAVM) alerts are received and acted on by all commands, agencies, and organizations within the DoD. The IAVM process provides notification of IA vulnerability alerts (IAVA) and requires that each organization take appropriate actions in accordance with the issued alert.

The DoD 8570 Directive provides guidance and procedures for the training, certification, and management of the DoD workforce conducting IA functions in assigned duty positions. The objective of the DoD IA workforce training and certification program is to establish a baseline of validated (tested) knowledge that is relevant, recognized, and accepted department-wide. This applies to technical and management categories and provides information and guidance on reporting metrics and the implementation schedule.

The National Information Assurance Partnership is a U.S. government initiative between the National Institute of Standards and Technology (NIST) and the National Security Agency (NSA). The goal of this partnership is to increase the level of consumer trust in information systems and networks.

In reality, many of the IA principles revolve around both smart data and data quality issues. Smart data is a group of processes and activities, which focus on meeting the executive's information needs, within the context of both security and privacy concerns. While much attention is given to improving infrastructure for processing data and communicating globally, deficient is attention to improving the data itself: data quality, data characteristics, data exchange, and data management and security. We advocate that strategic focus be given to data as it will lead executives and their enterprises to more precise solutions and to optimized performance.

For example, in a healthcare setting, the goal of smart data is to collect, manage, use, and store information that will improve organizational scientific/technical, patient-care, customer satisfaction, and administrative processes. We are contributing to optimizing enterprise performance with a strategy that will vastly improve enterprise data resource management reflected in more efficient and cost effective IT support that produces high impact results for executive and operations management users.

"Smart data" describes data that has been engineered to have certain superior characteristics and that is a product of state-of-the-art data engineering disciplines such that it is interoperable and readily exchangeable among qualified members of an enterprise user community. "Smart data" is an executive strategy and tool for exacting higher performance enterprise-wide. It operates by leveraging state-of-the-art infrastructure and enterprise services.

Joint Commission on Accreditation of Healthcare Organization (JCAHO) Perspective

Managed smart data processes may be affected by the IT infrastructure; however, it is important to first understand and define smart data from the Joint Commission on Accreditation of Healthcare Organization (JCAHO) perspective. The JCAHO provides a useful basis for conceptualizing smart data management in a healthcare unit [23].

Management of information plays a supportive role in quality performance of smart data. This role is a common thread throughout the TQM literature. Information and analysis are core components of the JCAHO. Management of information, data quality, and information assurance play supportive roles in making smart data processes more efficient and effective.

The healthcare sector is an information-laden industry. However, today's healthcare information requirements differ from those of the past, in terms of both the internal and external reporting needs of healthcare departments. Today, physicians, quality managers, TQM teams, marketing staff, financial managers, regulators, insurance plans, accreditation agencies, purchasers, coalitions, and other customers generate greater demands for scientific/technical, patient-care, customer satisfaction, and administrative information that follow the smart data paradigm. We will investigate each of these information types in the following sections.

Scientific/Technical Information Scientific/technical information is important for maintaining the organization's knowledge base. Knowledge about cutting edge techniques and procedures in the medical literature is essential to delivering quality healthcare from a smart data perspective.

Scientific/technical information provides the knowledge base for identifying, organizing, retrieving, analyzing, delivering, and reporting clinical and managerial journal literature, reference information, and research data for use in designing, managing, and improving patient-specific and departmental processes. This scientific/technical information may come from either the general environment or the healthcare environment. For example, laser technology was developed for uses outside the healthcare industry. Today, there is a continuing environmental slip in which uses for laser technology are being applied within both the healthcare and general environments.

Information flows are created through normal operations of individuals, departments, and organizations in both the general environment and the healthcare environment. However, some scientific/technical information may have more significance than others. For example, competitive advantage may be gained by learning about new diagnostic technologies such as magnetic resonance imaging (MRI). Organizations within the healthcare facility, which follow the smart data paradigm, must be able to access new information in order to respond properly to changes in the healthcare environment.

Patient-Care Information Specific smart data and information on patients are essential for maintaining accurate medical records of the patient's medical history and physical examinations. In addition, patient-specific smart data and information are critical to tracking all diagnostic and therapeutic procedures and tests. Maintaining accurate information about patient-care results and discharges is also imperative to delivering quality healthcare.

Communication of smart data within and across healthcare departments is particularly important for managing patient-care information. Coordination of patient-care information within departments helps guarantee quality healthcare delivery. However, linkages between departments require both sophisticated information technology and the strategic management of information, utilizing the smart data paradigm.

Management of information and information technology played a large part in changing healthcare delivery. *Medical informatics* is the term used to describe the management of information that supports the delivery of patient care. For example, some states have legal requirements that hospital departments provide patient-care information on admissions, discharges, and lengths of stay to the public.

In addition, state and federal governments specify the length of time that records must be stored.

The sheer quantity of patient-care information that is generated and stored can be overwhelming to a healthcare organization. Healthcare organizations are faced with maintaining records that document the patient's ailment, the physician's diagnosis, and the recommended course of treatment. Insurance companies, federal and state governments, and individuals use this information for reimbursement and tax purposes. In addition, this smart data may be used in malpractice cases, outcome measures, and research studies.

Customer Satisfaction Information   Customer satisfaction information may be gathered on external customers such as patients and their family and friends. In addition, customer satisfaction information is necessary for internal customers as well. Internal customers include physicians and employees. Surveys designed to gather information on patient satisfaction should take into account sociodemographic characteristics, physical and psychological status, attitudes and expectations concerning medical care, the outcome of treatment, and the healthcare setting. A smart data perspective is quick to point out that any of these factors may have a positive or negative influence on satisfaction.

It may be argued that the patient's family and friends may be more difficult to satisfy than the patient. However, satisfaction of this customer group is rarely evaluated. Family and friends are valuable sources of information and can be very influential in perceptions of quality performance and valuable input for smart data outcomes.

Physician satisfaction information is essential to any quality improvement effort. The *quality improvement strategy* (QIS) has been used to gather information on physician satisfaction with hospital services. The QIS method links quality performance with physician satisfaction.

Information on employee satisfaction is important for understanding the smart data approach to the overall healthcare work process. Employee satisfaction information provides a horizontal view of quality performance. Customer satisfaction information is necessary to understand results and trends, which are based on the customer's behavior and beliefs. Information on customer satisfaction is important in determining customer preferences and expectations about how to effectively deliver organizational healthcare services. Collection, use, and storage of customer satisfaction information are integral to instituting a listening and learning attitude by the organization in order to adopt a smart data approach.

Customer satisfaction information can be used to improve organizational quality performance by both formal and informal measures. For example, information about patient complaints can be used in the smart data paradigm, to provide better service and to institute new services when appropriate. Information should be accurate, accessible, timely, complete, secure, and bias free so that patient complaints can easily be resolved. Customer satisfaction information may be used to better identify individual patient needs, and to determine if there are any differences in the organization's approach to different customer segments.

Comparison of organizational customer satisfaction with other institutions is also important in the smart data approach to the management of healthcare information.

Administrative Information Administrative information is essential for formulating and implementing effective policies at both the organizational and departmental levels. For example, administrative information is necessary to determine the degree of risk involved in financing expansion of services. In a smart data approach, both strategic and operational strategies require administrative information before they can be implemented. Smart data and associated strategy are strategic to supporting these pursuits because nothing gets done without actionable data. Trying to get things done without the proper data slows progress down and makes it much less precise than it needs to be when resources are so constrained.

Traditionally, the CEO has been responsible for strategic management and the role of IT in this process. More recently, executives, managers, and employees have become more involved with the use of IT in managing administrative information. A smart data approach argues that IT can be a competitive weapon that can change an industry's structure, alter key competitive forces, and affect a company's choice of strategy. In order to meet the challenges of these new roles of IT, changes must occur in how IT activities are organized. The smart data approach also suggests that organizations adopt a contingency approach to organizing IT activities into responsible IT groups by exploring the relationships between certain characteristics of the organizational structure and the effectiveness of support provided by IT groups. Therefore executives, managers, and employees today have become more responsible for scanning and influencing the environment, managing key constituencies, and developing adaptive strategies.

Administrative information is required for strategy implementation. It is a key tool used by upper management to provide leadership and commitment to the organizational mission and vision. Administrative information is especially important in a smart data approach to keeping the quality objective before all employees and quality performance has become an important commitment for employees at all department levels. Top level managers are in the best position to use administrative information to encourage and reinforce this commitment. Administrative information is also a key ingredient in reengineering the organization. Managers must make the organizational vision meaningful to healthcare professionals by utilizing administrative information to appeal to the social and economic motives important for the operational success of departments and individuals.

In the smart data paradigm, information must be collected, interpreted, and stored in order to effectively manage the organization. The standardized definition, accuracy, analysis, quality, and reporting of information needs to be aggregated to support managerial decisions and operations, and to improve performance activities. Administrative information must be collected, interpreted, and stored in order to effectively schedule departmental processes, pinpoint errors, archive improved performance, and continuously improve financial services. In addition, administrative information is necessary for improving services by tracking customer needs, scheduling efficient methods of delivery, and coordinating financial data within the organization. Through accessible, accurate, timely, complete, secure, unbiased, and high quality administrative information, financial processes can be maintained and improved in order to achieve increased performance.

Management of Information and Healthcare   Management of information is a group of processes and activities that focus on meeting the organization's information needs. The goal of information management is to collect, manage, use, and store information that will improve organizational scientific/technical, patient-care, customer satisfaction, and administrative processes, by following the smart data paradigm. Managed information processes may be affected by the IT infrastructure; however, it is important to first understand and define management of information in terms of smart data processes. The JCAHO provides a useful basis for conceptualizing information management in a healthcare unit, from a smart data perspective.

JCAHO Objectives   The JCAHO has a set of standards to "describe a vision of effective and continuously improving information management in health care organizations." The objectives related to achieving this vision are more timely and easy access to complete information throughout the organization; improved data accuracy; demonstrated balance of proper levels of security versus ease of access; use of aggregate data, along with external knowledge bases and comparative data, to pursue opportunities for improvement; redesign of important information-related processes to improve efficiency; and greater collaboration and information sharing to enhance patient care" [24].

We have drawn on the JCAHO objectives for continuously improving information management as an input to our smart data paradigm. These JCAHO objectives provide a useful base but do not reflect an adequate list of attributes. Therefore we have expanded on these items to fill this gap at least partially. The smart data management of scientific/technical, patient-care, customer satisfaction, and administrative information is composed of several attributes. These attributes are now discussed.

Accessibility is an important attribute in the management of healthcare smart data. Often, healthcare data that reside in secondary storage are composed of records—patient records, records of procedures and tests, records of scheduled operations, and so on. After scientific/technical, patient-care, customer satisfaction, and administrative records are processed, they need to be accessed, either sequentially or directly. Accessibility of smart data is necessary for information assurance quality.

Accurate information is defined as the degree to which smart data are free of errors or mistakes. In some cases, inaccurate healthcare information occurs at the collection point. There have been numerous problems with inaccurate data stored in hospital information systems. Wrong limbs have been surgically removed as a result of faulty information. People have been denied access to healthcare because of inaccurate data. In some cases, inaccurate information can threaten the life of patients. Accurate smart data is the solution to these problems.

Most healthcare departments develop *standardized definitions* and policies for end-users and healthcare workers. The smart data paradigm dictates that there must be agreements on information formats and procedures. Scientific/technical, patient-care, customer satisfaction, and administrative information cannot maintain integrity unless these standards and policies are maintained.

Timeliness is an important attribute in the smart data management of healthcare information. Timely patient-care information requires a just-in-time basis. Knowing last week's patient test results is not as important as knowing what the results are today. Timing is also crucial for customer satisfaction and administrative applications, including inventory control, trends and forecasting, and cash-flow projections. In medical emergencies, fast processing of documents or reports can literally affect human life.

Complete information contains all the important smart data facts. A patient chart, for example, that does not include all important tests and procedures performed is not complete. Complete information is also important for administrative purposes. If hospital administrators file an investment report that does not include all important costs it is not complete.

Security is the protection of information from intentional or unintentional destruction, modification, or disclosure. Data administrators of scientific/technical, patient-care, customer satisfaction, and administrative information at the departmental level are responsible for accessing, updating, and protecting smart data. If proper controls are not applied, the database is vulnerable to security breaches because a large user community is sharing a common resource.

Analysis of smart data is an important component of both the JCAHO and the Baldridge Award Criteria. Scientific/technical, patient-care, customer satisfaction, and administrative information must be analyzed to find out how quality, customers, operational performance, and relevant financial data can be integrated to support departmental review, action, and planning.

Information can be sorted, classified, and presented as neatly formatted reports. *Reporting* of scientific/technical, patient-care, customer satisfaction, and administrative data in a pleasing and easily understood format is essential for the smart data approach to the management of healthcare information.

The *quality* of scientific/technical, patient-care, customer satisfaction, and administrative information should be consistently high according to the smart data paradigm. Physician reports, for instance, that are cluttered with unnecessary information or relevant smart data that is not in a sequential format could lead to mistakes and life-threatening errors.

A comprehensive list of smart data quality attributes has been constructed and presented. The attributes of accessibility, accuracy, standardized definition, timeliness, completeness, security, analysis, reporting, and quality have their origins in the JCAHO. Accessibility was included in our list because items could be adapted directly from the JCAHO. The same is true for the attributes of accuracy, timeliness, analysis, and quality. Standardized definitions, completeness, security, and reporting were included in the smart data requirements because of the identification of their importance.

And so we have explored cloud computing and introduced IA concepts into our smart data paradigm. And we arrive back where we started, which is enterprise-wide adoption of data-aligned methods and algorithms, optimization, and interoperability. And how do we accomplish these tasks? We accomplish these tasks through smart data, which is a consumer-driven, constituent-driven, investor-driven demand for higher enterprise performance from executives to use data smartly to manage more effectively; to make data smart through data engineering; to make enterprises smarter by adopting smart data strategy; and to make infrastructure more responsive by employing smart data technology in an enterprise context—IT infrastructure such as distributed computing, SOA, SOE, and cloud. Let's take a closer look at IT infrastructure from a healthcare perspective.

Information Technology Infrastructure

Traditionally, IT infrastructures closely paralleled the healthcare organization's rigid functional hierarchy. Hardware, software, databases, and networks were rarely integrated, standardized, or sophisticated. As a result of these fragmented and nonstandard systems, information flows were uncoordinated, unreliable, and disjointed.

Today, IT infrastructures are increasingly challenged in terms of responsiveness to scientific/technical, patient-care, customer satisfaction, and administrative information needs. To achieve requisite responsiveness, the IT infrastructure is evolving in its ability to connect professionals to one another, and in the delivery of timely information within and between functional departments enterprise-wide.

Healthcare IT infrastructures enable healthcare professionals in all of the functional departments within the organization to perform their work. Various healthcare teams and senior executives draw upon the IT infrastructure in order to gather, coordinate, and deliver scientific/technical, patient-care, customer satisfaction, and administrative information.

However, whether you lead a commercial enterprise or a government enterprise, there is one activity that is at the top of the hierarchy in importance for which all CEOs and senior executives are responsible and accountable, and that is to optimize performance. Optimizing performance means applying scarce resources to business processes under constraint, and transforming them into highest yield and best use outcomes by managing people and enabling technical mechanisms within the IT infrastructure.

IT infrastructure can be defined by the standardization and integration of its constituent components.

Components of the Healthcare IT Infrastructure The IT infrastructure of healthcare information systems is composed of four base technologies including hardware, software, databases, and networks. This is a highly technical area that is evolving rapidly. Let us look at each of these components individually.

Hardware ranges from centralized mainframe computers to decentralized microcomputers. A hardware trend is to move processing power out of the central site. This trend is picking up speed because today's desktop and portable workstations have more memory and are faster than mainframes of the 1980s. There is also a strong trend toward cooperative processing. As a result of computers working together in networks, processing power has been distributed away from the mainframe, and toward a client–server environment.

Software applications are used to support care-related operational processes and to improve the effectiveness and efficiency of these processes. However, the relative costs of software compared with hardware applications have increased dramatically. Software has become more important and expensive as a function of total systems costs. Software costs were a small percentage of total costs in the 1950s. Now, software can be 75% or more of the total cost of a present-day healthcare department's computer system.

There are several reasons for this trend of greatly increased expenditures in software. One reason is that there have been reduced hardware costs as a result of advances in hardware technology. A second reason for increased software costs is that the applications are becoming increasingly more complex and therefore require more time and money to develop. Finally, the salaries of software developers have increased because there is a greater demand for these individuals' skills. Software costs are expected to comprise an even greater portion of the cost of the overall computer system in the future.

Telecommunications networks range from private to public, narrowband to broadband. The driving force behind the emergence and evolution of networks is the need to support and access databases. While networks have experienced considerable change, the future promises even more drastic changes. Today, local area networks (LANs) connected to wide area networks (WANs) are leading to increased computer connectivity among information workers. This network infrastructure growth will hasten the transformation from a mainframe environment to a workstation centered, distributed computing one. In the future, wireless communication technologies such as spread spectrum and diffuse infrared will receive broader acceptance and allow people to do their jobs anytime or anyplace.

Databases are composed of stored information or classes of data, which may be centrally located or distributed in various departments and locations. Databases have evolved from file management systems to database management systems. Today, the trend is toward distributed smart data, and the focus has changed from data resources to information resources, both internal and external to healthcare departments. Data management is concerned mainly with internal facts organized in data records. Information resources also include data from external sources.

Integration of Components   Integration is defined as making the separate components of a technology base or business service work together and share resources. Healthcare information systems and comparative databases draw on data generated by diverse providers and other health professionals. These smart data and the applications that access the smart data must be integrated if a composite picture of the scientific/technical, patient-care, customer satisfaction, and administrative information is to be achieved. Connectivity is another way of saying that the hardware, software, databases, and networks are integrated.

Integration requires proprietary differentiation or unique ways of accomplishing things. This is achieved through a variety of means that begins with creative leadership: selecting the right customers to service, attending all constituents and specific needs (e.g., government), and selecting the right things to do by organizing activities and designing work. Proprietary differentiation also means integration of attributing activities with application of a superior mix of people and technology, applying scarce resources in an optimal manner, structuring the balance of consequences such that doing the right things the right way is rewarded and deviations are dissuaded, ensuring that customers receive valuable results, and assuring stakeholders that the enterprise is performing optimally.

The technological infrastructure of organizations is becoming increasingly complex. More and more, IT is being used to improve coordination of activities both within and across organizations. It can be argued that internal integration across value-added functions is a key to interorganizational information systems implementation. For example, computers and video networks are providing long-distance healthcare through medical connectivity.

This is a rich and sophisticated medium. Doctors can interact with each other and ancillary medical personnel not only through e-mail text, but also via video and audio means. A difficult patient case in a rural area can be given expert specialist attention simply by using "distance" medicine. Not only can patient records, text, and documents be transmitted instantaneously via electronic means, but live video, X-rays, and other diagnostic parameters can be discussed in an interactive manner with live discussions weaving a web of competency never before witnessed in medical history. This high bandwidth connectivity can enable real-time interaction in a healthcare context.

One of the most innovative integrated and recent developments in the management of healthcare information is the concept of *shared care*. This philosophy takes the idea of distributed computing and information technology to new heights. After a primary hospital stay, necessary with severe illness, an alternative to further hospitalization will be shared care. Shared care comprises the continued and coordinated and integrated activities of different people from different institutions applying different methods in different time frames, all in a combined effort to aid patients medically, psychologically, and socially in the most beneficial ways.

Standardization of Components   Standardization is defined as agreements on formats, procedures, and interfaces that permit users of hardware, software, databases, and networks to deploy products and systems independent of one another with the assurance that they will be compatible with any other product or system that adheres to the same standards. Mature organizations comprise a collection of processes, people (organizations a.k.a. bureaucracy), and technologies otherwise known as systems that evolved to their present state of standardization.

The collection of historical elements is referred to as *legacy*. Legacy assets have varying degrees of usefulness. Some legacy assets have long life cycles, while others have short ones. Life cycles may be extended through upgrade and change. Legacy consumes resources to sustain and maintain them. Legacy elements whose usefulness is depleted may represent targets for retirement or for footprint reduction because they no longer fit the standards.

Healthcare organizations can expect to see increasing degrees of connectivity given the widespread nature of networks and the standardization of connections. Steadily, healthcare organizations are moving toward the point where anyone, anywhere can connect to anyone else in order to provide shared care. Previously, the concept of connectivity was enabled by electronic data interchange (EDI). Today, the concept of connectivity has taken on a broader meaning, especially in the context of medical connectivity.

The medium is much richer today in terms of interactivity, temporal constraints, and bandwidth. As a consequence, healthcare enterprises today are swamped in a sea of data and their vision and goal attainment is impaired by a fog of unwanted and unnecessary costs. In addition, often infrastructure—processing and communication infrastructure—cannibalizes operational budgets such that more resources are consumed by IT than by the direct mission. Furthermore, in order for hospitals to realize shared care at an interorganizational level, they must first standardize the connections of IT between departments and improve enterprise management performance through adoption of a smart data strategy and improvements in IT support performance.

Sophistication of Components Technological sophistication refers to the recentness or currentness of the technology. In the early 1980s, the term "computers" was used to describe information processing. Information technology (IT) has become the generally accepted umbrella term for a rapidly expanding range of sophisticated equipment, applications, services, and basic technologies. Today, service-oriented enterprise implies that enterprises are about the business of delivering services (including products) that satisfy customers' or constituents' needs. Service-oriented enterprise is enabled by sophisticated smart technologies.

Advanced IT potential provides major changes in health service. Virtual surgery may become a reality. Surgeons may be able to congregate in healthcare facilities located in metropolitan areas and perform surgery anywhere in the world using ultra high bandwidth techniques. With advances in robotics and electronics, this surgery will be even more precise than present human capabilities allow. Diagnostics will not be tied to only one or two opinions, but a whole team of specialists can examine the patient through nanotechnological advances. Sophisticated virtual reality computers will allow surgeons and other medical personnel to practice their techniques not on live patients, but virtual patients, which cannot die or feel pain from human mistakes.

Nanotechnology, or remaking the world molecule by molecule, is an emerging science on the cutting edge of IT sophistication. Miniaturization is a major scientific field that will affect healthcare. Already the Japanese are working on developing a sophisticated technology that will tour the body and report back information. A working motor has been created by scientists from Cal Tec, which is only 100 microns in size and has been combined with a spot of light created by Bell Labs, which is only two millionths of an inch wide. Through miniaturization and distance medicine, future diagnosis of any area of the body can be done at any location in the world that has ultrahigh bandwidth connectivity capabilities. Sophisticated IT will enable team decisions to be made in the best interests of the patient, and ultrahigh bandwidth connectivity will be the medium that will make distance medicine an "actual" reality.

Integration, Standardization, and Sophistication of Information Systems   Systems integration, standardization, and sophistication are additional ingredients necessary for coordinating the data repositories of scientific/technical, patient-care, customer satisfaction, and administrative information systems. As the number and kind of healthcare workers using computers as a part of their daily activities increases, it is becoming evident that an integrated, standardized, and sophisticated IT infrastructure is essential for maintaining and improving quality performance.

IT infrastructure is an important differentiating factor between high- and low-quality performance in healthcare departments. Hardware, software, networks, and databases are the major components of IT that gather, store, and disseminate information about quality. The integration, standardization, and sophistication of these components significantly improve the capabilities of organizational information systems.

Conclusions   IT infrastructure moderates the effect of management of healthcare information on the quality performance in a healthcare department [25]. This follows from the concept that management of information, IT infrastructure, and quality performance go hand in hand. However, in order for healthcare organizations to meet the demands of the future, a radically different approach to information management is proposed.

The success of information systems (ITs) has been an elusive one to define. One dimension of IT success may be organizational impact. Productivity gains may be a measure of IT impact on the organization, with impacts of IT success not only on firm performance but also on industry structure. The impacts of IT on firm and industry-level performance may be viewed by measuring the effects of interorganizational systems on reduction of overhead, increases in customer switching costs, barriers to new firm entry, and product differentiation.

Many reasons exist for IT success and the sharing of information to develop quality care and to improve quality performance throughout a healthcare facility. Prevailing systems are driven by financial data; however, future systems must also consider clinical smart data. The basis for quality healthcare will be effective information systems.

Integration, standardization, and sophistication of all four IT infrastructure components are necessary for the effective and efficient operation of the scientific/technical, patient-care, customer satisfaction, and administrative information systems of healthcare departments. Currently, many information systems at the functional level are not designed in accordance with the TQM philosophy. Often departmental information systems cannot be integrated and standardized with other functional areas, and the results of this lack of sophistication are less than satisfactory. New systems that do not meet user demands are underutilized or not used at all. In addition, the management of healthcare suffers because scientific/technical, patient-care, customer satisfaction, and administrative information systems cannot be effectively employed to improve organizational quality performance.

To improve quality performance, the IT infrastructure must be able to support integrated, standardized, and sophisticated hardware, software, databases, and networks. Development of the IT infrastructure involves integration, standardization, and sophistication of technical analyzers, diagnostic equipment, LANs, WANs, order entry capabilities, external communication reports, physician-specific practice profiles, workstations, emergency backup units, and external and internal databases. The standardization, integration, and sophistication of distributed IT hardware and client–server architecture along with the various types of human–machine interfaces and processing modes are also important.

The healthcare industry is very information intensive. Scientific/technical, patient-care, customer satisfaction, and administrative data are required for decision making at all levels of the healthcare facility. Management of healthcare information and management of the organization are, in essence, very closely related. In order to be successful, a healthcare organization must be able to use health data from both the internal and external environments.

Management of healthcare information is a concept that views scientific/technical, patient-care, customer satisfaction, and administrative information as major resources. The management of healthcare information has lagged behind other industries in automating scientific/technical, patient-care, customer satisfaction, and administrative information. Scientific/technical, patient-care, customer satisfaction, and administrative information that is not effectively and efficiently used severely limits the healthcare manager's ability to handle decision making responsibilities. For example, the use of knowledge-based information by physicians for clinical problem solving is very important. MEDLINE is a bibliographic database containing more than 7 million citations, most with abstracts, from over 3500 biomedical journals and covering the period from 1966 to the present. The information obtained from MEDLINE had an impact on clinical problem solving, choosing the appropriate diagnostic tests, and making the diagnosis. There are many examples of inefficient use of information, including collecting information that is not needed, storing information after it is needed, disseminating more information than is necessary, inadequate means of collecting, analyzing, retrieving, and storing information, and difficulty in giving users access to relevant information.

Technological sophistication refers to the latest, most up-to-date technology being used by the organization. The level of integration is the extent to which various IT components are linked together in a strategic and economic manner. Standardization of IT refers to the extent to which the hardware, software, databases, and networks are compatible with each other.

A major impediment to the development of a computer-based clinical record system has been the lack of agreement on standards both for the clinical terminology to be used and for the computer technology. In order for computer-based record systems to be effective, they must replace the paper record completely. There is a collaboration between the Health Level Seven and the American Society for Testing and Materials (ASTM) to implement a standard for the protocol to be used in the communication of laboratory data in an electronic format. The lack of standards makes it difficult to properly measure treatment cost-effectiveness and to learn how patient outcomes are affected by clinical decisions.

In order to provide a bridge between the vision of an integrated, standardized, and sophisticated vision of the IT infrastructure and the development of a systems architecture to support that vision, a set of policies and standards are necessary to ensure connectivity and compatibility in the IT infrastructure. For example, international standards such as IBM's systems network architecture (SNA) and the open systems interconnection (OSI) reference model can be used to link the healthcare departments to the outside world. Standards for hardware acquisition, software adoption, and database and network implementation are also integral parts of developing the overall architecture.

An open systems architecture offers a solution for organizations that are sometimes locked into the technologies of the past. An open system architecture provides flexibility to take advantage of both existing systems and new technologies. While an open systems solution may be difficult to obtain, it is a viable alternative to retaining systems that are poorly integrated, difficult to maintain, costly, and hard to change.

Data integration is desirable and leads to increased organizational performance. Many organizations cannot make coordinated organization-wide responses to business problems because of data integration problems. Smart data can be used to improve quality. Healthcare employers are finding that there are many advantages to being equipped with specific smart data on the performance of their healthcare plans. Cost control and better healthcare for their clients are among the benefits that were realized from analyzing claims and utilizing smart data.

There have been many attempts by researchers to increase data integration. Integration of existing databases is critical to the development of the nationwide computer-based patient record required for healthcare in the 21st century. System integration can allow one hospital to perform the same procedure with the same results as another hospital for 75% less cost. Theoretical schemas can be developed for disparate databases, including entity relationships to conceptualize data integration and the use of information engineering for smart data integration.

Many integration efforts fail. One reason for this failure may be due to lack of user involvement and failure to secure top management support. Our paradigm stresses the importance of smart data that staff and management can access and apply to both strategic planning and day-to-day issues. We also stress that information management professionals should be involved in the development of a healthcare facility's information management plan. There may be a lack of integration between corporate and IS planning. Smart data integration is contingent upon organizational context that depends on three factors: degree of interoperability, degree of optimization, and the need for enterprise-wide adoption of data-aligned methods and algorithms.

Quality management activities are dependent on valid and reliable smart data about healthcare processes and patient outcomes. Information management professionals can use continuous quality improvement techniques to meet the challenge of providing scientific/technical, patient-care, customer satisfaction, and administrative information. Health information departments can begin to design appropriate integrated, standardized, and sophisticated systems that support quality management activities.

In summary, we suggest that the healthcare user community, probably led by the top executives, collaborate with representative members of the healthcare facilities to determine a smart way to exchange data between and among healthcare organizations. Our model suggests that (1) the user community will determine an open and interoperable standard that will serve as a neutral model for exchange, (2) data exchange servers either centralized or dispersed among the community will enable data exchange that is integrated and independent from legacy applications and infrastructure, (3) , the user community will employ semantic mediation technology to accept diverse sophisticated lexicons, (4) a common standardized and integrated system of configuration management and collaboration will be employed to support the user community, and (5) based on a value set by the head of an enterprise there will be sophisticated governance control (all or shared) over the technology enabler.

We don't suggest a particular set of vendors to support the concept. It is possible that government customer-led virtual enterprises might determine that a commercial and government utility is most desirable for managing healthcare data. A commercial services utility might emerge to sell data exchange services to various healthcare enterprises and commercial and government stakeholders. Most important is that the solution should evolve from the patients' requirements versus vendor or administrator push.

We also champion data governance in order to improve quality control. Data governance is an emerging discipline that embraces the smart data paradigm. The discipline embodies many of the topics discussed in this book, including data quality, data management, business process management, and risk management that we recommend for facilitating the smart data paradigm in an organization. Through data governance, organizations are looking to control the smart data related processes and methods used by end-users.

There are some commonly cited vendor definitions for data governance. IBM defines data governance as a quality control discipline for assessing, managing, using, improving, monitoring, maintaining, and protecting organizational information. It is a system of decision rights and accountabilities for information-related processes, executed according to agreed-upon models that describe who can take what actions with what information, and when, under what circumstances, using what methods.

While data governance initiatives can be driven by an internal desire to improve data quality within the organization, they are more often driven by external regulations. Examples of these regulations include Sarbanes–Oxley, Basel I, Basel II, and HIPAA. To achieve compliance with these regulations, business processes and controls require formal management processes to govern the smart data that is subject to these regulations.

Modern theories of quality performance have provided the healthcare industry with a conceptual framework. This conceptual framework can be used to link the Donabedian trilogy of structure, process, and outcome of care into a cohesive system for evaluating quality of healthcare performance at the organizational level. Opportunities for improving quality performance are typically stimulated by an examination of the structures or processes of care that affect outcomes [26].

The disadvantages of attempts to improve quality performance driven by management of smart data focused on only one of these three elements are legion. For example, medical records reviews may have a false-positive rate as high as 95% if they are based on generic screening criteria. In addition, their usefulness as a primary source of data for quality performance assessment and improvement has been questioned. Even checklist self-reporting by healthcare providers may be unsatisfactory if the reporting is inconsistent. Furthermore, completing the checklist may be of a low priority because it is not related to patient care.

Much like the Donabedian trilogy proposed for healthcare, we propose a Georgian trilogy for smart data management of the related Donabedian processes, structures, and outcomes. We propose that the adoption of a smart data paradigm of interoperability, optimization, and implementation of appropriate methods and algorithms is essential to managing organizational structures, processes, and outcomes.

How to Manage Data Across Agencies Using the Federal Enterprise Architecture Data Reference Model

Another topic featured on the Federal CIO Council website is called "How to Manage Data Across Agencies Using the Federal Enterprise Architecture Data Reference Model." Two co-chairs were identified presumably as authors of this briefing: Bryan Aucoin from the Office of Director National Intelligence, and Suzanne Acar from the Office of Occupational Health and Safety. The date on the briefing is October 2006.

The FEA DRM concept (Figure 4.15) includes three primary elements: (1) data description, (2) data context (taxonomies), and (3) data sharing (query points and exchange packages). These elements address three questions: (1) What does the data mean? (2) How do I find the data and access it? (3) How do I exchange the data?

This is followed by consideration about data structure (Figure 4.16). While there are more details, we observe two things: (1) the strategy is based on standardization, and (2) details about data exchange are not actionable.

While the picture is labeled "data strategy," we don't see one. We do see an enterprise context.

DoD CIO

The Honorable John Grimes is the current DoD CIO and his website is hosted under the National Information Infrastructure (NII) banner. The DoD CIO website features the following:

- A vision statement featuring "Delivering the Power of Information" to an agile enterprise

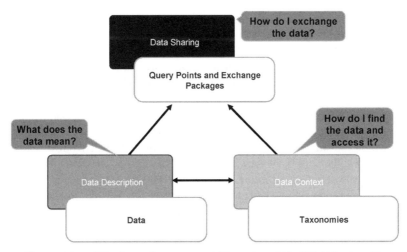

Figure 4.15 FEA data reference model based on FEA DRM Version 20.

- A mission for "Enabling Net-Centric Operations"
- Goals featuring "Information on Demand" from building, populating, operating, and protecting the net
- Outcomes including access, sharing, and collaboration

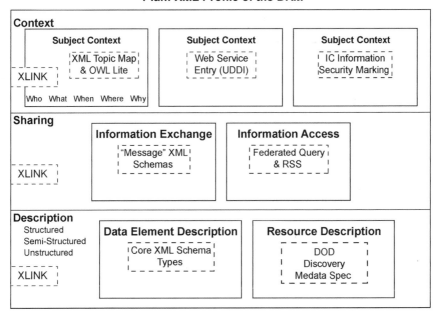

Figure 4.16 FEA DRM Structure.

We advise the executive customer, in this case Secretary of Defense Gates, to send this synopsis back to the drawing board. Surely this is a high-level snapshot, though it does not contain sufficient attention to data in a strategic sense.

Beginning with outcomes, the website message just doesn't address specific needs of the DoD and its military commanders. Surely information access, sharing, and collaboration are functions needed from the IT operation, but these are not outcomes. Complete dependence on something as nebulous as "the net" is just meaningless in the wake of detailed and robust requirements demand.

Posted on the site is a (now ancient) directive from Mr. Paul Wolfowitz that was updated April 23, 2007: Directive Number 8320.02 ASD (NII)/DoD CIO. The contents of this directive constitute the substantiation for smart data and smart data strategy. Attention to tagging and metadata discovery is consistent with the ideas that we advance in smart data and are consistent with the vision for the semantic web, for instance.

Note that specific DoD nomenclature used for numbering is omitted for simplification and clarity, although we preserve the DoD regulation structure for authenticity.

SUBJECT: Data Sharing in a Net-Centric Department of Defense

POLICY: It is DoD policy that:

- Data is an essential enabler of network-centric warfare (NCW) and shall be made visible, accessible, and understandable to any potential user in the Department of Defense as early as possible in the life cycle to support mission objectives.
- Data assets shall be made visible by creating and associating metadata ("tagging"), including discovery metadata, for each asset. Discovery metadata shall conform to the Department of Defense Discovery Metadata Specification (reference (d)). DoD metadata standards shall comply with applicable national and international consensus standards for metadata exchange whenever possible. All metadata shall be discoverable, searchable, and retrievable using DoD-wide capabilities.
- Data assets shall be made accessible by making data available in shared spaces. All data assets shall be accessible to all users in the Department of Defense except where limited by law, policy, or security classification. Data that is accessible to all users in the Department of Defense shall conform to DoD-specified data publication methods that are consistent with GIG enterprise and user technologies.
- Data assets shall be made understandable by publishing associated semantic and structural metadata in a federated DoD metadata registry.
- To enable trust, data assets shall have associated information assurance and security metadata, and an authoritative source for the data shall be identified when appropriate.
- Data interoperability shall be supported by making data assets understandable and by enabling business and mission processes to be reused where possible.
- Semantic and structural agreements for data sharing shall be promoted through communities [e.g., communities of interest (COIs)], consisting of data users (producers and consumers) and system developers, in accordance with reference (b).
- Data sharing concepts and practices shall be incorporated into education and awareness training and appropriate DoD processes.

RESPONSIBILITIES

- The Assistant Secretary of Defense for Networks and Information Integration/ Department of Defense Chief Information Officer (ASD(NII)/DoD CIO) shall:
- Guide and oversee matters relating to Net-Centric data sharing in support of the DoD Components, COIs, Domains, and Mission Areas by:

    Developing, maintaining, and enforcing enterprise metadata direction that uses existing Government and industry metadata standards when possible.

    Developing and maintaining direction on, and enabling use of, federated enterprise capabilities to publish metadata and to locate, search, and retrieve metadata and data. Federated enterprise capabilities shall include the Intelligence Community (IC).

    Develop the policies and procedures to protect Net-Centric data while enabling data sharing across security domains and with multinational partners, other Federal Agencies, and State and local governments in accordance with law, policy, and security classification, in coordination with the Under Secretary of Defense for Intelligence (USD(I)) and the Under Secretary of Defense for Policy (USD(P)).

    In accordance with the Deputy Secretary of Defense Memorandum (reference(e)), ensure that Domains within the Enterprise Information Environment Mission Area promote Net-Centric data sharing and effectively enable COIs, including adjudicating conflicts in metadata agreements and identifying authoritative sources.

    As an element of Information Technology (IT) portfolio reviews, and in accordance with reference (e), provide guidance to evaluate and measure:

    The status of the DoD Components in achieving Net-Centric data sharing.

    The degree to which Mission Area and Domain portfolios provide the capabilities needed to share data.

The Under Secretary of Defense for Acquisition, Technology, and Logistics, in coordination with the ASD(NII)/DoD CIO, shall:

- Ensure that the Defense Acquisition Management System policies and procedures incorporate the policies herein, and provide guidance for Milestone Decision Authorities to evaluate and approve system or program satisfaction of data sharing practices, in accordance with reference (e).
- Ensure that the Defense Acquisition University develops education and training programs to advocate Net-Centric data sharing in the Department of Defense based on policies herein.

The Under Secretary of Defense for Policy shall collaborate with the ASD (NII)/DoD CIO and the USD(I) to develop the policies and procedures to protect Net-Centric data while enabling data sharing across different security classifications and with multinational partners, other Federal Agencies, and State and local governments, in accordance with law, policy, and security classification.

The Under Secretary of Defense (Comptroller)/Chief Financial Officer, in accordance with reference (e), shall ensure that Domains within the Business Mission Area promote Net-Centric data sharing and effectively enable COIs, including adjudicating conflicts in metadata agreements and identifying authoritative sources.

The Under Secretary of Defense for Intelligence shall:

- Collaborate with the ASD(NII)/DoD CIO, the USD(P), and the IC CIO in developing policies and procedures to protect Net-Centric data while enabling data sharing across different security classifications, and between the Department of Defense, the IC, and multinational partners, in accordance with policies herein and consistent with Director of Central Intelligence Directive 8/1 (reference (f)).
- In accordance with reference (e), ensure that Defense Intelligence Activities within the Domains of the National Intelligence Mission Area promote Net-Centric data sharing and effectively enable COIs, including adjudicating conflicts in metadata agreements and identifying authoritative sources.
- Ensure counterintelligence and security support to network-centric operations.

The Heads of the DoD Components shall:

- Ensure implementation of Net-Centric data sharing, including establishing appropriate plans, programs, policies, processes, and procedures consistent with policies herein.
- Ensure that all current and future data assets are made consistent with policies herein.

Support Mission Areas and Domains by taking an active role in COIs. The Chairman of the Joint Chiefs of Staff shall:

- In coordination with the ASD(NII)/DoD CIO, ensure the policies herein are incorporated into the Joint Capabilities Integration and Development System and the procedures of the IT and National Security Systems' (NSS) interoperability and supportability certification and test processes.
- In coordination with the ASD (NII)/DoD CIO, direct the National Defense University to develop education and training programs to advocate Net-Centric data sharing in the Department of Defense based on policies herein.
- Ensure that Domains within the Warfighter Mission Area promote Net-Centric data sharing and effectively enable COIs, including adjudicating conflicts in metadata agreements and identifying authoritative sources.

The above directive and policy is surely consistent with what we advocate although there is something significant that is missing. The policy applies to enterprise data sharing but lacks focus on enterprise performance optimization and executive user accountability.

Smart Data Subjects and Subtopics in the News

What is the most significant enterprise engineering topic in the news on this day of writing? It is the U.S. government hands-on management of General Motors and the directed firing of the CEO. There is nothing more dramatic of an event than that in American and global commerce. Combine that with the U.S. government overseeing a merger between private company Chrysler and the Italian company Fiat.

No matter how large the company, and no matter how long the enterprise has been in business, if leadership executives take their eyes off the customers, their competition, and the trends of technology and economy that are the data facts, then the game may well end.

In searching subjects that indicate attention to the enterprise context, there are many glancing blows that are contained in popular technology topics that we have discussed. Missing is the notion that enterprise engineering must move beyond systems engineering. Missing too is an understanding about what it means to be outcome focused—not just the attributes but the real outcomes and measures for success in all process domains and in the aggregate enterprise rollup.

Enterprise Engineering Based on Architecture and Ontology

There are a host of bloggers addressing our subjects and social media are definitely a way to engage our topics more fully with peers. We went to the international community searching for content that would support our enterprise engineering ideas. We did this because, in our experience, Europeans tend to lead in standards development and collaboration. After all, the Norwegians pressured Lockheed and the JSF Program to attend ISO 10303 PLCS, for instance.

We are not qualified to interpret why this might be true, but we went fishing for ideas and found some good material beginning in The Netherlands.

We discovered an interesting and useful blog post entitled "Enterprise Engineering Based on Architecture and Ontology" [28] by Johan den Haan, Head of Research & Development at Mendix, and a firm in The Netherlands. He posed some interesting discussion that is pegged on ideas originating from Jan Hoogervorst, who is the IT Director at Royal Dutch Airlines. Author George was familiar with Hoogervorst from a conference hosted by the Dutch Airlines for the Air Transport Association addressing electronic commerce and enterprise integration in the past.

According to Johan den Haan,

> In a lot of enterprises IT is blamed for the lack of business–IT alignment, but why? Isn't IT just a tool to reach the enterprise goals? If we buy a bike to improve our traveling, can we blame it for non-functioning if we just walk next to it? IT should function as part of an Enterprise Architecture and it should be an integral part of the enterprise strategy. The emerging field of Enterprise Engineering takes care of the integral design of an enterprise, but what exactly is Enterprise Engineering and where does it come from? [27].

Why Enterprise Engineering?   According to Johan den Haan [27]:

> The successful implementation of strategic initiatives in Enterprises is very difficult. According to Hoogervorst [28] numerous studies show that less than ten percent of strategic initiatives are implemented successfully. Hoogervorst also refers to a lot of studies showing that a lack of coherence and integration is the most important reason for this high failure rate. All elements (resources, business and organization) should be aligned with one another for making a great corporate strategy. The failing introduction of IT has also a lot to do with this lack of coherence and integration.
>
> According to Hoogervorst (based on several studies), the power to do a coherent implementation of strategic choices is more important than the quality of these strategic

choices itself. So an integral design of the enterprise is essential for being successful [28]. In this way competences for realizing integral designs are even more important than competences for formulating corporate strategy!

Problems in enterprises are often addressed with use of black box thinking based knowledge, i.e., knowledge concerning the function and the behavior of enterprises. Such knowledge is totally inadequate for changing an enterprise to meet its goals. For these purposes white-box based knowledge is needed, i.e., knowledge concerning the construction and the operation of enterprises.

According to Dietz and Hoogervorst [29] developing and applying such knowledge requires no less than a paradigm shift in our thinking about enterprises, since the traditional organizational sciences are not able to ensure that enterprises are coherently and consistently integrated wholes. They call for a new point of view: "The needed new point of view is that enterprises are purposefully designed, engineered, and implemented systems. The needed new skill is to (re) design, (re) engineer, and (re) implement an enterprise in a comprehensive, coherent and consistent way (such that it operates as an integrated whole), and to be able to do this whenever it is needed"

Roots of Enterprise Engineering   Enterprise engineering (Figure 4.17) is a new field, emerging from information systems engineering and the organizational sciences. From top to bottom the evolution of thinking within the field of computer science is shown. It consists of two parts. The first revolution, around the 1970s, was the recognition of the difference between form and content. Data systems engineering became information systems engineering.

At the moment the second evolution/revolution is going on. It consists of recognizing two different parts in communication: proposition and intention. Proposition can be seen as content while intention expresses the relation between the communicator (by example, the speaker) and the proposition. Examples of intention are requesting, promising, stating and accepting. Like the content of communication was put on top of its form in the 1970s, the intention of communication is now put on top of its content. It explains and clarifies the organizational notions of collaboration and cooperation, as well as notions like authority and responsibility. For a more comprehensive explanation of the concept of intention I'd like to recommend the book *Enterprise Ontology* [30].

This current revolution in the information systems sciences marks the transition from the era of information systems engineering to the era of enterprise engineering. However, Enterprise Engineering is closely related to organizations, so it focuses on

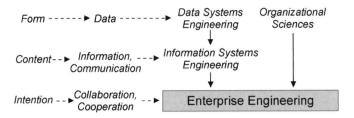

Figure 4.17   The roots of enterprise engineering [29].

a socio-technical system. Therefore it adopts also knowledge of the organizational sciences. Enterprise Engineering is about engineering enterprises in the sense of focusing on the construction, the white-box view, of enterprises. It doesn't forget that an enterprise is a system also consisting of people [27].

The Core Notions of Enterprise Engineering

As said before, the basic condition of Enterprise Engineering is that an enterprise is designed and engineered coherently and consistently. To make that happen two core notions are needed: ontology and architecture.

The ontology of a system is theoretically defined as the understanding of its construction and operation in a fully implementation independent way. Practically, it is the highest-level constructional model of a system, the implementation model being the lowest one. Compared to its implementation model, the ontological model of an enterprise offers a reduction of complexity of over 90% [30]. This reduction of complexity makes an organization for a manager intellectually manageable and transparent. It also shows the coherence between all fields within the enterprise, like business processes, workflow, organization structure, etc.

While ontology describes the system itself, architecture describes how to design a system. We adopt here a prescriptive notion on architecture [27].

Conclusion Johan den Haan concludes [27]: "We think Enterprise Engineering is a nice, and needed, point of view for designing and implementing enterprises. The important notions of ontology and architecture are very useful in reducing complexity and in giving more insight in the construction (white box view) of an enterprise"

Present and Near-Term Thinking

This section is intended to address the next 5-year situation as far as we can determine. As near term as this window is, the picture remains foggy due to a number of factors:

- The U.S. economy is in a state of correction for which the outcome is uncertain.
- A new political administration is still getting established while dealing with crises on an unprecedented scale.
- Changes are being made and the environment is fluid.

Furthermore, at this writing in 2009, the fifth year from now is 2014, two years into a second term for the Obama administration or into a new administration altogether.

Will there be an American enterprise strategy? Will there be smart data strategy in government and commercial enterprise? Will the notion of enterprise performance optimization support systems take shape?

The probability of these things happening will be based on success in securing attraction to these ideas through publishing, academic work, and public organizing through professional associations like the AFEI and political engineering.

There is an immediate development that serves as an indication of what is to come. While only recently appointed, the U.S. Chief Information Officer has made an announcement as was expected and it delivers on a promise from the president that is profound because it demonstrates how rapidly the CIO has had to hit the ground running. According to a report from *Information Week*, June 4, 2009:

> The federal government's "Data.gov site launched last month with a measly 47 data sets, but it's on track to make 100,000 data feeds available to the public in programmable form by the middle of June.
>
> The upgrades to the site, which will be available in a few months, will feature new ways to find and use data, including the ability to tag data sets. The government also will begin to make some data available in more finished forms, especially if data sets are rated poorly by the public.
>
> Once we've got the raw format, we're going to be going back and seeing data sets that haven't been rated at all, or data that has low ratings. We're going to be saying, what do we need to do to make it more usable? Kundra said.
>
> Data.gov isn't just transparency for transparency's sake. The goal is to let the public use government-generated data in new ways. We don't believe we have a monopoly on best approaches," Kundra said, pointing to moneymaking innovations made possible by government release of genomic and GPS data, as well as to an application development challenge by the Sunlight Foundation, an open government advocacy group.
>
> The site makes "raw" data available in a variety of formats, including XML and text. The data can be searched, extracted, and analyzed using a catalog of tools. Kundra hasn't prescribed specific data formats, but has made it a guiding principle that the government move toward open data standards and make data available in multiple machine-readable formats.
>
> Data.gov, which Kundra has made one of his and the agency CIOs' highest priorities, will require upgrades to serve up more data and improve data quality, but the investment will eliminate duplicate data collection and cut costs over the longer term, Kundra says. For example, the Office of Management and Budget will look at reducing duplicate map data.
>
> A second iteration of Data.gov will be available over the coming months and will include new ways for the public to use and find the data, including the ability to tag data sets. By the end of June, Kundra also will launch an IT dashboard to measure spending and effectiveness of federal IT projects, and make that data available on data.gov. [31].

There is no doubt that sharing government data assets and encouraging the public to exploit the information is innovative and delivers on the promise to make government more transparent. However, the highest priority from a citizen's view and executive view is to make certain that government enterprise executives have the data they need to optimize performance that is delivered through an effective management system that today simply does not exist as it should.

4.4 TEN YEARS FROM NOW

Where will smart data be in year 2020? Our vision begins by promoting smart data and smart data strategy to support enterprise performance optimization in 2010 such that the president and department heads and Congress agree to the development of a government enterprise performance optimization system (GEPOS), the first of its kind.

Development of GEPOS requires rapid planning and development following concurrent development of smart data and smart data strategy. It is a government executive control system for the United States government. Implementation of the idea will track the natural progression of Web-based technologies and international open standards for interoperability.

The greatest threats to global government and commercial automation and enterprise performance optimization will continue to be in the realm of cyber security and environmental calamities that can wreck the current computer systems. Priority attention must be given to securing the electronic environment throughout the world.

REFERENCES

1. J. A. Rodger and P. C. Pendharkar, "A Field Study of the Impact of Gender and User's Technical Experience on the Performance of Voice-Activated Medical Tracking Application," *International Journal of Human Computer Interactions*, **60**, 529–544, 2004.

2. D.L. Goodhue and R.L. Thompson, "Task-Technology Fit and Individual Performance," *MIS Quarterly*, **19** (2), 213–236, 1995.

3. J. A. Rodger and P. C. Pendharkar "A Field Study of Database Communication Issues Peculiar to Users of a Voice Activated Medical Tracking Application," *Decision Support Systems*, **43** (2), 168–180, 2007.

4. J. A. Rodger, T. V. Trank, and P. C. Pendharkar, "Military Applications of Natural Language Processing and Software," *Annals of Cases in Information Technology*, **4**, 12–28, 2002.

5. Barry W. Boehm, *Software Engineering Economics*, Prentice-Hall Professional Technical Reference, New York, 1981.

6. DOD Data Strategy, John Stenbit, DOD CIO, http://www.defenselink.mil/cio-nii/docs/Net-Centric-Data-Strategy-2003-05-092.pdf.

7. http://www.iso.org/iso/home.htm.

8. http://www.iso.org/iso/catalogue_detail.htm?csnumber=39369.

9. M.G. Vincent Boles, "Enterprise Approach to Supporting Military Operations in the War on Terror," 2008 Logistics Symposium, Richmond, Virginia, May 14, 2008.

10. J. A. Rodger, P. Pankaj, and M. Hyde, "An Empirical Study of the Determinants of Backorder Aging: The Case of the Missing National Item Identification Numbers (NIINS)," *Journal of Global Management Research*, **5** (1), 35–43, 2009.

11. J. A. Rodger, *An Empirical Study of the Determinants of Backorder Aging for the Battlefield Breakout Backorder Initiative (B3I)*, Midwest Decision Sciences Institute, Erie, PA, 2008.

12. Defense Logistics Agency (DLA), About, http://www.dla.mil/about_dla.aspx.

13. Scott Wilson and Robert O'Harrow Jr., "President Orders Review of Federal Contracting System – More Competition, Accountability for Procurement Sought," *The Washington Post*, March 5, 2009 http://www.washingtonpost.com/wp-dyn/content/article/2009/03/04/AR2009030401690.html.

14. http://en.wikipedia.org/kiki/Government_Accountability_Office.

15. Anna Mulrine, "On Pentagon Waste Just Give it to us Straight," March 3, 2009, *US News and World Report*, http://www.usnews.com/articles/new/2009/03/03/on-pentagon-waste-just-give-it-to-us-straight.html.

16. Mark Fox and Michael Gruninger, "Enterprise Modeling," *AI Magazine*, **19** (3), 1998.

17. Tom Gerber, "Ontology" in Ling Liu and M. Tamer Ozsu (Eds.), *Encyclopedia of Database Systems*, Springer-Verlag, New York, 2008.

18. http://tomgruber.org/writing/ontology-definition-2007.htm.

19. N. R. Nagel and R. Dovel, Twenty-First Century Manufacturing Enterprise Strategy: An Industry Lead View, Tech. Report, Iacocca Institute, Lehigh University, 1991.

20. P. C. Pendharkar and J. A. Rodger, "Data Mining Using Client/Server Systems," *Journal of Systems and Information Technology*, **4**(2), 71–81, 2000.

21. P. C. Pendharkar, J. A. Rodger, G. Yaverbaum, N. Herman, and M. Benner, "Association, Statistical, Mathematical and Neural Approaches for Mining Breast Cancer Patterns," *Expert Systems with Applications*, **17**, 223–232, 1999.

22. http://wikipedia.org/wiki/Information_assurance.

23. J. A. Rodger, P. C. Pendharkar, and D. J. Paper, "Management of Information Technology and Quality Performance in Health Care Facilities," *International Journal of Applied Quality Management*, **2** (2), 251–269, 1999.

24. Joint Commission on Accreditation of Hospitals, 2010 Comprehensive Accreditation Manual for Hospitals: The Official Handbook (CAMH), Joint Commission Resources, New York, 2010.

25. J. A. Rodger, D. J. Paper, and P. C. Pendharkar, "An Empirical Study for Measuring Operating Room Quality Performance Attributes," *Journal of High Technology Management Research*, **9** (1), 131–156, 1998.

26. P. C. Pendharkar, J. A. Rodger, D. J. Paper, and L. B. Burky "Organizational Status of Quality Assurance and End User Perceptions of Information Technology for Information Management," *Topics in Health Information Management*, **21** (1), 35–44, 2000.

27. http://www.theenterprisearchitect.eu/.

28. J. A. P. Hoogervorst,"Enterprise Governance & Architectuur, Corporate, IT, Enterprise governance in samehangend perspectief," Academic Service, 2007.

29. J. L. G. Dietz and J. A. P. Hoogervorst, "Enterprise Ontology and Enterprise Architecture–How to Let Them Evolve into Effective Complementary Notions," *GEAO Journal of Enterprise Architecture*, **2** (1), 2007.

30. J. L. G. Dietz, *Enterprise Ontology—Theory and Methodology*, Springer-Verlag, New York, 2006.

31. J. Nicholas Hoover,"Federal CIO Kundra Looks Forward To Data. Gov 2.0," *Information Week*, June 4, 2009.

Chapter 5

CEO's Smart Data Handbook

It is the CEO's and enterprise smart data strategy, not the CIO's or CTO's smart data strategy.

5.1 STRATEGY

Enterprise executives of government and commercial business and industry need and deserve improved support for their primary responsibility, which is to optimize enterprise performance. We have studied this and believe that executive enterprise performance optimization systems can be developed employing the notion of smart data and smart data strategy that will vastly improve enterprise executives' ability to:

- Anticipate and predict
- Plan
- Decide
- Solve problems
- Make sense
- Model and simulate

Notice that we list "anticipate and predict" first because therein is the strategic path toward leadership and competitive superiority, which is a product of innovation. Too often, an enterprise's future is data deprived by deficient support. Data support

Smart Data: Enterprise Performance Optimization Strategy, by James A. George and James A. Rodger
Copyright © 2010 John Wiley & Sons, Inc.

includes automated analytical capacity to process actionable data in anticipation of enterprise needs and executive requirements. Improving this, executives' capacity can be increased to accommodate the explosion of demand from operating in a global environment. Increased and improved support of the type we envision will permit more timely attention to operational and tactical needs while increasing the range in anticipating and envisioning the future.

Both government and commercial enterprise operations are made exceedingly more complex from the impact of being in the global marketplace. Old tools and systems are inadequate to support management responsibilities. We want to equip executives with words that will help them communicate their needs, requirements, and expectations.

Since so much focus has been given to President Obama and his dealing with commercial enterprise to right the U.S. economy, imagine his requesting a U.S. government enterprise performance optimization system (GEPOS) that can be employed to monitor and evaluate departments' and agencies' performance toward achieving specific outcomes aligned with the administration's strategy. His capacity to govern can be greatly increased from development and implementation of a modern support system designed under contemporary requirements. Such a system would support future presidents as it would continuously improve.

Imagine Fritz Henderson, the new GM CEO, having a commercial enterprise performance optimization system (CEPOS) that greatly improves his visibility about new markets and division performance and enables testing what-if scenarios as well as supporting immediate operational decisions in a more robust and timely manner.

Imagine Secretary of Defense Gates having specific data to support his reinventing the department's procurement system with specific visibility and transparency for congressional actions that often undermine fresh thinking that is vitally important in the new global economic environment. Envision a system that preserves the wisdom of senior leadership such as this from administration to administration.

Smart data is addressed in three dimensions:

1. Enterprise context—comprehensive, holistic, not stovepiped
2. Interoperability technology—metadata management, semantic mediation, openly exchangeable
3. Data-aligned methods and algorithms—decision support accompanying actionable data

To begin this chapter, we pursue the idea that began with the intention of improving executives' ability to optimize enterprise performance by focusing on data necessary to accomplish this. The idea was sparked by a Department of Defense customer who was exasperated by what appeared to him to be a great disconnect between what executives and managers need to perform and what information technologists actually deliver. We all observe mountains of nonactionable data that is the symptom of a problem. Problems arise and mistakes are made without transparent reasons.

In addition, we should note that while accountants serve up the daily spreadsheets that reflect a view of business performance, the balance sheet is far short of what is needed to lead and guide the enterprise to higher performance and self-renewal. Needed is data that permits you to envision the future, to model and test ideas, and to plausibly extrapolate changes from here to there.

Given the present state of affairs in commercial and government enterprise, operations can be trapped by present conditions into starving the future. Executives can unwittingly fail to make sense of circumstances and fail to accurately anticipate emerging opportunities for development.

Problem statement. Enterprises do not have a data focus. They lack smart data and a smart data strategy that is essential for optimizing enterprise performance.

Solution. There are a host of details to consider in solving the problem, but we suggest that executives should request an enterprise performance optimization system (EPOS) to support them that incorporates enterprise context, interoperability technology, and data-aligned methods and algorithms. Turning ideas into something like a support system serves to make them real and tangible.

Of course, other things are real and tangible such as auditing data throughout the enterprise, and knowing how data is input into processes and how it manifests as outcomes. It is the act of analysis that stimulates fresh thinking in the search of better ideas and strategies and that achieves higher performance.

We know that there are better ways to exchange data and to conduct business transactions that are made possible by today's standards and technologies as well as emerging technologies associated with the semantic web and cloud computing. Employing a data focus to produce better direct support to executives to improve enterprise performance optimization is a critical path initiative demanded by today's economic and business environment.

The authors observed the same deficiencies for many years while supporting government and commercial customers and have had several opportunities to think about the best way to inject ideas that can make a difference, as enterprises seem to be trapped by their legacy behavior.

The first thing is to get executives on board with our intention to improve support in ways that they perceive to be significant.

Want to learn what's right and wrong in your enterprise? Follow the data.

There is no more vivid example of the consequences of not having a smart data strategy and data-focused management than the Wall Street debacle of 2008. Data assets and value-tracking were deficient at every step in the banking enterprise, which could have been prevented with smart data strategy. While the president and congressional leadership pursue new policies and controls, inclusion of an overarching smart data strategy for all government and private enterprise is encouraged.

There are three aspects of smart data strategy: (1) smart data itself has characteristics that are different from dumb data; (2) smart data is a product of methods, engineering disciplines, and technology that is a markedly different approach from former practices; and (3) data-focused strategy is embraced by executives as a primary value in pursuing enterprise performance improvement.

This chapter presents the executive pathway to understanding data and defines how you may develop and adopt smart data strategy to optimize enterprise performance. Even if you do not read Chapters 1 to 4, this chapter is designed as a standalone and practically beneficial document. To gain more detailed insight about how we arrived at these contents, please read the rest of the book.

This book is not about technical aspects of data modeling and database management, as those subjects are for information technologists. This book is about what executives should expect from data and what you can do to gain better leverage from smart data practices.

Executive focus on data and how it is used by the enterprise is a catalyst for change and improvement. Because data touches every process, it provides a comprehensive vehicle for examining associated performance. Smart data strategy aims at eliminating inefficient technologies and methods and replaces them with a solution that leapfrogs former barriers to higher performance. This happens only with executive direction, persistence, and reinforcement through constant use.

While the potential cost savings cannot be underestimated, the advantage of a smart data focused approach to enterprise performance optimization cannot be overestimated.

Here is some terminology to review before we delve into data as a pristine element of our focus.

Business data describes everything used to operate the enterprise. *Business intelligence* is a term invented by Howard Dresner of the Gartner Group describing concepts and methods to improve executive decision making. *Business model* is how enterprise architects depict processes and data, often shown as the current state and a future state. *Business transactions* is a unit of work acted on by the system to manage business data documenting and recording the facts describing individual business events. *Business rules* govern data and impose standards, often a legal test, to ensure proper use and agreement.

Smart Data Strategy

Smart data strategy is needed by every enterprise, commercial and government. Reasons why?

1. *All enterprise assets are represented by data and CEOs and executive management cannot optimally control and protect assets without explicit data strategy.*

   How much do executives need to know about data? Certainly, data, its descriptions and representations, user dynamics, associated functionality, governance, enabling technologies, and supporting professional skills have complex dimension. Yet, because executives depend on it as if their enterprises' livelihoods do, it begs for more in-depth understanding and hands-on management.

The Wall Street calamity illustrates that investors and the public in general did not have adequate visibility into data affecting them. Management may or may not have had proper data, and determining fraud and manipulation will be based on who had the data. Business rules did not levy the proper balance of consequences. Multiple institutional processes were noninteroperable, and all of these things led to errors in oversight.

Federal Reserve Chairman Ben Bernanke said Tuesday March 10, 2009, "We must have a strategy that regulates the financial system as a whole, in a holistic way, not just its individual components," addressing the Council on Foreign Relations.

Throughout this discussion, we remind ourselves that data is a very practical thing. It is the way we communicate and our day-to-day language. However, when it is processed and exchanged in high states of automation, our systems must have greater capacities than we do as individuals. They must work smarter, faster, better, and cheaper. They are products of human ingenuity, but are intended to extend our capacity to become super human. Achieving this is nontrivial.

Often, we drag down our best invention and subordinate it to our inferior processing ability and therefore undermine the payoff. Applying creative interpretation to achieve innovation is encouraged, but it should be fact-packed, data-based, and autonomic.

When the task of oversight and regulation enforcement seems foreboding, smart data strategy with supporting automation makes effective management possible. Concepts such as automated contracting environment and automated regulatory environment become a practical reality. Envision contracts and regulations being delivered in executable code such that your internal systems can automate large portions of your process auditing routines, reducing time and distraction while increasing accuracy and preventing and mitigating problems in real time.

In an article reported by MSNBC by AP economics writer Jeannie Aversa entitled "Federal Reserve Chairman Bernanke Says Regulatory Overhaul Needed," she observes that "revamping the U.S. financial rule book—a patchwork that dates to the Civil War—is a complex task" [1]. We extend the observation that the "rule book" is the governance over our automated systems as we will discuss.

Our computer systems have evolved to their present state, not so much driven by strategy as much as they are developed in response to changing requirements. Information technology advances rapidly while developing systems to satisfy needs is like a greyhound chasing a robot rabbit in a race it cannot win. In older organizations, layer upon layer of systems exist. Some get retired; more get added on and modified. A common thread is data – publishing data, processing data, exchanging data, analyzing data. Oddly, data is rarely addressed strategically.

2. *CEOs are responsible for maximizing return on assets whereby all assets are represented as data.*

Do CEOs truly have a grasp for understanding data at the level required in our highly automated world? We advise that for an enterprise to gain competitive superiority, increasing knowledge and competence in data management from the top down is essential.

Understanding data is an integral part of process engineering that produces the building blocks for enterprise engineering that is aimed at satisfying certain needs among targeted populations of customers and constituents. Assets include materials, capital, people, processes and products, systems and enabling technology, customers and constituents, stakeholders, and trading partners. (Note that all of these "things" require unique descriptions and are subject to different standards and influences for accurate representation.)

Smart data is different from data that is not engineered for smartness. How will your organization measure data smartness? This guide will identify criteria that your organization may customize for your situation. Your organization can perform a self-assessment employing these criteria.

Smart data strategy implies having smart data and hosting a specific enterprise strategy that exploits unique benefits. How will the enterprise track the advantages and corresponding measurable benefits? This guide will suggest a business case approach to tracking benefits and assuring expected return.

Employing engineering discipline to thinking about the enterprise and data is a path toward making the enterprise smarter about data, assets, and performance management. Making data management more explicit at the executive level of the organization is critical to improving enterprise performance.

We want to answer the following questions.??

- What are the attributes of smart data?

Data is data and may not appear to the executive or user any differently than in the past. The difference is that smart data will be more reliable, timely, accurate, and complete. Therefore smart data is actionable. It will be the product of a more cost-effective approach to delivering higher quality information.

- How is smart data engineered?

Here lies the critical difference. Smart data is engineered for high states of interoperability among disparate users that includes semantic mediation that translates meaning from one enterprise lexicon to another despite differences. This is essential when conducting commerce or delivering critical human services, for instance, where inaccuracies are unacceptable and subject to legal liability.

To understand the difference requires (1) knowing how information is exchanged by the enterprise today, including the support footprint and costs, and (2) knowing how information is engineered in a smart data paradigm by contrast. Answering this question consumes the intent of this book. We want to provide executives with a glimpse about what they should expect from technologists to support them in their pursuit. We want executives to challenge the status quo as it has surely failed business and government in ways and with consequences that are still unfolding.

Automation can be a friend in sorting out the truth. However, automation can propagate damage when it is not managed properly.

Information technologists will point to the middleware and enterprise services as being different between what we suggest as smart data versus the norm. Smart data leverages metadata and open standards and introduces semantic mediation to produce an interoperable mechanism for data exchange that is independent from hard coded and brittle standard interfaces. Smart data advances more cost-effective data exchange and business transacting as a result.

- Who would resist change?

The answer is that most certainly the vendors that have vested interest in the status quo, and they are considerable in size and number. Strong customer enterprise executive leadership is essential to free the grip from legacy practices and infrastructure.

- What does a smart data environment look like? (Can one tell the difference?)

The information systems skill set and infrastructure for a smart data environment is different from the as-is state. The differences will appear in the organization chart, job skills, and corresponding spreadsheet as well as the introduction of performance standards that reinforce proactive support for executives. A principal analyst function will accompany the support system to provide customized response capability.

- How does it perform differently from other data?

Smart data strategy enables accurate data exchange and robust actionable data to support executive management. It does this in an open and interoperable environment characterized by high collaboration among participants.

- At what does the smart data strategy take aim?

Otherwise, what is the enemy? Can the enemy be the absence of data strategy? Can the enemy be deficient data strategy and technically deficient methodologies for data management and exchange? Before we direct the fight, we have much more to discuss.

Our proposed technical solution builds on the IEEE description below; however, leaping to the technical approach requires much greater understanding by senior executives about needs, problems, and requirements as well as the business case. Therefore tuck this away as it is intended for executive note and review (but for the exceptional executive in pursuit of technological advantage a deeper understanding may be desired).

Observe also that this early description appeared in 1991. There has been much iteration during the 19 years since, which indicates that the approach and smart data strategy required significant development and maturity before only now being employed as a practical and much needed solution.

It is in the search engines that we see rigorous application of semantic mediation, for instance, in what is called natural language processing. Google, Lexxe, and Powerset (a Microsoft acquisition) are examples. They address unstructured data interpretation problems for which applications are here and now.

From a referenced abstract from IEEE, "smart data" is defined as "a hybrid approach taken for the interoperability of heterogeneous information management systems. The authors combine the federated (unified/global) view approach with the multidatabase approach to provide access to relational databases, object-oriented databases, and spatial databases. Case-based reasoning techniques are explored to semi-automate the schema integration process.... In addition, the incompatibilities among the information systems that cannot be transparently supported by the federated approach are resolved using specified procedures" [2].

"Interoperability of heterogeneous information management systems" is a phrase worth understanding.

The University at Oldenberg, Germany, conducts a workshop on the subject and describes the following:

Interacting and exchanging information with internal or external partners is a key issue. If companies want to survive in the current market situation, a tight collaboration is needed. A seamless collaboration is possible, only if all participating information systems are able to semantically exchange information. The system topographies in today's companies are characterized by a high heterogeneity linked with various data models and formats. Interoperability between systems is therefore a major challenge for enterprises because of the high heterogeneity.

Today there are several approaches for creating a unified possibility of accessing heterogeneous information systems. Possible solutions are, for example, found in the domain of merging ontologies, linking or mapping different standards, or using model interoperability technologies like model morphing or transformation [3].

Alright, we are addressing the need for interacting and exchanging information with internal or external partners as a critical success factor. Seamless collaboration is a key attribute of the solution, as is semantic exchange of information. The technical terminology used here is what we are talking about as well.

From U.S. Patent 7072900, we discovered the following definition of "system topography."

A software system is designed using a layered approach that provides a topography that is suitable to a particular management philosophy or particular customer requirements. The topography can be viewed as a fabric that provides an infrastructure that supports the customer's management philosophy and other requirements. The topography addresses deployment mechanisms, such as interfaces between applications and users, security infrastructure, such as what control is asserted and maintained for the topography, component interaction defining how components installed on the topography interact with one another, and operation conduits that determine where and how processing is performed within the infrastructure. These same capabilities are addressed by other topographies that are developed. Common topography-neutral application components are designed and built to be installed on any topography" [4].

Witness that systems engineers and scientists are addressing the technical solution, and these are cursory artifacts offered as mere evidence. Before any of this makes sense, and it surely will to advanced information technologists, we must pave the way with executive education where plain business English is spoken, accompanied by popular science.

Management scientists and IT gurus have long sought scientific principles on which to guide and evaluate enterprise performance optimization. As Arthur Koeslter defined in *The Ghost in the Machine*, data are holons, something that is simultaneously a part and a whole. In fact, an enterprise is a holon. Daniel S. Appleton pushed the adoption of holonics as a way to understand enterprises in a more scientific manner. His influence can be seen in the Wikipedia definition as follows, which is given because it helps us to understand enterprise dynamic complexity.

> Since a holon is embedded in larger wholes, it is influenced by and influences these larger wholes. And since a holon also contains subsystems, or parts, it is similarly influenced by and influences these parts. Information flows bidirectionally between smaller and larger systems as well as rhizomatic contagion. When this bidirectionality of information flow and understanding of role is compromised, for whatever reason, the system begins to break down: wholes no longer recognize their dependence on their subsidiary parts, and parts no longer recognize the organizing authority of the wholes [5]. Appleton said an enterprise approaches entropy under these circumstances.

Today, we witness government enterprise approaching entropy, although government may not be aware that it is. There are at least three reasons for this: (1) enterprise leadership changes too frequently for them to be aware and to be accountable, (2) enterprise bureaucrats have vested interest in keeping things the same without the balance of consequences to motivate improvement or to punish poor performance, and (3) constituents do not have adequate visibility into government performance as it lacks needed transparency.

What does this have to do with data? The answer is that executives must be aware of the environment and, no matter what their duration or tenure, must understand that the fast track to producing extraordinary results is with a data focus. Enterprise stakeholders—shareholders and constituents—can serve their own interests by expecting executives to address performance and results with a smart data focus.

To be fair, some government organizations are pioneering solutions in the direction of what we envision. The Army is driven to do this to provide warfighters with what they need. The Defense Logistics Agency and U.S. Transportation Command (TRANSCOM) are collaborating to produce a common integrated data environment, for example. Necessity is the mother of invention, it seems.

We suggest that organizations with executive-driven data focus will set goals for data clarity, currency, quality, efficiency, rate, volume, frequency, cost, and timeliness related to specific support for planning, problem solving, decision making, sense making, and predicting. Goals will be set in context of enterprise parts that may be functionally defined; however, goals will also be set for shared capabilities that cut horizontally across vertical or functional disciplines.

To understand this requires some detailed discussion about data.

In practical terms, subordinate organizations perform autonomously in an enterprise whereby they generate and transact data, some of which is uniquely valuable and some of which may be redundant or deficient to such an extent that it is noise. Without a strategy that engineers data for optimal exposure to qualified users, including senior management, data remains in isolated pockets without proper configuration management and protection.

For example, there was an initiative in the Department of Defense to improve the tracking of material assets that range from a complete weapon system like a helicopter to electronic components comprising communications gear. Too much material was getting lost and the history of many items was incomplete, inaccurate, or otherwise deficiently managed. Repair and maintenance costs are extraordinarily high because sometimes systems get repaired or maintained whether they need it or not. At direction from Congress, property managers were required to place bar codes on all items of a certain minimum value. Having unique identification was the first step to asset visibility.

Bar codes contain data describing the item for identification, storage, use, and maintenance purposes including traceability to the original equipment manufacturer. While extraordinary effort was put into establishing standards and conventions for labeling items, including procuring readers to interpret the bar codes, planners did not allocate sufficient funding to capture, store, and manage the configuration of data. Information was captured electronically and sent to a repository that was not prepared for configuration management. Information piled in but was unusable. Therefore the exercise was a bureaucratic dead end.

In the haste to label items, some things were mislabeled, leading to shipments of the wrong items and items to the wrong places. This is a simplified description of a very large and complex problem. There was an instance of nuclear fuses sent to China that were thought to have been batteries!

Why do these things happen? This is an instance of an incomplete smart data strategy that should be a prerequisite to systems engineering in an enterprise context. We refer to this as the service-oriented enterprise (SOE) to distinguish it from traditional systems engineering and systems-of-systems engineering that is often performed out of enterprise context. Traditional systems engineering produces stovepipes, whereas SOE produces integrated and harmonious functionality. Smart data strategy is aimed at producing integrated and harmonious functionality.

Figure 5.1 illustrates that a larger enterprise comprises subordinate enterprises beginning with one that has a prime management role. The prime organization may have subordinate organizations within it. Some enterprise data is confidential and business sensitive and must be protected as internal data assets. Examples of prime organizations include Ford Motor Company and Boeing Corporation, each of which have thousands of subcontractors and suppliers participating as trading partners in what Michael Porter called "value chains."

Prime organizations have relationships with other subordinate enterprises, and these enterprises engage other smaller enterprises as suppliers, partners, and vendors. Some data is useful to many members of the enterprise and should be shared freely among qualified users with proper credentials and privileges. Some data is proprietary and private, subject to strict regulation.

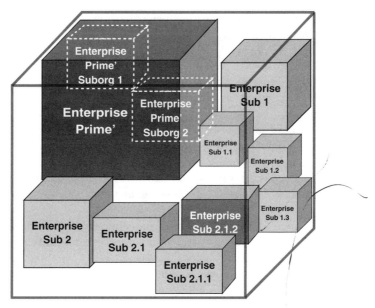

Figure 5.1 Enterprise boundaries.

Enterprise data is subject to governance that is manifest in business rules, policies, laws, and regulations. Relationships among enterprise entities are contractual, and contracts govern data and its uses. Contracts and other sources of governance vary with regard to time constraints and conditions making data management a complex and dynamic process. Such a circumstance begs for automated assistance that must be a part of a smart data strategy. Powerful tools are needed for use by the enterprise to manage complexity and the good news is that they are available and affordable.

Since much of the enterprise engineering is dependent on collaboration among holonic entities, communities of participants need to consider what standards, methods, and tools they need to ensure optimal and harmonious results. Some consideration should be given to establishing an enterprise community utility for data engineering and enterprise engineering. An enterprise engineering utility for communities of participants can help contain costs while achieving needed efficiencies and effectiveness.

One measure of enterprise performance is efficiency and effectiveness in data transaction and management. The cost and efficiency of enterprise transaction and management aggregate to the prime enterprise. Subordinate enterprises may serve more than one prime enterprise, some of which may be competing. All participants want to share equitably in the benefit.

The extent to which a prime enterprise can solicit contribution to the overarching enterprise smart data strategy leads to performance optimization. Subordinate organizations must serve many more customer enterprises than a single prime enterprise. For this reason, strategies that are based on open interoperability are better for them than those based on rigid standardization and proprietary applications. Flexibility is better than brittle rigidity.

Latent sources of data may be lost as unaccounted assets that absorb resources and create risk and exposure. They consume resources in varying amounts. They are like a light bulb left burning whether anyone is using the light or not. They are like a back door left unlocked, giving easy entry to unwanted intruders.

Often, organization executives want to make decisions that should be based on certain dependable data. They respond by directing analysts to go find and interpret data. Commonly, this is a costly wild goose chase because data needed was an afterthought. Users may stumble upon some useful data, make an effort to apply it and then abandon usage for lack of resources to support it. Data forensics reveals what appears at first to be promising, and then data fragments turn out to be useless.

Too often a good deal of time and resources are consumed trying to manipulate data fragments into something useful without successful results. Enterprises are staffed excessively with analysts for this reason.

Enterprises and their data can be modeled, which is synonymous with being engineered. Enterprises are complex and modeling demands engineering disciplines and professional tools. Executives should know that today's tools and technologies enable the complete modeling of the enterprise, such that they can see performance in real time and can plan and test strategies, situations, and ideas in simulation. Smarter organizations are able to do this to increasing levels of detail, proficiency, efficiency, accuracy, and predictability.

Herein lies the potential for competitive advantage. Having the ability to model enterprise performance and associated data supports agile planning, problem solving, decision making, sense making, and prediction. It helps organizations define their capacity for change and improvement.

Enterprise executives need a cockpit or command center from which to control and regulate the enterprise in real time, and today's technologies permit such creation and support. We use the U.S. Army and Department of Defense as examples in this discussion for several reasons: (1) these government organizations are very large and complex and are spearheading the best application of technology to catch up with commercial enterprise and global best practices; (2) their examples provide lessons learned that include good and bad instances for thoughtful consideration; and (3) access to information is publicly available in many cases.

At a breakfast meeting, the U.S. Army's CIO LTG Jeffrey Sorenson described the notion that warfighters, soldiers in the field, is where data strategy begins. Similarly, might a corporate executive say that employees closest to the customer are where data strategy should begin? Might the Secretary of Health and Human Services say that constituent citizens are where data strategy should begin?

In fact, only in the last half of the last administration did the U.S. government begin to address data requirements for citizens' healthcare management. Whether or not the strategy is a smart one is to be determined.

General Sorenson suggested that individual soldiers are from the computer generation and their higher computer literacy demands that the Army keep them equipped with state of-the-art technology to which they have been accustomed at home, in school, and in their commercial experience. Cannot the same be said about new employees in a commercial enterprise? Cannot the same be said about many American constituents of all government departments and agencies?

Then the general talked about the lag in the Army's supporting infrastructure, which is a product of a budgeting and funding process that is woefully untimely and deficient. It is constrained by business rules that inhibit discretion needed for agile enterprise management.

Some commercial enterprises might question the responsiveness of their annual forecasting, budgeting, and funding process. However, commercial enterprises are much more responsive to the needs of the business, keeping pace with demand, than government enterprise. Government enterprise is metaphorically hardwired by legislation, whereas commercial enterprise is more adaptive within constraints of public and private financing. These observations must be tempered by recent actions that blur the lines between government and commercial enterprise even further.

Referring again to the immediate situation in which Wall Street and government are converging, this commercial and government collaboration puts the emerging smart data utility on a critical path for business performance.

The Army general described the enormous challenge of accomplishing simple things like tracking soldiers in dynamic deployment, ensuring that they have an electronic signature, e-mail, and communications access that travels with them. He described the challenge to ensure that infrastructure keeps pace with the commercial rate of change and improvement. He described the circumstance whereby legacy systems and information technology infrastructure cannibalizes the Army IT budget, such that it is impossible for his organization to keep pace. The Army's state of the art is in a perpetual case of falling behind as budget and resources, and enabling processes, are deficient.

Distinguished leaders like Dr. Paul Strassmann, the Department of Defense's first CIO and now distinguished professor at George Mason University, continues to warn how infrastructure and legacy systems undermine government performance by consuming an inordinate share of resources just to stay even. Staying even is truly falling behind the commercial rate of technology advance.

We witness and hear these complaints and observe that in an economy already exacerbated by competing demands, it is imperative to think outside traditional approaches. Technology supports smart data strategy whereby we assist executives to leap beyond legacy constraints and engineer data such that it hurdles barriers of past performance.

One measure of success will be to reduce expenditures in legacy maintenance while directing increased investment in smart data engineering and management with corresponding improvements in support to management. To accomplish this requires clear articulation of a new strategy. It requires investment in information engineering while halting the practice of throwing good money after bad on antiquated approaches.

3. *A primary responsibility therefore is to know your assets and to know how your assets are represented and transacted as data.*

The Department of Defense devoted considerable resources to modeling the enterprise, creating an enterprise architecture expressed as process models. The technique allows for tracking data as inputs and outputs, controls, and mechanisms. The effort must be continuous, as organizations are in a constant state of improvement and, sometimes, degradation.

When the military goes to war for extended periods, or when the Department of Homeland Security and FEMA manage multiple hurricanes, their equipment wears out and supplies run short. People in the field identify new requirements that create new priorities. New data flows through the system.

Historically, it is known that the data and associated data models for an enterprise are inherently more stable than the process models. That is, the core data needed to manage the business does not change as variably as the enterprise activities.

One can imagine the impact of announcing the integration of 22 disparate departments and agencies into the Department of Homeland Security. The intent was enterprise integration, but what resources would be required to address the order of magnitude change? Was there a data strategy? There was not, and surely there are significant opportunities to improve performance by addressing this requirement.

In the instance of the Department of Defense enterprise and its subordinate military service enterprises—Army, Navy, Air Force, and Marines, where Marines are a subordinate element of the Navy—there is a unique relationship with weapon systems products. Weapon systems products are assets that are represented as data with corresponding models.

Weapon systems products are intellectual property whereby ownership is variable based on unique contracts between the prime producer and the government customer. For instance, the Army's Future Combat Systems (FCS) is based on the integration of 44 unique technologies, all under development at one time and being completed at various stages. One product of this is a lethal mechanized vehicle that carries troops and ammunition as well as serving as an integrated communications technology platform. The Army's concept of operation is designed to leverage this piece of hardware.

A component of the Army enterprise is therefore designed around the availability of weapon systems technologies, in this case produced by Boeing Corporation as a prime contractor. The operational deployment of the Army warfighting enterprise is dependent on the critical relationship with the prime contractor and its suppliers.

The same can be said for the Joint Strike Fighter (JSF) that is managed by Lockheed Corporation. Interesting dynamics are revealed when considering who owns the weapon systems data. The U.S. Department of Defense might claim to own the JSF weapons system. However, the Lockheed contract specifies that data ownership is with Lockheed and data is behind the Lockheed corporate firewall.

The JSF product is being produced by an international team of contractors under Lockheed's leadership. The JSF product is sold or licensed to a host of international customers. The Lockheed supply chain includes international suppliers. Customers and suppliers need and want access to JSF data. Who get what? Credentialing and privileging data access is a complex process.

The advanced weapon system is covered under the U.S. Arms Export Control Act and other laws and regulations and therefore is subject to governance by the United States Customs and Department of State.

Therefore when the DoD contracted for the JSF under a performance-based service agreement, for instance, who has access to the product data and under what circumstances? The autonomic logistics support system developed by LMCO is a Lockheed asset, as is the data inside it. This means that the U.S. government does not have free use of the technology that it paid to have developed, as it is outside the contractual boundary.

These examples just begin to indicate the complexity of issues related to knowing your data. Being left in the fog without an explicit data strategy does not disguise the responsibility and associated risks. Smart data strategy is essential to concise and appropriate management visibility and accountability.

A data repository is where data are stored. The enterprise may have a data repository that is centralized or dispersed. Certainly data that is outside the organizational boundary are dispersed. Locations to where the enterprises send data are dispersed. There are various strategies about pushing and pulling data to and from suppliers and customers. Critical to developing any data management strategy is having a smart data strategy and making data explicit. Dispersed data begs for supporting a virtual data environment.

The historic progression describing data connectivity is as follows:

- Point-to-point
- Hub-and-spoke
- Virtual

Point-to-point is the most expensive type of interface to maintain because it requires constant maintenance and is most difficult to change. Hub-and-spoke is better and less costly but requires central control that runs counter to enterprises that desire greater freedom, innovation, and agility. Smart data strategy supports a virtual enterprise environment. Virtual enterprise is dependent on a federated data strategy, whereby members employ a self-regulated approach. Smart data strategy is a federated data strategy.

Enterprise data is made explicit by modeling in accordance with international standards that address syntax and semantics. Syntax refers to data formatting and semantics refers to how meaning in language is created by the use and interrelationships of words, phrases, and sentences.

Information technologists devoted considerable effort to data formatting, although now technology accommodates diversity with greater ease. There is less need for structured formatting that leads to brittle and costly interfaces, for instance.

Enterprise communities have devoted considerable resources to harmonize data dictionaries and implementation conventions for the purpose of ensuring precise meaning. Today's technologies make it easier to achieve precise meaning while being more accommodating of diversity in the use of languages and terminology.

When communities such as industry associations elect to simplify through harmonization strategies, with today's technology they can do so without dumbing down the enterprise lexicon that may support competitive advantage. Dumbing down data was the old paradigm that restricted pursuit of best practices and innovation.

Popular previously were rigid standardization approaches that are counter-intuitive to enterprise needs for agility. Legacy systems remain prevalent today such as electronic data interchange (EDI) based transactions that represent technology solutions that are over 30 years old and fail to leverage today's advantages. They are the enemy.

Smart data strategy promotes accommodation of enterprise diversity in both syntax and semantics. Some correlate accommodation of diversity with entrepreneurial freedom and innovation.

Smart data strategy views investment in rigid standardization of terms as a resource drain when employing advanced technology is a more efficient and effective approach.

Corporations and government organizations invest considerably in industry working groups and standards bodies to harmonize the use of standards and implementation conventions. Standards that govern data transactions must be maintained. If user organizations approach this activity passively, then they do so at the will of others.

Smaller organizations do not have the resources to participate in the host of standards initiatives affecting them; therefore some participate in associations where industry and professional delegates represent their interests. These activities often involve global participation and governments may manipulate or influence policies favorable to their position. Enterprise executives need to be aware of the politics of standardization relating to their businesses because the consequences have significant associated costs.

Smart data strategy addresses this circumstance by advocating executive action as follows.

- Restrict participation and use of standards to those that are most beneficial to your organization and community of participation.
- Favor standards that are open and interoperable and nonproprietary.
- Employ technology to engineer data such that open interoperability is supported, implemented, and managed automatically.
- To measure the benefit from the strategic element, audit your organization's current use and support for standards, and define their characteristics. Compare and monitor the difference when adopting an automated approach that is a result from smart data engineering.
- Select vendors as allies in implementing smart data strategy: reject proprietary solutions and accept open and interoperable technologies.

4. *Data assets may be inputs into your enterprise processes; they most certainly are outputs from enterprise processes, as they are enabling mechanisms.*

   Needed is a way to model and track each of these elements in relation with one another. There are modeling techniques that are supported by software tools that capture these relationships. Some techniques are quite mature, including the IDEF family as well as the UML family among others. It is not our purpose to teach these techniques in this book; however, we use them and we refer to government examples where they are employed to produce enterprise architectures for government organizations.

   Tools that information technologists find useful for programming may not answer questions that executives need answers for in order to optimize performance in the enterprise. Executives need special methods and tools as do information technologists. Needs and tools overlap and enabling technologies exchange information between them. Seamless data exchange is a part of smart data strategy.

   Certain terms are shared among all members of the enterprise. For instance, *inputs* are things such as capital and materials that are transformed by *activities* into higher yield and higher value *outputs*. Outputs may be products, also called solutions, as well as their corresponding outcomes.

   Activities are the lowest order elements comprising *processes*. Processes and their activities are work that is performed by people and combinations of people and technologies—hardware, equipment, and software whereby people and technologies are *mechanisms*. Their interaction is sometimes called *systems*.

   Mechanisms actually perform the work defined as processes under "constraints" such as business rules, laws and regulations, policies, budgets, and schedules that are also called "controls." Mechanisms consume capital and materials as well as time to complete activities.

   Herein, executives take note. The consumption of enterprise capital and time can be managed to the lowest level of activity. The data, and information reporting and describing outputs from activities, can be tracked as value-added. Value additions can be tracked to the point where they become tangible assets. At last, this is true asset and resource management.

   The discipline employed to design work is systems engineering, which may involve various engineering and other professional specialties. Systems engineering adds such considerations as workflow and rate, volume, frequency, cost, time, and quality.

   Systems engineering is a relatively recent invention in human history, invented at Bell Labs in the 1940s and 1950s. The editor of this series of Wiley books, Dr. Andrew Sage, Professor Emeritus, George Mason University, is a pioneer of this discipline. Enterprise engineering is systems engineering in an enterprise context. This is an important distinction because systems engineering has been performed to produce products and outcomes in accordance with specific functional requirements.

   Functional requirements generally correspond with certain processes that are aligned with professional disciplines or professions. This is sometimes described as vertical alignment or stovepipes. Systems engineering typically produces solutions that are specified by functional domain and associated organizational domains. This circumstance is problematic because enabling technologies and people tend to become stovepiped. Stovepiping is counterintuitive to enterprise integration, whereby the enterprise seeks seamless performance without islands of automation or noninteroperable functional domains.

Computer technology is an even newer invention. During the past 40 years, business and government have developed computer systems, layering them on top of one another. Systems get replaced and modernized for different reasons and on different schedules. Without a common enterprise strategy in approaching the need to change and improve, the result is chaotic and entropic. Focusing on data is one way to address the need for modernization without getting hung up in the infrastructure.

Since all of these things are rooted in data, it is important to discuss them as we try to define how to engineer enterprises such that they can be managed for optimal outcomes.

Some believe that the most important assets are people and therefore we need to talk about them with special consideration. AT&T once used the term "personnel support subsystems" to emphasize that some systems are subordinate to people, although that is not always the case.

Some may argue that people are the most important aspect of an enterprise, where people are employees, customers, and stakeholders of a variety of types.

Recently, in America, the healthcare records of each citizen have been considered in the context of developing a standard that will make medical histories digitally portable. Nothing is more personally touching than that as an example. Yet our health records are just a beginning. We have banking records too and a host of records describing where we live, how we live, and even what we eat.

In the Army, food and caloric intake is planned as part of a warfighter's logistics footprint. Civilians are also preoccupied by what they eat, though with much less precision than warfighters by contrast.

In considering smart data strategy, we must be specific.

For commercial enterprise, "customers" are people who may be end-user consumers, or they may represent an organization that pays for products and services, whereas organizations may be public or private organizations, corporations, and government entities. In business-to-business and business-to-government relationships, people have roles to play in representing their affiliate organizations.

Richard Beckhard, the "father of organization development," from MIT described the idea that when businesses purchase something, a pattern appears as follows.

- Responsibility: One and only one person has the "R" for buying something.
- Approval: Typically the "R" must get approval from finance and from one higher authority.

- Support: The staff of "R" and "A" typically support the process with as many as five people involved.
- Information: There are usually a couple of people in other organizations, inside and outside the buying organization, that need to be informed.

When you add up the number of persons it is at least 10 people involved in the decision. These people are considered influencers of various types. They represent people who need access to data to support their roles in the process. Not every person should have equal access to all of the data, as individuals have qualified privileges based on their credentials and access privileges.

Employees are people who perform specific functional duties in organizations interacting with supervision and with others.

Citizens are the ultimate constituents for government as customers are the ultimate targets for commercial business.

All people have logistics footprints comprising all of the things a person needs to live and execute their responsibilities. They have personal footprints in their private lives at home. They have footprints within the domain of organizations in which they belong. The footprints are expressed and described by data.

Our lives depend on our logistics footprints and the data therein. For instance, a warfighter cannot survive long without ammunition. A sick person may not survive long without medication. When the data links are broken, people may become stranded and vulnerable.

An employee of a commercial enterprise has a different contractual relationship with an employer than a warfighter has with his/her military service government employer. Yet, both have logistics footprints comprising capital, materials, other people, and technologies: equipment, machines, software, real estate, infrastructure, and so on.

Interesting is the relationship between people and machines in the context of a system. People may be supported by machines. Machines may be operated by people. Machines may perform work formerly performed by people, and there are economic decisions to be made about which is better—people or machines performing certain work. Making these decisions is influenced by the environment and by economics, which are represented by data.

In the military enterprise, systems are often developed around weapons systems. In government enterprise, planning and problem solving often begins with a propensity to preserve the bureaucracy or organization to which people belong. In part, this is because people are leaders of the organization and participants who want to preserve their jobs first.

We report this observation because when smart data strategy is considered as a catalyst for change and improvement, it is imperative that it be done from the top down such that it does not get compromised by subjective considerations that may resist change.

Yet, as we complete our work today, congressional leaders are second guessing automakers closing dealerships, otherwise compromising objectivity in the process of changing and improving.

Defense Secretary Gates completed a review of the type we advocate, which broke from the tradition of permitting bureaucracies to influence the analysis; thus many more opportunities to reduce expenses and to reallocate resources became apparent.

On the other hand, when the Health and Human Services leadership developed their strategic plan, they socialized the effort to include representatives in workshops from all aspects of their community.

These are two striking examples of very different management approaches to planning and strategizing.

5. *Some data assets are shared with others in the enterprise community who are sometimes called trading partners, customers, constituents, and stakeholders.*

From an executive perspective, here are some considerations. Participating in a roundtable discussion hosted by the Association for Enterprise Integration (AFEI), an affiliate of the National Defense Industry Association (NDIA), the topic was "Social Media." Attendees included representatives from the defense industry, and many more representatives from federal government agencies and their CIOs: Defense, Homeland Security, and Environmental Protection Agency, among others.

What do "social media" have to do with smart data?

George proposed a definition to the group in the form of a whitepaper: Social media is a utility enabling individuals and organizations (government and commercial) to accelerate their information and communications needs for actionable data among their communities of interest.

At the time of this writing Dr. Brand Neiemann, EPA Sr. Architect, Office of CIO, EPA who was in attendance, said that he had been tasked by the Federal CIO Council to develop a draft policy for what the Obama administration is calling "Data.gov." Data.gov delivers on the promise to share with the American people everything about the government in as complete transparency as possible.

- *Data entering your enterprise that you want.* Driven by your balanced scorecard, for instance, what information do you need and want? Information that you need is represented in the data model for your enterprise. Your enterprise data model depicts all of the data needed by your processes to achieve prescribed results. It should include identifying sources of data, internal and external.

  When performing a data audit, you may determine that some external sources are not providing data that you need. When these external sources are subcontractors and suppliers, you may have leverage to improve the situation. When these external sources are customers and competitors, you may discover different means for improving data accessibility.

- *Data entering your enterprise that you don't want.* Cyber security restricts most intruders. Data that is inaccurate for any reason is information that you do not want. There are other criteria for unwanted data and that may be worth brainstorming.

Data that is redundant, the same data stored in many places, is an example of unwanted data because it is excessive and demands configuration management. Data that lacks configuration management where it is unreliable is unwanted.

- *Data that you buy.* Data that you buy may be an asset with limited shelf life. That is often the case. Nonetheless, it is an asset as long as it is useful. Sometimes, data is maintained longer than its useful life and that absorbs extra cost.

  Disposition of data assets deserves deliberation because sometimes data may lose its immediate value but its value may resurface as a matter of historical reference and for trend analysis and prediction.

- *Data generated within your enterprise that stays within your enterprise.* Some data is proprietary and must be protected with high security, accessible with credentialing and privileging rules. How is this data identified and protected?

- *Data generated within your enterprise that is shared with others outside for free.* Some data should be shared freely with partners and customers. Some of this needs protection from unauthorized users, such as competitors and enemies. Careful consideration as a matter of routine will maximize advantage from sharing while limiting risk.

- *Data generated within your enterprise that is shared with others outside for a fee for limited use.* Some information is so valuable and proprietary that it is worth charging for use under licensing and other contractual terms.

- *Data generated within your enterprise that is shared with others outside for a fee for unrestricted use.* Some data may be licensed or shared under unrestricted use as prescribed in agreements. This circumstance is where data use contributes to strengthening brand awareness or market development, or otherwise strengthens the contribution to the federated way of doing business.

  From work on an advanced fighter program involving a multitude of customers and suppliers with varying contractual privileges, we understand the necessity to be deliberate and thorough in addressing these conditions describing data.

6. *Data are governed by business rules that may be contractual and regulatory.*

   Some rules are inherited or imposed, while others are invented by management. Rules constrain processes, and they constrain how and where data assets are used or otherwise shared.

   Anticipated are trends toward automated regulatory environment and automated contractual environments. This is where legal and regulatory code is delivered to govern systems replacing paper documentation. Anticipated is that there will be combinations of paper instructions, electronic words, and actual code for automating compliance, validation, and verification. These things are possible with smart data.

7. *Creating, receiving, processing, publishing, accessing, and storing data assets consumes resources expressed by labor to support data management as well as essential enabling technology.*

   From the latest experience, we have learned that it is essential to make certain that the organization has the right skill sets to support smart data strategy to perform all of the data processing and management routines. It is essential to employ resources that have up-to-date certifications and knowledge about essential enabling technology.

   It is fine to leverage the current workforce; however, it is essential that they are equipped with the latest knowledge and with the most current tools and technologies to create a smart data environment. Failure to keep pace is the source of excess cost and uncompetitive performance.

8. *Data must be engineered for smartness so that it is interoperable and secure.*

   From experience in supporting advanced weapon systems programs and logistics systems modernization for the Department of Defense, we identified special skills, knowledge, and proficiency needed to implement smart data strategy. In addition to routine skills such as program and project management there are additional skills needed:

   - Integration and interoperability architect
   - Transformation/mapping and data architect
   - Process modeler
   - Adapter developer
   - Technical infrastructure services architect
   - Librarian taxonomist
   - Repository specialist
   - Content editor
   - Business case analyst

   Also, having appropriate domain expertise is essential for collaboration with the customer in requirements planning.

   Figure 5.2 illustrates the operational solution environment. Three engineering disciplines are present in the illustration: process engineering, grid engineering, and data engineering. Smart services, smart grid, and smart data are products from these disciplines that are created and managed with an enterprise view.

   The SOE paradigm was originally developed as part of the DoD Electronic Logistics Information Trading Exchange (ELITE) program. The intent is to bring together the DoD's net-centric strategy and commercial service-oriented architecture (SOA) web services technology. Smart data is the missing link in the DoD and government strategies, and that is how we discovered the topic. It is not our intention to discuss the detailed merit of DoD strategies, although it is important to note that considerable effort is made today at implementing part of this approach. Addressing the omission will help improve its success and effectiveness.

   Here is a description of the critical terminology introduced in the figure that deserves understanding by executives because they are elements that

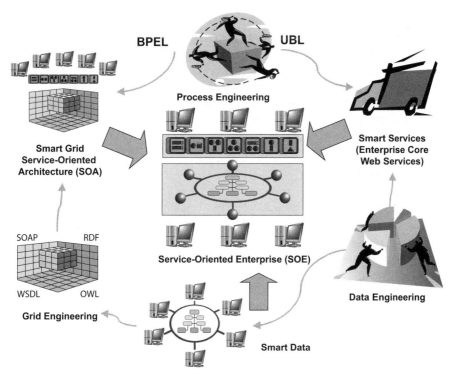

Figure 5.2 Service-oriented enterprise model.

your enterprise will need to develop under the guidance of smart data strategy.

- Service-oriented enterprise (SOE) icon illustrates that data models are exchanged via a plug and play capability that accommodates a variety of plugs. These things are a product of smart data engineering.
- Metadata engineering and metadata management are integral elements of smart SOE and smart data where metadata is data that describes data.
- Smart grid describes the use of SOA that provides infrastructure enablement. This is a product of grid engineering, not to be confused with the idea of the same name applied to power grids.
- Process engineering and business processes are the practical and functional context enabled by advanced standards and technologies among which are BPEL and UBL. BPEL is an XML-based language for the formal specification of business processes and business interaction protocols. According to OASIS, "UBL is Universal Business Language that is the product of an international effort to define a royalty-free library of standard electronic XML business documents such as purchase orders and invoices."
- Smart services refer to enterprise web services that are possible through the combination of SOA and smart data operating under an SOE paradigm and

smart data strategy. Smart services are enterprise services, with functionality that services the enterprise as a common utility.

9. *Smart data strategy represents the best attempt at defining governance, rules, and the means for interoperable data exchange and usage by all members of the user community based on specified credentialing and privileges.*

Interoperability is enabled by applications of open source and international interoperability standards as discussed.

Sources of Governance and Rules.

- *Enterprise Internal.* Internal rules and constraints include all self-imposed controls. They may appear in leaders' value statements, annual reports, policies and plans, contracts, procedures, and those that are embedded in software that has been adopted by the enterprise for certain uses. Budgets and schedules are also sources of governance and guidance.

- *Enterprise External.* Laws and regulations, customer-imposed policies, procedures, and contracts are enterprise external controls. Suppliers too may impose external requirements under different circumstances that are contractual and that may come about indirectly when the suppliers cannot comply with their customers' requirements.

The notion of credentialing and privileging is most commonly associated with the healthcare industry, although it is directly applicable to establishing advanced smart data environments. Every user has a profile whereby individuals require certain skills, knowledge, and proficiency, and whereby their roles and responsibilities require the granting of access privileges by participating enterprises. While credentialing and priviledging is accomplished today on an ad hoc or enterprise-by-enterprise basis, there is room for improvement as smart data ideas mature.

10. *Smart data strategy is measured by attainment of outcomes with corresponding metrics that represent the most efficient and effective use and protection of data assets.*

Performance improvement goals from smart data strategy will target primary areas, including:

- Infrastructure cost and operations improvement
- Enterprise shared services as a means of retiring ad hoc, redundant, and dispersed services
- Proprietary and standards-based functionality to be replaced by open standards and interoperable solutions
- Associated information technology labor for maintaining rigid application interfaces

Identifying and quantifying opportunities for significant improvement is the product of a smart data strategy audit that will track data use and exchange throughout the enterprise, quantifying associated labor and resources and associating this with enterprise performance and functionality that executives can understand.

11. *Data assets are owned by the enterprise and controlled by CEOs, enabled and supported by information technologists in accordance with smart data strategy.*

    Does the enterprise have a policy and strategy for smart data management that addresses the subject from the CEO to the enterprise? It begins with plain business language. While our book is careful not endorse any vendor, in researching enterprise data policy we discovered an announcement from Tizor Systems Corporation, a leading provider of enterprise database monitoring and protection solutions for the data center. On September 9, 2008 it was announced that the United States Patent and Trademark Office awarded the company a new patent (U.S. Patent Number 7,415,719). "The patent is for the first comprehensive data auditing policy framework that integrates easy to understand English-like rules and templates with behavioral analytics to detect and prevent data theft and data breach in real time." While this technology is more narrowly focused than a broad enterprise policy, it serves to note that technology firms are addressing the topic in practical ways.

12. *Smart data strategy is unique to the enterprise as it is a matter of degree.*

    Smart data strategy has the potential for competitive advantage, and conversely, it may be a factor in competitive disadvantage.

    Because smart data strategy is rooted in leveraging open standards, the competitive baseline is neutral. However, the degree to which an enterprise pursues strategy implementation among its trading partners provides a measure of competitive advantage with regard to depth and breadth of strategy implementation.

13. *Problems and symptoms are visible that can be mitigated through smart data strategy, such as enterprise data being contained in a myriad of places and systems.*

    Data analysts are assigned to obtain and interpret the data and this takes a significant investment of time and resources. No one is really sure about the ratio of results to investment. Improvement comes from becoming more precise and specific about how data impacts and is used by the enterprise.

    Business intelligence (BI) is a term that has been in use at least since the late 1950s, referring to the basis for decision support. Today, BI is associated with decision support systems (DSSs) that are, of course, data driven. This subject includes data warehouses and data marts as the place where data resides. However, the smart data paradigm connotes more dispersed data and the technological means of pulling data from distributed places for the purpose of answering questions in real time. Performance management and optimization require real-time responsiveness combined with enabling modeling, simulation, and prediction. Furthermore, BI is the data store for benchmark performance data.

14. *An organization can have an excessively large IT footprint, though no one knows for certain.*

    Indications are that IT footprints for mature organizations are too large or excessive. The reasons for this come from the fact that systems have evolved and legacy infrastructure and data have accrued. We have already reported how some large organizations have identified this as a significant problem as operations budgets consume resources for maintenance that could otherwise be used for performance improvement through advancing technology.

Breaking this cycle is a huge target for government performance improvement, but it comes at the expense of reducing expenditures to some systems integrators that profit from inefficiency.

Here is an interesting observation. When the GEIA electronics industry association conducts their annual review of government expenditures, they identify operations and maintenance budget items as the onerous enemy because increases in operations and maintenance (O&M) come at the expense of investment in new electronic technologies. Product sales suffer when service labor sales increase.

On the other hand, when aerospace defense contractors who are in the systems integration business meet, they see operations and maintenance increases as a positive development because it means more contracting for professional services.

Smart data strategy is an objective means of identifying opportunities for performance improvement and therefore should be conducted by third-party objective professionals. This also means that objectivity must be preserved for internal and external analyst and auditor participants.

15. *It is too costly to maintain interfaces to a host of legacy systems inside the organization and among enterprise constituents.*

A legacy system is a computer system or application program that continues to be used because the user does not want to replace or redesign it. Legacy systems performance is often deficient, slower performing and more costly to maintain. Often, their use is so high and critical they cannot be taken offline. Redesign is often costly due to their being large and complex and often documentation needed to understand the code and how the system works is deficient.

For these reasons, there needs to be a way to exchange data with and otherwise transact with these systems while injecting a degree of agility, adaptability, and flexibility. Smart data is the technological means for achieving this.

16. *It is also too costly and impractical to change to one perfect solution.*

Smart data is the means to overcome legacy system barriers. It requires stopping investments in maintenance and diverting resources toward new solutions.

17. *Often, the data needed is found in fragmented and nonconfigured states and therefore is incomplete and unreliable.*

This circumstance can be eliminated by beginning with knowing what decisions the enterprise needs to support and providing configured data for reporting, decision making, sense making, and predicting.

18. *Incoming data takes extraordinary effort to translate into usable format and syntax for internal systems.*

Smart data strategy places the burden of deviation on the source of the deviation. That is, smart data strategy provides a universal basis for data exchange that is addressable via mapping from the enterprise to the universal translator and from there to diverse users. Any enterprise that desires to participate in the community is responsible for exchanging data through the agreed upon utility.

19. *Outgoing data may or may not satisfy user needs.*

    Data that is processed through the enterprise utility solution that is a product of smart data strategy will satisfy end-user needs.

    - *Data is overly redundant.* Data redundancy excesses are eliminated through this strategy.

    - *Data is overly costly.* Smart data strategy aims at providing the most cost-effective means for enterprise participants to exchange data. It also contributes to overall reduction in operations and maintenance costs.

    - *Data is not actionable.* Data is not actionable if its source is unaccredited or unknown. Processing data through an enterprise with smart data strategy provides data that is actionable. When deficient for whatever reasons, the caveats are identified and attributed to the source or cause.

20. *Decision support is deficient because needed data is absent, incomplete, and lacking in quality.*

    The best way to ensure that decisions are supported with adequate data is to identify decision requirements as part of the smart data strategy.

21. *Sense making and prediction are deficient for the same reasons, and so is data for planning and problem solving.*

    Requirements for sense making and prediction must be anticipated. When they are not, the smart data system will support ad hoc queries with reliable data and with known caveats.

22. *Data storage, access, and processing are excessively expensive, or the enterprise does not have a good metric for knowing whether data management is effective and efficient or not.*

    Smart data strategy advocates an audit that will produce a baseline from which to improve performance.

23. *Infrastructure costs cannibalize resources, leaving little for improvement and discretion needed for agile and adaptive performance.*

    Having the discipline to halt the cause of deficient performance is made easier by having a detailed plan to mitigate risk from change and improvement.

24. *Enterprise performance is affected by incoming data.*

Data from Customers   Data from customers identifies and describes them, including demographic and psychographic data. Customer profiles can be detailed and private. Customers may place restrictions on using their data.

Some customer classifications include retail and wholesale, for instance. In the Department of Defense, the lines between retail and wholesale are blurring as a result of seeing cost reductions and performance gains similar to those discovered by Wal-Mart, for instance. Also, the lines are blurring between logistics and transportation.

Data from customers includes their orders for products and services. For large enterprises this process is massively complex.

Distinctions are made between customers for consumables and customers for capital equipment, for instance. Customer requirements vary widely with regard to what they are buying and consuming.

In the case of government customers, the process in accessing them is through an acquisition and procurement regime that can be long and complicated. It is governed by the Federal Acquisition Regulation (FAR).

There are information services that provide data about government customers and their procurement plans and status that includes identifying contracting officers and the like. However, as in the case with commercial business, direct customer relationships make the difference as well as knowing details about customers' needs and organization missions and priorities.

Customers come in different types depending on your enterprise view:

- Clients
- Commercial customers
- Constituents
- Consumers
- Decision makers
- End users
- External customers
- Government customers
- Influencers
- Internal customers
- Product customers
- Service customers
- Warfighters

Inc. magazine posted information on its website addressing this subject.

Inc. says that "cynicism about the use and misuse of customer data" underscores the importance of having a "clearly written privacy policy." The magazine also suggests that you might have to provide incentive to customers to provide needed information. Part of the incentive is for service providers simply being able to understand customers better so that their needs can be better serviced.

What is the technology component of smart data strategy that is used to gather and protect customer data? Customers may be asked to provide requirements. From research we discovered that the Army AcqBusiness organization is established to provide web service supplied data and analysis to Army customers from more than 40 organizations. They have a website soliciting requirements from customers. The Army established a committee to review and prioritize the requirements from which some will be selected for action.

The trouble with this approach is that it leads customers to believe that their needs will be addressed, although some may lose in the evaluation process and not get attended to in a timely manner. It is important that customers understand the rules and intent of service providers such that their expectations are properly managed. That is part of managing customer data.

Websites and traditional survey methods are employed to gather information proactively, although other information comes from analyzing business transactions. The data is collected; users and customers must understand the policy governing this information. Both commercial and government enterprise must "control the need to know." We call this credentialing and privileging. RSS software is employed to gather intelligence on clients and competitors.

Many organizations simply have too much data and are unable to act upon it because there is no smart data strategy. Ultimately, customer data must be integrated into internal systems. Many enterprises have adopted the term customer relationship management (CRM) as the rallying point for pulling in the data and making it actionable.

Smart data strategy addresses the technical mechanism for translating data from customers' environments into your environment automatically.

There are reports about redundant customer data and difficulty in managing it. We discovered a company called Purisma that uses fuzzy logic to sort things out. Apparently they were on the right track as they were acquired by Dun & Bradstreet [6]:

> Purisma's unique self-learning matching system progressively learns from contributing data sources, external reference data and data steward actions to ensure continuous improvement in match quality. While some meta data management (MDM) solutions throw away key knowledge about data correlations and hierarchical relationships, Purisma maintains complex relationships between elements by retaining associations with data entries such as alternative spellings, abbreviations, misspellings or data entry mistakes. As a result, the Purisma Data Hub™ allows you to:

- Gain the highest match accuracy
- Continuously improve accuracy over time
- Improve match quality by learning from data stewards
- Improve match rates based on both internal and external reference data

This is one example of how technology is keeping pace with technical requirements for smart data strategy.

Data from Suppliers We discovered a company called Austin-Tetra with a product called Supplier Data Management™ that "creates a consolidated view of enterprise supply base from fragmented, disparate and inconsistent data silos" [7]. We use this company's product as a basis for comparing and contrasting smart data strategy.

The solution "leverages best-in-class content, extensive knowledge base, and proprietary technology to support better business decisions, reducing costs and risk" [7].

By contrast, the proprietary aspect is external to smart data strategy that leverages open and interoperable standards instead.

The Austin-Tetra product addresses data quality requirements: data cleansing (eliminating duplicates and fixing errors), standardizing data formats, configuration management (ensuring data currency), and maintenance. These functionalities are essential to smart data strategy.

By contrast, smart data strategy does not standardize data formats; it maps and translates, providing data exchange based on open and interoperable standards for both syntax and semantics.

Other Austin-Tetra capability includes risk assessment, analyzing possible disruption in the supply chain, which is an application of decision support. The Austin-Tetra product addresses "process improvement" that is a theme for smart data strategy. In the supplier data context, data is used to profile and evaluate suppliers, for prequalification, and for workflow and monitoring that is a part of policy compliance.

In another example, "The Digital Entertainment Group and consultants Capgemini and Teradata finished the initial phase of an industry-first supply chain study analyzing merchandising execution at what's known as the 'the last 100 feet' of retail, from the stockroom to the store floor. What they found was significant room for improvement in data synchronization between suppliers and retailers" [8].

Large scale industry and government initiatives were initiated in the early 1990s to attend to supplier data as electronic commerce took off. These efforts are the bureaucratic-intensive initiatives that consumed many resources to harmonize the use of standards and implementation conventions. While such efforts were useful in focusing attention on supplier and prime producer interaction, they had difficulty accommodating advancing technology as the perception was that it was easier to "standardize" in rigid electronic data interchange technology than to accommodate the emerging service-oriented architecture and web services.

Looking at Boeing, for instance, the company governs supplier relationships with a standard trading partner agreement.

> Boeing employs a "Trading Partner Agreement for Electronic Commerce" that is a document containing legal verbage describing at a high level, the potential risks and liabilities inherent when using the internet to transmit data between companies. It is generally a "for your information" type document to remind suppliers that this is typically a reliable process but there may be unexpected delays to deal with [9].

Like many large companies, Boeing has a supplier portal that is password protected providing "instant access to information by anyone from anywhere in the world" [9]. Boeing says that "the key characteristics of the Supplier Portal are that it provides: 'One Company Image to our Suppliers'" [9].

Boeing employs the term "single point of entry (SPOE)" for Boeing simplified access to complex information links to news and events pertaining to supplier activities across Boeing.

Smart data strategy addresses the details of how suppliers and prime contractors/producers exchange data in a manner that is open and interoperable.

When supporting the Lockheed Martin Joint Strike Fighter SPOE program, Lockheed and the government customer had certain objectives and requirements.

Technical Objectives.

- Virtual exchanges instead of fixed integration points
- Not platform dependent
- Not site dependent
- Can accommodate multiple co-op/security strategies
- Incremental integration, not big bang style solution
- New integrations do not disturb legacy integrations
- Synchronization of budgets and schedules not required
- Enterprise context instead of data stovepiping
- Local information models are maintained as interoperability is achieved
- Reduction of glue code over point-to-point integration
- Mediation requires less custom code than point-to-point integration
- SPOE reduces interface adaptor requirements
- Service component architecture
- Service-oriented solution with coarse (not chatty) service exchanges
- Best practices and patterns for SOA design
- SOA component guidelines—universal SOA deployment
- Pluggable services that can be leveraged alongside many COTS middleware providers
- Source or custom deployments as needed by SPOE consumers

Data About Competitors or Enemies. On the verge of a new era in transparency, much more information about your competitors will be in the sunshine, so to speak. If your enterprise is commercial and you address the needs of a broad consumer public, the chances are you will discover more about your competitors than if you are tightly focused on a niche. Even then, there are advantages about operating in a smaller space, although customers and competitors tend to be more tight-lipped.

In the government space, information about competitors is widely available from both direct and indirect sources. The question is, what is being done about it? Is it prepared such that it is actionable? Often, it is treated more tactically than strategically. Decisions are often made based on old beliefs more so than facts.

There are layers and layers of information about enemies and competitors in which to delve for better decision support. The most efficient and effective way to leverage data resources that answer difficult questions is through smart data.

Data About Economic Environment. We are now overwhelmed with information about the economic problems of the world. Uncertainty about the effectiveness of government intervention and mitigation strategies makes risks apparent. To break the cycle requires (1) solving the enterprise's need for capital and (2) organizing the market such that it is capable of quid pro quo sustainment. These challenges are best addressed with smart data and smart data strategy.

25. *Enterprise performance is affected by outgoing data.*

Data to Customers. When we talk about data to customers and attempt to do this from the viewpoints of government and commercial enterprise, the subject can become confusing. The government can be a customer. The government can fill the need as a trading partner.

Prime producers of goods and services can be customers to suppliers and vendors, both of which can serve government customers. Data sharing among them is symbiotic and smart data strategy serves all.

"Toyota Joins American Rivals, Urges Aid to U.S. Parts Makers," is a *Washington Post, march 12, 2009* headline by Kendra Marr. The story is that all prime auto producers are concerned about getting capital to suppliers to sustain their viability: 20–30 suppliers can bring down an industry unless they have sufficient capital. Auto suppliers don't get paid for 40–60 days on average.

So, in this example, the prime producers are the customers and they typically inform suppliers well in advance about their production schedules via an enterprise resource planning mechanism.

Suppliers likewise inform prime producers about their status and in this instance they are saying they don't have sufficient capital to stay afloat to support future orders that are coming in reduced numbers. In today's environment, prime producers are informing government about the threat to their viability such that the government now becomes a factor as a trading partner with industry. The ultimate customer, auto consumers, are hearing this scenario and this fuels their concerns: they may perceive greater risk associated with a major capital purchase at a time when capital is king.

Of course, in rosier times, data goes to commercial consumers from marketers via multimedia channels, increasingly digital media. Consumers are buying more via electronic commerce and are receiving more information via directed channels over which consumers have increasing control. Google permits customers to tailor their advertising channel such that they receive and have access to only what they want— pull versus push. Smart data strategy in the advancing new digital economy will become an increasingly important success factor.

Data to Suppliers Prime producers provide information to suppliers. Government customers provide data to supply chains. Industrial viability is increasingly dependent on governments' industrial policies—formal or defacto.

Data to Regulators or Legislators New transparency will likely result in greater demand for data from regulators; trouble in the past was that data calls are costly, redundant, or unreliable and not actionable. Smart data will enable automated regulation and governance.

Data to Stakeholders Stakeholders of all kinds have greater access to data, some that is pushed to them and some that is available to pull. Smart data applied to communities of stakeholders will likely strengthen relationships.

Data to Partners and Allies Data to partners and allies will be managed via credentialing and privileging relationships. Smart data is an enabler to trusted relationships.

Smart Data Criteria

In Chapter 4 we described how smart data has three dimensions. Table 5.1 is a checklist summarizing the evidence of smart data. There are two basic elements: concept of operations (CONOPS) and operational architecture (OA).

Concept of Operations (CONOPS) Enterprise context describes the CONOPS. By this we mean the executive will lead the charge to optimize performance in an enterprise context, subordinating traditional stovepiped systems engineering approaches to those that consider enterprise-wide performance and impact.

Enterprise scope for performance optimization will be documented in policy that is unique to the enterprise and will likely append the balanced scorecard or equivalent. It most surely will be augmented by six sigma and other quantitative initiatives. Smart data strategy greatly improves data actionability and transparency and revolutionizes performance management in both operational and technical dimensions.

Executive users are prime smart data customers and their visible use and dependency on supporting the enterprise performance optimization system (EPOS) is statement enough about commitment. However, system management and governance requires hands on management.

The enterprise performance management process is proprietary to the enterprise and is the source of competitive advantage and superior performance. Nothing else in the enterprise trumps it as a central source of enterprise knowledge. Fully attributed process models are evidence of completion. An operational system is evidence of practice. The fully attributed process is modeled with contributions from functional management and subject matter experts facilitated by qualified modelers and technical specialists.

Operational Architecture (OA) OA is a description of the technical elements needed to implement and operate the CONOPS. There are four categories of technology enablement for smart data: (1) data interoperability technology, (2) data-aligned methods and algorithms, (3) cloud computing, and (4) semantic web.

Data Interoperable Data Technology Smart data is differentiated with model-driven data exchange. This requires the ability to model processes and data, and to strategically select standards for neutral data exchange. Such selection is performed in an enterprise context and therefore requires communication with all of the critical enterprise touch points from customers, suppliers, and other appropriate stakeholders. Some describe this as a harmonization activity.

TABLE 5.1 Smart Data Criteria Checklist

Smart Data	Three Dimensions	Criteria	Evidence
Concept of operations (CONOPS)	1. Enterprise context	1.1. Not traditional systems engineering by stovepipes or vertical	1.1. Executive- directed strategy
		1.2. Enterprise-wide scope	1.2. Defined in policy
		1.3. Executive user leadership	1.3. Executive chaired meetings
		1.4. Enterprise performance optimization process	1.4. Process model
			2.1. Staffed
Operational architecture (OA)	2. Interoperability technology	2.1. Data engineering	
		2.2. Model-driven data exchange	2.2. Implemented replacing brittle standards and interfaces
		2.3. Standards	2.3. Standards selection
		2.4. Semantic mediation	2.4. Present to support diverse lexicons and typographies
		25. Metadata management	
		2.6. Automated mapping tools	2.5. Competently staffed and enabled
		2.7. Service-oriented enterprise paradigm that includes smart data, smart grid, and smart services	2.6. A part of the toolkit
			2.7. Embraced and implemented via SOA and other specific enablement
		2.8. Credentialing and privileging	2.8. Part of security enablement
		2.9. Enterprise topology	2.9. Operational architecture
	3. Data-aligned methods and algorithms	3.1. Data with pointers for best practices	3.1. Commitment to identifying best practices and incorporating methods and algorithms with data
		3.2. Data with built-in intelligence	3.2. Present
		3.3. Autonomics	3.3. Present
		3.4. Automated regulatory environment (ARE) and automated contracting environment (ACE)	3.4. Present

Old harmonization activities focused on using brittle standards that required much work to achieve synchronization within the user community. Today's smart data technology leapfrogs the legacy by achieving data exchange based on open and interoperable standards and implementation conventions.

Service-oriented architecture describes the infrastructure, accompanied by data exchange servers that some call smart grid. Enterprise services are common among the user community as a part of a shared services strategy and utility service concept.

Work requires metadata management and data mapping that employs specific tools.

The comprehensive environment is called the service-oriented enterprise (SOE), described in earlier chapters. In an exceedingly open and transparent operating environment, it is imperative to manage credentialing and privileging to ensure flawless and secure data exchange, access, and use.

Data-Aligned Methods and Algorithms   Data is associated with methods and algorithms represented by best practices and preferred uses for analysis. This is accomplished via tagging.

Aligning methods and algorithms with data is aimed at ensuring executives have the best practices available to support planning, problem solving, decision making, sense making, and predicting. Employing contemporary technology, executives can expect data to be actionable and available immediately to support analysis.

The smart data paradigm aims at building in analytical intelligence such that the executive depends more on automated systems and less on a large staff of analysts. In this regard the smart data system is autonomic.

As an extended application of smart data autonomics, automated regulatory environment (ARE) and automated contracting environment (ACE) are offshoot applications.

Cloud Computing   "A computing paradigm in which tasks are assigned to a combination of connections, software and services accessed over the Internet. This network of servers and connections is collectively known as 'the cloud.' Computing at the scale of the cloud allows users to access supercomputer-level power. Using a thin client or other access point, like an iPhone, BlackBerry or laptop, users can reach into the cloud for resources as they need them" [10].

Semantic Web   "The Web of data with meaning in the sense that a computer program can learn enough about what the data means to process it." [11].

People are a strategic element, especially those possessing data Engineering disciplines.

Interoperable data technology includes the data engineering disciplines that are possessed by qualified staff combined with enabling technologies — software tools and hardware.

Strategy Elements

Applying Daniel Hunter's suggestions about four aspects of effective strategy, a smart data strategy template begins as follows.

1. *Set Direction.* Describe your starting position. What are the current CONOPS and OA for enterprise performance optimization employing data from the senior executive perspective? Model the as-is situation.

Given an introduction to smart data concepts and characteristics, where do you want to be? Model the to-be situation. What performance outcomes do you want? What do you want to change?

What is the current state of your data and associated assets? Have data specialists perform an objective assessment.

2. *Concentrate Resources.* The chances are that much of your IT support resources are skewed to infrastructure maintenance and management. This skew must be shifted to direct more resources directly to smart data vision development and direct and immediate executive support for planning, problem solving, decision making, sense making, and predicting.

3. *Maintain Consistency.* Develop a CONOPS and OA that will ensure commitment and focus on processes and mechanisms that will ensure maximum executive support for optimizing enterprise performance.

4. *Retain Flexibility.* Technology and ideas about smart data are advancing. Accommodate the advance by adopting spiral development and iterative reviews. Your smart data strategy must be customized to your unique enterprise and to your unique position in the total enterprise context.

Technical Solution and Enablement

Selecting standards that are most relevant to your enterprise and ones that satisfy the criteria for openness and interoperable data exchange is a technical requirement. Qualified staffing and profession development are critical success factors in breaking from the cycle imposed by layers of legacy barriers. Having strong command of modeling techniques and analytics are critical ingredients. Enterprise performance is optimized by smart data strategy through policy, goals, and actions.

5.2 POLICY

There are many different types of policies, and even multiple policies regarding data as we have reviewed examples. We are addressing a specific policy advocating smart data ideas applied specifically in support of executive enterprise performance optimization. The intent of the policy is to address every member of the user community with regard to their credentials and privileges and rights to access, publish, distribute, use, repackage, and exchange information.

1. *Policy is sponsored by the CEO and owned by the enterprise.* You will have to tailor your policy to the needs of your executive and the needs of the enterprise. When searching for examples of smart data policy, we really didn't find many good ones.

We appreciated the order of the Open Group Model Geographic Data Distribution Policy and will share the items that appear in the index for discussion.

- Definitions
- Assumptions
- Support
- Public Data
- Legal
  - Data Recipients
  - Control & Security
  - Copyright and Notices
  - Indemnification
  - Disclaimer
  - Privacy & Restrictions
  - Positive ID
  - Data Dictionary Service
  - Metadata Management and Maintenance
  - Data Correction and Updates
  - Data Redistribution & Third Party License
  - Derivative Data or Products
  - Value-Added Services

2. *Policy is developed and maintained by the controller or legal department.* We appreciate the sentiments of James Kavanaugh, Controller and Treasurer of Parker Thompson, published in an article entitled "Writing a Happy Ending," in *BizTech. com* in 2006. Kavanaugh said, "Ask yourself which of your corporate assets you could least afford to lose: a) your company's strategic plan; b) client list; c) new product marketing campaign; your pricing matrix?"

Of course the answer is you do not want to lose any asset. Yet, most companies don't have an explicit policy. Our focus is even more specific than a general data policy as we attend the specific data the enterprise executives need to optimize performance. Is there any difference? Probably not.

Typically, enterprise policies are given to a custodian once they are developed, and often the custodian is in the legal department and/or finance and accounting. Smart data policy will have legal ramifications and that is where it should be managed.

Observe in the previous section that the intent was on data sharing and not just protecting assets. In the smart data environment, you do not want to hunker down to the extent that you miss the point of credentialed and privileged data exchange.

3. *Policy is improved continuously.* Smart data policy is complementary with continuous improvement culture, although by spearheading continuous improvement with a data-focus makes it more effective.

4. *Policy is enforced by balance of consequences.* Make clear what are desired uses and what is prohibited. Make certain that participants understand the value of model-driven data exchange that is a level of performance independent of proprietary and legacy environments. Establish an enterprise-wide value statement.

5.3   ORGANIZATION

- *Management Continuity.* Policy as discussed above should reinforce the value of ensuring continuity of smart data strategy from administration to administration.
- *Government Cycles and Seasons.* Smart data use spans government cycles for planning, budgeting, funding, and authorization.
- *Commercial Cycles and Seasons.* Smart data use spans commercial cycles for planning, budgeting, funding, and authorization. Note that commercial enterprise that conducts business with government has a symbiotic relationship with customers whereby commercial planning may precede government planning, though the funding and authorization follow government.
- *Industry Associations, Standards Bodies, and Professional Organizations.* Many industry associations, standards bodies, and professional organizations are active participants in what we are calling smart data initiatives. Enterprise executives must observe that some of these external bodies have earned fees for brittle standards maintenance as a primary source of revenue. Therefore open and interoperable model-driven data exchange poses a threat to some of these, and you may experience resistance for that reason. They may have a conflict of interest.

5.4   ACTIONS

Here is an example of smart data and smart data strategy at work.

From an interview focused on data strategy with Michael E. Krieger, director of information policy for the Department of Defense, by Ben Bradley for *CIO* magazine, we learned that he is responsible for providing policy and guidance for implementing the DoD's net-centric data strategy and enabling the transition to an enterprise service-oriented architecture (SOA).

The DoD is separating data from applications, decoupling data, because systems that are too tightly coupled to the data result in high maintenance expense and other problems. The data strategy gives priority to visibility, accessibility, and understandability over standardization.

When asked by Bradley what the DoD's Achilles' heel is Krieger replied: "the transformation and change required to get communities together to address information-sharing challenges by agreeing on shared vocabularies and exposing and sharing data as a service."

Krieger said that "industry understands the agility and power that separating data from applications represents. Consider the Google Maps service and the Google Earth application. It is based on a community vocabulary for modeling and storing geographic data called KML (keyhole markup language). By publishing data as a service in KML, the Google Maps service or the Google Earth application seamlessly plots the data on a map or a globe."

On metadata, Krieger explained how it makes data assets visible or discoverable. "For example, in digital cameras, where the data is the photographic image, metadata would typically include the date the photograph was taken and details of the camera settings. The Department of Defense Discovery Metadata Specification (DDMS) specifies how to use metadata to make data assets visible to the enterprise. It describes how developers, engineers and users should advertise data assets posted to shared spaces so that others can discover them. We didn't make up our metadata standard. The DDMS is based on the industry Dublin Core standard, and we added a security leg to it. This approach moves people away from hoarding data and increases data visibility and sharing."

Asked by Bradley how he handles resistance to the information standard, Krieger replied: "It helps when the community develops the vocabulary or information exchange semantics. This helps both data consumers and producers. One-way translation from the community standard at the producer or consumer location is the only requirement. Technologies like XML easily enable data wrapping for one-way translations. Resistance is reduced because you only need to fix one side of any legacy system. This is how you can republish data from a legacy system to unanticipated users without rebuilding legacy applications."

Smart Data Strategy Work

The work begins with an assessment of the current situation. Are your primary processes modeled and fully attributed as described in the book? Have you, as executive, clearly inventoried the questions that you need answering to optimize performance in the enterprise and requirements for planning, decision making, problem solving, sense making, and predicting?

Following these prerequisites, your management team, including information technologists, can address the enterprise performance optimization system CONOPS and OA. Then you will implement with system development and operational improvements in executive support.

Smart Data Strategy Staffing

Is your organization properly staffed for the initiative? Review the skills, knowledge, and experience of your staff to ensure they have what it takes to implement the strategy.

TABLE 5.2   Twenty Outcomes to Expect from Smart Data Strategy

Outcomes	Measures
1. Enterprise performance optimized	• Key performance indicators and metrics commercial enterprise • Key performance indicators and metrics government enterprise • Adoption of artificial intelligence methods and algorithms • Data mining of business intelligence to produce smart data
2. Executive planning, problem solving, decision making, sense making, and predicting optimized	• Actionable data support key responsibilities in a more timely, accurate, and complete manner, at an optimal rate, frequency, cost, and quality • Real time versus batch processing: Business at the speed of thought
3. Enterprise resource management optimized	• As a subset of #1 • Utilize cost accounting and activity-based costing to outperform traditional accounting methods
4. Stakeholder relationships optimized	• Consistently positive feedback from stakeholders for enterprise management • Utilize negative feedback as a means for continuous improvement and BPR
5. Enterprise processes optimized by design	• Significantly improved metrics from aggregate and granular performance as reflected in cost and time, among other process-specific indicators • Follow an iterative spiral development process for design by focusing on planning and analyzing both the physical and conceptual design schemas necessary for implementation and maintenance
6. Regulatory, legal, and business rule compliance improved	• Deviations minimized, cost to comply lowered • Eternal vigilance by gathering business intelligence to reengineer outdated processes and implement an SOE using an SOA
7. Performance visibility improved	• Executives see real-time results and can simulate scenarios • Adoption of AI methods, improved algorithms and modeling and simulation techniques such as Monte Carlo applications
8. Quantitative methods more readily applied	• Actionable data is available on which to apply improved methods and algorithms

TABLE 5.2 (*Continued*)

Outcomes	Measures
	• Apply a data mining mentality to produce actionable smart data to optimize performance
9. Enterprise managed by "how to" not just by "what"	• Smart data-driven process management that is supported by AI and improved algorithms
10. Enterprise assets fully accounted	• One hundred percent • Quality is an enterprise-wide undertaking
11. More resources available for IT executive service support with corresponding reduction in legacy infrastructure maintenance and management	• Make 30% more available for reassignment to better uses • Adopt AI methods for ITCB such as SA and replace traditional outdated A-Rank and D-Rank methods that are not as optimal
12. Advantages and benefits from SOA maximized	• Advantages from openness and interoperability enterprise-wide • Lower total cost of ownership (TCO) in IT through consolidation opportunities and more standards-based integration • Agility for faster implementation and changes in IT through reuse, modeling, and composite application development techniques • Better alignment through business processes and their IT realization through more transparent process modeling and monitoring based on service
13. Higher state of enterprise integration with associated performance advantages from automation achieved	• Fifty percent higher states of automated support to executive responsibilities • Bring together business semantics, process content, and delivery of applications through leveraging SOA
14. Operational results from specific applications increased	• Domain specific • Smart data improves results at the operational, tactical, and operational levels
15. Relationships between government and commercial enterprise improved	• Seamless • Standardization such as XML PSLC 10300 allows for improved communication between vendors and suppliers and government
16. Performance management precision increased	• Higher alignment accuracy among planning, anticipation, prediction, and results • Adoption of smart technologies such as AI and data mining improve predictive capabilities

(*continued*)

TABLE 5.2   **Twenty Outcomes to Expect from Smart Data Strategy**   *(Continued)*

Outcomes	Measures
17. Enterprise agility, flexibility, and adaptability increased	• Higher states • Adoption of dashboards and other business intelligence techniques improves fluidity and capability to predict change
18. Accurate anticipation and prediction increased	• Greater accuracy with more reliable predictive results by adopting smart data mentality, techniques, and methods
19. Alignment between enterprise performance data and executive users increased	• Higher executive satisfaction • Better stakeholder and customer satisfaction
20. Smart data-focused management culture created	• Such as six sigma, CMMI, ISO-9000, PSLC 10300, and others

5.5 TIMING

It is reasonable to expect to take a year to institute smart data and associated strategy in the enterprise.

5.6 FUNDING AND COSTING VARIABLES

There are two types of funding requirements: (1) development and implementation of smart data strategy, and (2) development and implementation of enterprise performance optimization system (EPOS).

Some investment may be needed to upgrade modeling and mapping tools. Some investment may be needed to develop data exchange servers. Investment is needed to work with the community to implement the strategy.

5.7 OUTCOMES AND MEASUREMENTS

Anticipate and plan for internal and external outcomes and associated measurements, including collaborative benefits that may be scaled with regard to proportional contribution. (see Table 5.2.)

REFERENCES

1. Jeannie Aversa, "Federal Reserve Chairman Bernanke Says Regulatory Overhaul Needed," AP/MSNBC, March 11, 2009.

2. S. Dao, D. M. Keirsey, R. Williamson, S. Goldman, and C. P. Dolan, *Interoperability in Multidatabase Systems*, 1991. IMS apos; 91. Proceedings. First International Workshop on Volume, Issue, 7–9 Apr 1991 Page(s): 88–91 Digital Object Identifier 10.1109/IMS.1991.153689.

3. *The 1st International ACM Workshop on Interoperability of Heterogeneous Information Systems—IHIS'05* is organized in conjunction with the *14th ACM Conference on Information and Knowledge Management (CIKM'05)*, Proceedings of the First International Workshop on Interoperability of Heterogeneous Information Systems, 2005.

4. U.S. Patent 7072900.

5. http://en.wikipedia.org/wiki/Holon_(philosophy).

6. http://www.purisma.com/.

7. http://www.austintetra.com/.

8. http://www.dvdinformation.com/GroupInfo/index.cfm.

9. http://www.boeing.com/companyoffices/doingbiz/tdi/partner.html.

10. http://saint-michael.trap17.com/blog/2009/05/flying-high-in-the-clouds/.

11. http://www.w3.org/People/Berners-Lee/Weaving/glossary.html.

Index

WILEY SERIES IN SYSTEMS ENGINEERING AND MANAGEMENT

Andrew P. Sage, Editor

YACOV Y. HAIMES
Risk Modeling, Assessment, and Management, Third Edition

DENNIS M. BUEDE
The Engineering Design of Systems: Models and Methods, Second Edition

ANDREW P. SAGE and JAMES E. ARMSTRONG, Jr.
Introduction to Systems Engineering

WILLIAM B. ROUSE
Essential Challenges of Strategic Management

YEFIM FASSER and DONALD BRETTNER
Management for Quality in High-Technology Enterprises

THOMAS B. SHERIDAN
Humans and Automation: System Design and Research Issues

ALEXANDER KOSSIAKOFF and WILLIAM N. SWEET
Systems Engineering Principles and Practice

HAROLD R. BOOHER
Handbook of Human Systems Integration

JEFFREY T. POLLOCK AND RALPH HODGSON
Adaptive Information: Improving Business Through Semantic Interoperability, Grid Computing, and Enterprise Integration

ALAN L. PORTER AND SCOTT W. CUNNINGHAM
Tech Mining: Exploiting New Technologies for Competitive Advantage

REX BROWN
Rational Choice and Judgment: Decision Analysis for the Decider

WILLIAM B. ROUSE AND KENNETH R. BOFF (editors)
Organizational Simulation

HOWARD EISNER
Managing Complex Systems: Thinking Outside the Box

STEVE BELL
Lean Enterprise Systems: Using IT for Continuous Improvement

J. JERRY KAUFMAN AND ROY WOODHEAD
Stimulating Innovation in Products and Services: With Function Analysis and Mapping

WILLIAM B. ROUSE
Enterprise Tranformation: Understanding and Enabling Fundamental Change

JOHN E. GIBSON, WILLIAM T. SCHERER, AND WILLAM F. GIBSON
How to Do Systems Analysis

WILLIAM F. CHRISTOPHER
Holistic Management: Managing What Matters for Company Success

WILLIAM B. ROUSE
People and Organizations: Explorations of Human-Centered Design

GREGORY S. PARNELL, PATRICK J. DRISCOLL, AND DALE L. HENDERSON
Decision Making in Systems Engineering and Management

MO JAMSHIDI
System of Systems Engineering: Innovations for the Twenty-First Century

ANDREW P. SAGE AND WILLIAM B. ROUSE
Handbook of Systems Engineering and Management, Second Edition

JOHN R. CLYMER
Simulation-Based Engineering of Complex Systems, Second Edition

KRAG BROTBY
Information Security Governance: A Practical Development and Implementation Approach

JULIAN TALBOT AND MILES JAKEMAN
Security Risk Management Body of Knowledge

SCOTT JACKSON
Architecting Resilient Systems: Accident Avoidance and Survival and Recovery from Disruptions

JAMES A. GEORGE AND JAMES A. RODGER
Smart Data: Enterprise Performance Optimization Strategy